PLASTICITY IN THE VISUAL SYSTEM:
From Genes to Circuits

PLASTICITY IN THE VISUAL SYSTEM:
From Genes to Circuits

Edited by

Raphael Pinaud, Ph.D.
OHSU, USA

Liisa A. Tremere, Ph.D.
OHSU, USA

and

Peter De Weerd, Ph.D.
University of Maastricht, The Netherlands

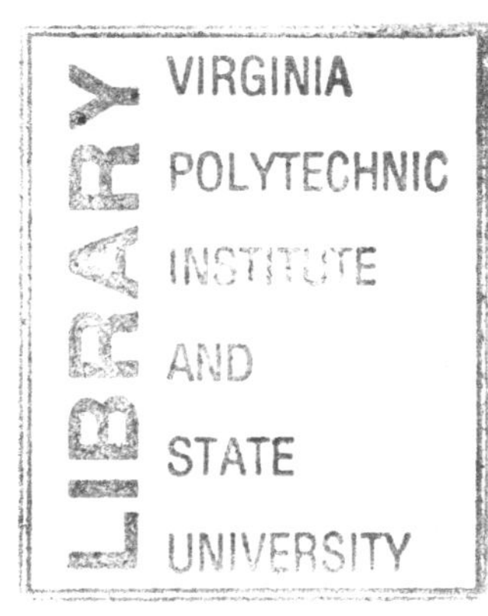

Library of Congress Cataloging-in-Publication Data

Plasticity in the visual system : from genes to circuits / edited by Raphael Pinaud, Liisa A. Tremere, Peter De Weerd.
    p. cm.
   Includes bibliographical references and index.
   ISBN-13: 978-0-387-28189-6
   ISBN-10: 0-387-28189-4
   ISBN-13: 978-0-387-28190-2 (e-book)
   ISBN-10: 0-387-28190-8
    1. Vision. 2. Neuroplasticity. I. Pinaud, Raphael. II. Tremere, Liisa A. III. De Weerd, Peter.

   QP475.P62 2005
   612.8'4—dc22

            2005051576

*This book is dedicated to
my family.*

To the origin: Luiz Fernando, Sandra and Rodolpho,
Vó Vilma, Vô Néo e Eunice; *To the present*, Liisa;
To the future, my Daniel.
(R.P.)

For my mentors, who have opened the doors that allowed
my entry into science. For my family, who has helped
me get through them . . . And for my husband, who lays
his coat down over the puddles along the way so, that
I may go the distance with a little more grace.
(L.A.T.)

To good friends and family.
(P.D.W.)

In Memoriam

After my colleagues and I secured a contract with Springer to publish this book, a large effort was put forward by us to recruit leading authors in the various fields tackled herein. In the "Plasticity in the Visual Thalamus" section, we extended an invitation to Dr. Bertram Payne, from Boston University. Dr. Payne had spent several decades investigating plasticity phenomena in the visual thalamus, in particular dynamic plasticity of thalamocortical relationships. Bertram Payne promptly accepted our invitation with enthusiasm and sent me a rough outline of his chapter. Two months before the deadline for submissions of the first drafts, Bertram contacted me asking whether I would have a problem with him including some novel ideas in his chapter concerning the effects of cooling in the lateral geniculate nucleus, which was the focus of his most recent research efforts. I obviously encouraged him strongly after reading some of his material. A month after this last exchange Peter contacted me and stated that Bertram had passed away. We recognize here the major contributions of Dr. Payne to the understanding of basic principles involved in visual system plasticity.

Raphael Pinaud

Contents

Preface

Light information represents the fastest and possibly the most physically complex source of physical energy processed by most mammals. All three editors share the bias that the visual system likely houses the most advanced evolutionary design for the encoding and efficient handling of sensory input. To understand the mechanisms of plasticity within the visual system, one must initially explore the extent to which reorganization at higher levels could be explained by changes that occur at earlier stations such as the retina. The idea of plasticity in the retina is one of the controversial issues in the field of CNS reorganization and neural plasticity. In the present book, several experts in the field of visual system plasticity describe and discuss the extent of detectable change and the mechanisms proposed to underlie neural plasticity in the retina, sub-cortical structures, and cortex. To facilitate cross-talk between researchers with different technical backgrounds and perspectives, who conduct research in the visual system, we invited a wide range of investigators to discuss their own work, and to offer ideas and interpretations going beyond their own field of expertise. All authors were encouraged to suggest future applications of research findings from neural plasticity in the visual system with an emphasis on potential clinical uses and engineering within the biomedical sciences.

This book was born well after midnight on a kitchen table in Arizona, amid bottles of good wine and fine Belgian chocolates. After having solved the world's other problems, the three editors fell upon a discussion of neural plasticity and its involvement in the complex activities of perception, learning and memory. With Raphael's formation in molecular biology and neural plasticity, Peter's background in psychophysics and mechanisms of attention and learning in the visual system, and Liisa's training in neural plasticity and inhibitory processing in sensory systems, it was amazing we could establish sufficient common ground to hold an informed discussion at all. It was then that we identified a contribution that we could make: We decided to repeat our kitchen table discussion, this time by pooling the findings and perspectives of leading scientists interested in neural plasticity, working at the level of molecular and cellular biology, systems neuroscience and most areas in-between. So, wherever you may fall on this spectrum of interests, we hope that you will enjoy reading "Plasticity in the Visual System: From Genes to Circuits".

An edited volume stands or falls with the chapters produced by its contributing authors. We feel lucky to have worked together with the present group of researchers, and we offer our sincerest thanks to all of the authors for their generous contributions and excellent chapters. We have greatly appreciated their preparedness to share their efforts and wisdom towards the completion of this project, and we thank them for their productive interactions and support throughout the editorial process.

Finally, we would also like to thank our partners at Springer: Joe Burns, Marcia Kidston, Sheri Campbell and Marc Palmer. Their enthusiastic support, advice and cooperation were indispensable in the creation of this work.

Raphael Pinaud, Portland, OR, USA
Liisa A. Tremere, Portland, OR, USA
Peter De Weerd, Maastricht, The Netherlands

Contributors

LUTGARDE ARCKENS
Laboratory of Neuroplasticity and Neuroproteomics
Katholieke Universiteit Leuven
Leuven, Belgium

GIUSEPPE BERTINI
Department of Morphological and Biomedical Sciences
University of Verona
Verona, Italy

CHRISTINE E. COLLINS
Department of Psychology
Vanderbilt University
Nashville, TN

YUZO M. CHINO
College of Optometry
University of Houston
Houston, TX

STEVEN K. FISHER
Neuroscience Research Institute
University of California Santa Barbara
Santa Barbara, CA

ANTONIO F. FORTES
Department of Neuroscience
University of Minnesota
Minnesota, MN

STEPHEN GROSSBERG
Department of Cognitive and Neural Systems and
Center for Adaptive Systems
Boston University
Boston, MA

BRYAN W. JONES
John A. Moran Eye Center
University of Utah School of Medicine
Salt Lake City, UT

JON H. KAAS
Department of Psychology
Vanderbilt University
Nashville, TN

GEOFFREY P. LEWIS
Neuroscience Research Institute
University of California Santa Barbara
Santa Barbara, CA

ROBERT E. MARC
John A. Moran Eye Center
University of Utah School of Medicine
Salt Lake City, UT

HUGO MERCHANT
Instituto de Neurobiología
UNAM, Campus Juriquilla
Queretaro Qro.
Mexico

VICENTE MONTERO
Department of Physiology and Waisman Center
University of Wisconsin
Madison, WI

BERTRAM R. PAYNE
Center for Advanced Biomedical Research
Department of Anatomy and Neurobiology
Boston University School of Medicine
Boston, MA

RAPHAEL PINAUD
Neurological Sciences Institute and Vollum Institute
Oregon Health & Science University
Portland, OR

BARRY J. RICHMOND
Laboratory of Neuropsychology
National Institute of Mental Health
National Institutes of Health
Bethesda, MD

THOMAS A. TERLEPH
Department of Psychology
Rutgers University
Piscataway, NJ

ELLEN TOWNES-ANDERSON
Department of Neurology and Neurosciences
New Jersey Medical School
University of Medicine and Dentistry of New Jersey
New Jersey, NJ

LIISA A. TREMERE
Center for Research on Occupational and Environmental Toxicology
Oregon Health & Science University
Portland, OR

CARL B. WATT
John A. Moran Eye Center
University of Utah School of Medicine
Salt Lake City, UT

PETER DE WEERD
Department of Psychology
University of Maastricht
Maastricht, The Netherlands

NAN ZHANG
Department of Neurology and Neurosciences
New Jersey Medical School
University of Medicine and Dentistry of New Jersey
New Jersey, NJ

1

Introduction to Plasticity in the Visual System: From Genes to Circuits

PETER DE WEERD[1], RAPHAEL PINAUD[2] AND LIISA A. TREMERE[3]

[1]Department of Psychology, University of Maastricht, Maastricht, The Netherlands,
[2]Neurological Sciences Institute, OHSU, Portland, OR, USA and
[3]CROET, OHSU, Portland, OR, USA

In this edited volume, the visual system is used as a model to study contemporary issues in central nervous system plasticity in the developing and adult organism. We evaluate plasticity induced by injury, sensory experience, and learning at all three anatomical levels of the ascending visual pathway: the retina, visual thalamic relays such as the lateral geniculate nucleus (LGN), and various visual cortical areas, with a particular focus on the striate cortex (V1) (Tessier-Lavigne, 2000; Wurtz & Kandel, 2000a,b; Lennie, 2000). One definition proposed for the term plasticity is the changing of connectivity between neurons (Hebb, 1949), which forms the basis for altered perception, cognition, and behavior. Plastic changes in neuronal activity and morphology in the visual system have been studied for decades in a variety of preparations using anatomical, electrophysiological and brain imaging techniques in both humans and animals. In parallel with advances in the neurosciences, research in the field of molecular biology has uncovered some principles underlying the physical basis for neuronal plasticity. These recent developments permit the realization of a conceptual bridge between plasticity viewed as changes in neuronal activity and morphology, and the molecular basis of these changes. The principal aim in generating this book, therefore, was to foster new perspectives on the topic of visual plasticity by integrating data obtained from systems neuroscience with insights from molecular and cellular biology. In a sense, this book will provide a zoom lens that will permit the reader to consider visual plasticity from the level of genes, proteins, and altered synaptic weights, to altered electrical activity measured in single neurons as well as in large neuronal circuits or populations.

The advent of modern visual neuroscience has been associated with studies in retinal ganglion cells by Kuffler (1953), who described the center-surround characteristics of retinal receptive fields (RFs) and a series of studies from

R. Pinaud, L. A. Tremere, P. De Weerd(Eds.), Plasticity in the Visual System:
From Genes to Circuits, 1-10, ©2006 Springer Science + Business Media, Inc.

Hubel and Wiesel (1959, 1962), who described the orientation tuning and other response properties of neurons in striate and extrastriate cortices. Their studies represent the start of a long tradition of single-unit recording studies that aimed to elucidate the response properties of (visual) neurons. While these studies were typically conducted in the anesthetized cat, the development of techniques to record single-neuron activity in awake, behaving primates (Wurtz, 1969; Evarts, 1968) soon allowed the study of perceptual, motor, and cognitive performance while neuronal activity was recorded simultaneously. Relating (changes in) perception, cognition and behavior with (changes in) neuronal activity has been one of the great accomplishments of modern neuroscience. A powerful example of this approach was the discovery by Fuster (1973) of neuronal activity in pre-frontal cortex that bridged a time interval between the presentation of a sample stimulus and a test stimulus that the animal had to judge as matching or non-matching. This type of 'delay activity' has been considered to be a behavioral correlate of working memory or attention. In the context of such studies, it was discovered that high-level cognitive factors such as attention (Moran & Desimone, 1985) and learning (Yang & Maunsell, 2004) could be read-out from neuronal activity within sensory areas, including primary visual cortex (Gilbert et al., 2000). In addition, it was discovered that injury to the retina could lead to profound reorganization of retinotopic maps in visual cortex (Kaas et al., 1990). The discovery that primary visual cortex remains plastic in adulthood went against long-held beliefs that cortical plasticity was limited to developmental plasticity occurring before closing of the critical period. In the present book, evidence will be presented that adult plasticity extends even to the retina itself.

The use of a single electrode to correlate neural activity with perception or cognition can lead to the fallacy of believing that the response properties of neurons observed at the tip of the electrode are generated locally. That the contrary is true has become evident from several lines of work. Anatomical studies have demonstrated the complexity and extent of the neural network that can influence the activity of any single neuron (Press et al., 2001). Neural network modeling studies have been instrumental in demonstrating how the properties of individual nodes are a product of interactions with many other nodes at a number of levels in the system (e.g., Rumelhardt, 1987; Tielscher & Neumann; Xing & Andersen, 2000). Finally, it has become clear that data from ensemble recordings are more readily correlated with behavior than with data from single neurons. The idea that stimulus values or movement parameters are encoded by invariant patterns of firing in large sets of neurons is also known as population coding, a concept championed by Georgopoulos (1986). Thus, when one is interested in learning behavior and associated neural plasticity, techniques that could evaluate plasticity in extended neural networks would be highly useful.

Functional magnetic resonance imaging (fMRI) and other imaging techniques such as positron emission imaging (PET) and optical imaging can reveal plastic changes in large neural networks at a macro scale. fMRI, pioneered by Ogawa et al. (1990), is an ideal technique to use in humans, because it can be used repeatedly and non-invasively. fMRI signals reflect increased blood flow in activated cortex, and research in monkeys has shown that the fMRI signal correlates only partly with spiking behavior (Logothetis & Wandell, 2004). Nevertheless, many findings on perception and learning originally done using extracellular recordings in animal studies have been replicated in humans with fMRI (Ungerleider, 1995). Among those findings are the time-dependent roles of a number of brain structures such as basal ganglia and cerebellum during skill learning (Doyon et al., 2003), and the function of hippocampus during episodic memory formation (Ryan et al., 2001). In addition, fMRI has permitted the study of the topography of human sensory and motor cortex (for review in visual cortex, see Orban et al., 2004), as well as plastic changes in topography induced by skill learning and peripheral injury, thereby corroborating many findings previously reported in animal studies.

Despite the explosion of knowledge in the visual neurosciences, there are two major limitations that characterize the majority of the studies conducted in this field. One limitation is the correlative nature of the data, which does not permit the establishment of causal relationships. The other limitation is the difficulty in studying cell functioning and plasticity of neural networks on a cell-by-cell basis.

It would do injustice to the field of the neurosciences to claim that it would not have provided its own answers to these two fundamental questions. Anatomical lesions (Sprague et al., 1985; Heywood & Cowey, 1987; De Weerd et al.,1999), pharmacological lesions by local injection of toxins, such as ibotenic acid (Murray & Mishkin, 1998), or NMDA (Hampton et al., 2004), and reversible lesions by cooling (Lomber et al., 1999) or pharmacological means (Logothetis, 2004) have directly and successfully tested whether correlative relationships between neuronal activity and behavior could be interpreted in a more causative manner. Furthermore, important microstimulation studies in behaving rhesus monkeys by Newsome and colleagues (Seideman et al., 1998) have shown that the electrical stimulation of clusters of neurons encoding a particular direction of motion biased movement perception towards the movement encoded by these neurons, and microstimulation has also been reported to influence other types of perception and cognition (Cohen & Newsome, 2004); such findings will remain essential in order to transcend what can be achieved based on correlative study alone.

The development of ensemble recording techniques (Nicolelis et al., 2003; Hoffman & McNaughton, 2002), in which multiple neurons can be recorded over a number of sessions, have permitted the correlation between neural activity in large populations of neurons with cognitive behavior and learning/plasticity. For example, ensemble recordings in the rat and monkey hippocampus have successfully demonstrated the encoding, consolidation or memory formation during sleep, and retrieval of different spatial environments (Hoffman & McNaughton, 2002; Wilson & McNaughton, 1993,

1994). Ensemble codes in the hippocampus have been used to predict the position of rats in their environment (Wilson & McNaughton, 1993), and ensemble codes in primary motor cortex have been used successfully to steer and control robotic arms (Carmena et al., 2003). Such findings indicate that initially correlative findings can acquire such a strong causative footing that they can become useful in the design of therapeutic intervention.

Many of the initial insights into the molecular basis of neuronal plasticity have come from the use of simple invertebrate systems, such as *Aplysia* and *Drosophila*, which were introduced in the late 1950's to analyze elementary aspects of behavior and learning at the cellular and molecular level. A series of studies from Eric Kandel's group (e.g., Castellucci et al., 1970, 1989) has shown that the long-term sensitization of the gill-withdrawal reflex in *Aplysia* requires *de novo* protein synthesis and, consequently, gene expression. A perceptive, early insight by Thorpe (1956) that learning and memory might be universal features of the nervous system is confirmed by current research. For example, synaptic enhancement described in *Aplysia* shares a number of features with synaptic enhancement observed in mammalian hippocampus (Bailey et al., 1996; Bailey et al., 2000; Kandel, 2001). Furthermore, the hours-long consolidation periods reported for both skill learning and episodic memory formation seems to reflect the time period during which genes are expressed that regulate plastic structural changes of synapses and neurons (Bailey & Chen, 1983; Bailey and Kandel, 1993), and, in line with the above-referred data in *Aplysia,* these types of memory formation can be blocked by protein synthesis inhibitors (Luft et al., 2004; Agnihotri et al., 2004; Fisher et al., 2004). A number of findings also show that several late-response genes, which putatively regulate structural changes associated with plastic changes are controlled by a class of fast-responding genes known as immediate early genes (IEGs). IEGs are activated within minutes of cell stimulation or learning experience. The expression of these IEGs in certain cells can thus be used both as a marker of activity, and many IEGs are currently under review as candidate-plasticity genes (see Pinaud, 2004, and Chapter 8).

In contrast to electrophysiological recording methods, which offer high temporal resolution in a limited part of the brain, the use of IEGs offers the possibility to mark activity on a cell-by-cell basis throughout the brain, at the cost of temporal resolution. IEG expression profiles can be used as a tool to localize the effects of recent sensory-driven activity throughout the brain. For example, Ribeiro and colleagues. (1998) have used stimulus-dependent activation of the IEG *zenk* to visualize tonotopic organization in the songbird auditory system. Guzowski and collaborators have monitored the distribution of the IEG *arc* to reveal different hippocampal ensemble activity while rats were exposed to different spatial environments (Vazdarjanova & Guzowski, 2004). These examples show that molecular tools can be used to image activity in large numbers of cells recruited during perception, with single-cell resolution, and during various forms of memory formation and learning paradigms (Kaczmarek and Robertson, 2002). By applying these techniques at different moments in a learning process, plastic changes in a neuronal network could be revealed directly. The main price to be paid for the high spatial resolution afforded by these techniques is the loss of certain time windows for studying plasticity related events, as changes occurring milliseconds or seconds after a manipulation cannot be captured histologically (Kaczmarek and Robertson, 2002). Furthermore, the above techniques have the disadvantage that animals must be sacrificed, and that particular IEGs can be expressed in more than a single anatomically defined cell type. New developments may soon enable the *in vivo* measurement of activity in neural networks of known cell type using sensors for membrane voltage, intracellular messengers and pH, which can be targeted to specific cell types by use of vectors such as viruses (Miesenbock et al., 1998; Bozza et al., 2004). These techniques applied at different learning stages, or during experience-dependent reorganization, could become invaluable tools to study neural network plasticity.

Knowledge of the molecular basis of neuronal plasticity also offers ways to directly test the causative status of links between manifestations of neuronal plasticity, and their perceptual, cognitive and behavioral correlates. Such tests avoid some of the limitations characteristic of the lesion and reversible deactivation approaches used in the field of the neurosciences, and they offer their own specific opportunities. In particular, a drawback of anatomical lesion methods and (reversible) deactivation methods is that they indiscriminately deactivate or destroy all neurons in a given region, even though different cell types in that region might contribute differently to the sensory or cognitive function being studied. An attractive alternative approach offered by molecular biology is to generate functional lesions by interfering with normal genetic contributions to a particular perceptual or cognitive function, though the delivery of antisense molecules (Hebb & Robertson, 1997; Lee et al., 2004). Antisense molecules bind to the mRNA encoded by a specific gene of interest and, its delivery to a particular cortical or brain region (iontophoretically or through other means, such as pressure injection) can thus block the contribution of a single gene to a particular cortical function, or to a particular perceptual or cognitive ability. Other approaches, such as viral-mediated gene targeting and conditional gene expression, have been used extensively more recently and enhance the potential of gene expression control in-vitro and in-vivo (Mansuy et al., 1998; Mower et al., 2002). Although it is an ambitious goal to start parsing the contribution of individual genes to neuronal plasticity (Goldberg & Weinberger, 2004), and associated cognitive behavior, these approaches are being successfully used in a number of animal models (Lee et al., 2004; Izquierdo & Cammarota, 2004; Ogawa & Pfaff, 1996; Liu et al., 2004). Studies of this type have contributed significantly to the generation of deep insights into the nature of learning and memory formation, and continue to hold tremendous promise for applied use. As a case in point, human gene therapy, in which viral vectors are used to insert genetic material into particular chromosomes of targeted cells to remediate specific illnesses (e.g., Amado et al.,

1999) plausibly could be expanded in the future to also treat sensory, motor, and cognitive deficits (Fink et al., 2003; Tinsley & Eriksson, 2004).

Much of the work on neural plasticity and associated forms of cognition that precedes any human applications remains to be done. In order to increase our understanding of plasticity in the human brain, it is necessary to develop model systems in higher mammals and primates. In this book, the visual system of higher mammals, including primates, is used to study plasticity from the molecular to the systems neuroscience level. We hope that the combination of different viewpoints and approaches of plasticity will contribute, ultimately, to a deeper understanding of higher order cognitive processes such as learning and memory formation. Briefly, the book is divided into three parts, with Part I (Chapters 2–6) devoted to retinal and thalamic plasticity, Part II (Chapters 7–14) to cortical plasticity, and Part III (Chapters 15–16) to some theoretical and integrative considerations. Chapters 2–5 will treat structural neurochemical and functional aspects of retinal plasticity induced by various types of damage and degeneration, as well as the experience-dependent expression of IEGs. In Chapter 6, Fos imaging is used to reveal cortico-thalamic networks recruited by attention. Chapters 7 and 11 discuss effects of neuromodulatory transmitters and inhibition on normal sensory processing and during experience-induced plasticity in primary visual cortex, while Chapters 8 and 9 focus on the molecular, biochemical and cellular mechanisms that participate in the control of plasticity in the visual cortex. The outcomes of plastic re-arrangement of circuitry in V1 and related lower-order visual cortex are discussed at the systems level in Chapters 10, 12 and 13. Chapter 14 is an example of the type of knowledge that emerges when high-level cognition is studied by a combination of systems neuroscience and molecular approaches. The final chapters aim to provide a theoretical framework for a number of topics covered in this volume, with Chapter 15 emphasizing a computational modeling approach at the systems level, and Chapter 16 providing an integrative view of molecular and systems approaches of neuronal plasticity.

References

Agnihotri NT, Hawkins RD, Kandel ER, Kentros C (2004) The long-term stability of new hippocampal place fields requires new protein synthesis. *Proc Natl Acad Sci USA* 101, 3656–3661.

Amado RG, Mitsuyasu RT, Zack JA (1999) Gene therapy for the treatment of AIDS: animal models and human clinical experience. *Front Biosci* 4, D468–475.

Bailey CH, Kandel ER (1993) Structural changes accompanying memory storage. *Annu Rev Physiol* 55:397–426.

Bailey CH, Bartsch D, Kandel ER (1996) Toward a molecular definition of long-term memory storage. *Proc Natl Acad Sci* U S A 93:13445–13452.

Bailey CH, Giustetto M, Huang YY, Hawkins RD, Kandel ER (2000) Is heterosynaptic modulation essential for stabilizing Hebbian plasticity and memory? *Nat Rev Neurosci* 1:11–20.

Bailey CH, Chen M (1983) Morphological basis of long-term habituation and sensitization in Aplysia. *Science* 220, 91–93.

Bozza T, McGann JP, Mombaerts P, Wachowiak M (2004) In vivo imaging of neuronal activity by targeted expression of a genetically encoded probe in the mouse. *Neuron* 42(1), 9–21.

Carmena JM, Lebedev MA, Crist RE, O'Doherty JE, Santucci DM, Dimitrov DF, Patil PG, Henriquez CS, Nicolelis MA (2003) Learning to control a brain-machine interface for reaching and grasping by primates. *PLoS Biol* 1(2):E42.

Castellucci VF, Pinsker H, Kupperman I, Kandel ER (1970) Neuronal mechanisms of habituation and dishabituation of the gill-wthdrawl reflex in Aplysia. *Science* 167, 1745–1748.

Castellucci VF, Blumenfeld H, Goelet P, Kandel ER (1989) Inhibitor of protein synthesis blocks long-term behavioral sensitization in the isolated gill wthdrawall reflex of Aplysia. *J Neurobiol* 20, 1–9.

Chawla MK, Lin G, Olson K, Vazdarjanova A, Burke SN, McNaughton BL, Worley PF, Guzowski JF, Roysam B, Barnes CA (2004) 3D-catFISH: a system for automated quantitative three-dimensional compartmental analysis of temporal gene transcription activity imaged by fluorescence in situ hybridization. *J Neurosci Methods* 139(1), 13–24.

Chew SJ, Mello C, Nottebohm F, Jarvis E, Vicario DS (1995) Decrements in auditory responses to a repeated conspecific song are long-lasting and require two periods of protein synthesis in the songbird forebrain. *Proc Natl Acad Sci USA* 92, 3406–3410.

Cohen MR, Newsome WT (2004) What electrical microstimulation has revealed about the neural basis of cognition. *Curr Opin Neurobiol* 14(2), 169–77.

Cole AJ, Saffen DW, Baraban JM, Worley PF (1989) Rapid increase of an immediate early gene messenger RNA in hippocampal neurons by synaptic NMDA receptor activation. *Nature* 340(6233), 474–476.

De Weerd P, Peralta MR 3rd, Desimone R, Ungerleider LG (1999) Loss of attentional stimulus selection after extrastriate cortical lesions in macaques. *Nat Neurosci* 2(8), 753–758.

Doyon J, Penhune V, Ungerleider LG. (2003). Distinct contribution of the cortico-striatal and cortico-cerebellar systems to motor skill learning. *Neuropsychologia* 41, 252–262.

Evarts EV (1968) A technique for recording activity of subcortical neurons in moving animals. *Electroencephalogr Clin Neurophysiol* 24(1), 83–86.

Fink DJ, Glorioso J, Mata M (2003) Therapeutic gene transfer with herpes-based vectors: studies in Parkinson's disease and motor nerve regeneration. *Exp Neurol* 184(S1), S19–24.

Fischer A, Sananbenesi F, Schrick C, Spiess J, Radulovic J (2004) Distinct roles of hippocampal de novo protein synthesis and actin rearrangement in extinction of contextual fear. *J Neurosci* 24, 1962–1966.

Fuster JM (1973) Unit activity in prefrontal cortex during delayed-response performance: neuronal correlates of transient memory. *J Neurophysiol* 36(1), 61–78.

Georgopoulos AP, Schwartz AB, Kettner RE. Neuronal population coding of movement direction. *Science* 233(4771):1416–1419.

Gilbert CD, Minami Ito M, Kapadia M, Westheimer G (2000) Interactions between attention, context and learning in primary visual cortex. *Vision Res* 40, 1217–1226.

Goldberg TE, Weinberger DR (2004) Genes and the parsing of cognitive processes. *TINS* 8, 325–334.

Guzowski JF, McNaughton BL, Barnes CA, Worley PF (1999) Environment-specific expression of the immediate-early gene Arc in hippocampal neuronal ensembles. *Nat Neurosci* 2(12), 1120–1124.

Guzowski JF, McNaughton BL, Barnes CA, Worley PF. (2001) Imaging neural activity with temporal and cellular resolution using FISH. *Curr Opin Neurobiol* 11(5), 579–584.

Hebb (1949) *The organization of behavior: A neuropsychological theory.* New York, John Wiley & Sons.

Hebb MO, Robertson HA (1997) End-capped antisense oligodeoxynucleotides effectively inhibit gene expression in vivo and offer a low-toxicity alternative to fully modified phosphorothioate oligodeoxynucleotides. *Brain Res Mol Brain Res* 47, 223–228.

Heywood CA, Cowey A (1987). On the role of cortical area V4 in the discrimination of hue and pattern in macaque monkeys. *J Neurosci* 7(9), 2601–2617.

Hoffman KL, McNaughton BL (2002) Coordinated reactivation of distributed memory traces in primate neocortex. *Science* 297(5589), 2070–2073.

Hubel DH, Wiesel TN (1959) Receptive fields of single neurons in the cat's striate cortex. *J Physiol (Lond)* 148: 574–591.

Hubel DH, Wiesel TN (1962) Receptive fields, binocular interaction and functional architecture in the cats's visual cortex. *J Physiol (Lond)* 160, 106–154.

Izquierdo I, Cammarota M (2004) Zif and the survival of memory. *Science* 304, 829–830.

Kaas JH, Krubitzer LA, Chino YM, Langston AL, Polley EH, Nlair N (1990) Reorganization of retinotopic cortical maps in adult mammals after lesions of the retina. *Science* 248, 229–231.

Kaczmarek L, Robertson HA (2002) *Immediate early genes and inducible transcription factors in mapping of the central nervous system function and dysfunction.* Amsterdam: Elsevier Science B.V.

Kandel ER (2001) The molecular biology of memory storage: a dialogue between genes and synapses. *Science* 294:1030–1038.

Kuffler SW (1953) Discharge patterns and functional organization of mammalian retina. *J Neurophysiol* 16(1), 37–68.

Lee JL, Everitt BJ, Thomas KL (2004) Independent cellular processes for hippocampal memory consolidation and reconsolidation. *Science* 7, 839–843.

Lennie P (2000) Color vision. In Kandell ER Schwartz JH, Jessell TM (Eds). *Principles of neural science.* NY, McGraw-Hill, pp 572–589.

Liu Z, Richmond BJ, Murray EA, Saunders RC, Steenrod S, Stubblefield BK, Montague DM, Ginns EI (2004) DNA targeting of rhinal cortex D2 receptor protein reversibly blocks learning of cues that predict reward. *Proc Natl Acad Sci USA* 101, 12336–12341.

Logothetis NK, Wandell BA (2004) Interpreting the BOLD signal. *Annu Rev Physiol* 66, 735–769.

Logothetis NK (2003) MR imaging in the non-human primate: studies of function and of dynamic connectivity. *Curr Opin Neurobiol* 13, 1–13.

Lomber SG, Payne BR, Horel JA (1999) The cryoloop: an adaptable reversible cooling deactivation method for behavioral or electrophysiological assessment of neural function. *J Neurosci Methods* 86(2), 179–194.

Luft AR, Buitrago MM, Ringer T, Dichgans J, Schulz JB (2004) Motor skill learning depends on protein synthesis in motor cortex after training. *J. Neurosci* 24, 6515–6520.

Mansuy IM, Winder DG, Moallem TM, Osman M, Mayford M, Hawkins RD, Kandel ER (1998) Inducible and reversible gene expression with the rtTA system for the study of memory. *Neuron* 21:257–265.

Miesenbock G (2004) Genetic methods for illuminating the function of neural circuits. *Curr Opin Neurobiol* 14(3), 395–402.

Moran J, Desimone R (1985). Selective attention gates visual processing in the extrastriate cortex. *Science* 229, 782–784.

Mower AF, Liao DS, Nestler EJ, Neve RL, Ramoa AS (2002) cAMP/Ca^{2+} response element-binding protein function is essential for ocular dominance plasticity. *J Neurosci* 22:2237–2245.

Murray EA, Mishkin M (1998) Object recognition and location memory in monkeys with excitotoxic lesions of the amygdala and hippocampus. *J Neurosci* 18: 6568–6582.

Nakazawa K, Quirk MC, Chitwood RA, Watanabe M, Yeckel MF, Sun LD, Kato A, Carr CA, Johnston D, Wilson MA, Tonegawa S. (2002) Requirement for hippocampal CA3 NMDA receptors in associative memory recall. *Science* 297, 211–218.

Nicolelis MA, Dimitrov D, Carmena JM, Crist R, Lehew G, Kralik JD, Wise SP (2003) Chronic, multisite, multielectrode recordings in macaque monkeys. *Proc Natl Acad Sci U S A* 100(19), 11041–11046.

Ogawa S, Lee TM, Nayak AS, Glynn P (1990) Oxygenation-sensitive contrast in magnetic resonance image of rodent brain at high magnetic fields. *Magn Reson Med* 14(1):68–78.

Ogawa S, Pfaff DW (1996) Application of antisense DNA method for the study of molecular bases of brain function and behavior. *Behav. Genet* 26, 279–292.

Orban GA, Van Essen D, Vanduffel W (2004) Comparative mapping of higher visual areas in monkeys and humans *Trends Cogn Sci* 8(7), 315–324.

Pinaud R (2004) Experience-Dependent Immediate Early Gene Expression in the Adult Central Nervous System: Evidence from Enriched-Environment Studies. *Int J Neurosci* 114, 321–333.

Press WA, Olshausen BA, Van Essen DC (2001) A graphical anatomical database of neural connectivity. *Philos Trans R Soc Lond B Biol Sci* 356(1412), 1147–1157.

Qian Z, Gilbert ME, Colicos MA, Kandel ER, Kuhl D (1993) Tissue-plasminogen activator is induced as an immediate-early gene during seizure, kindling and long-term potentiation. *Nature* 361(6411), 453–457.

Ramirez-Amaya V, Vazdarjanova A, Mikhael D, Rosi S, Worley PF, Barnes CA (2005) Spatial exploration-induced arc mRNA and protein expression: evidence for selective, network-specific reactivation. *J Neurosci* 25(7), 1761–1768.

Ribeiro S, Cecchi GA, Magnasco MO, Mello CV (1998) Toward a song code: evidence for a syllabic representation in the canary brain. *Neuron* 21(2), 359–371.

Rumelhart DE (1987) *Parallel Distributed processing.* MIT Press.

Ryan L, Nadel L, Keil K, Putnam K, Schnyer D, Trouard T, Moscovitch M (2001). Hippocampal complex and retrieval of recent and very remote autobiographical memories: evidence from functional magnetic resonance imaging in neurologically intact people. *Hippocampus* 11(6):707–714.

Seidemann E, Zohary E, Newsome WT (1998) Temporal gating of neural signals during performance of a visual discrimination task. *Nature* 394(6688), 72–75.

Sprague JM, Berlucchi G, Antonini A (1985) Immediate postoperative retention of visual discriminations following selective cortical lesions in the cat. *Behav Brain Res* 17(2):145–162.

Tessier-Lavigne M (2000) Signal processing by the retina. In Kandell, ER, Schwartz JH, Jessell TM (Eds). *Principles of neural science.* NY, McGraw-Hill, pp 507–522.

Thielscher A, Neumann H (2005) Neural mechanisms of human texture processing: texture boundary detection and visual search. *Spat Vis* 18(2):227–257.

Thorpe WH (1956) *Learning and instincts in animals*. Cambridge, Massachussetts, Harvard University.

Tinsley R, Eriksson P (2004) Use of gene therapy in central nervous system repair. *Acta Neurol Scand* 109, 1–8.

Ungerleider LG (1995) Functional brain imaging studies of cortical mechanisms for memory. *Science* 270(5237):769–775.

Vazdarjanova A, Guzowski JF (2004) Differences in hippocampal neuronal population responses to modifications of an environmental context: evidence for distinct, yet complementary, functions of CA3 and CA1 ensembles. *J Neurosci* 24:6489–6496.

Wilson MA, McNaughton BL (1993) Dynamics of the hippocampal ensemble code for space. *Science* 261(5124), 1055–1058.

Wilson MA, McNaughton BL (1994) Reactivation of hippocampal ensemble memories during sleep. *Science* 265(5172), 676–769.

Wisden W, Errington ML, Williams S, Dunnett SB, Waters C, Hitchcock D, Evan G, Bliss TV, and Hunt SP (1990) Differential expression of immediate early genes in the hippocampus and spinal cord. *Neuron* 4, 603–614.

Wurtz RH (1969) Visual receptive fields of striate cortex neurons in awake monkeys. *J Neurophysiol* 32(5), 727–742.

Wurtz RH, Kandel ER (2000) Central visual pathways. In Kandell ER Schwartz JH, Jessell TM (Eds). *Principles of neural science*. NY, McGraw-Hill, pp 523–547.

Wurtz RH, Kandel ER (2000) Perception of motion, depth, and form. Kandell ER Schwartz JH, Jessell TM (Eds). *Principles of neural science*. NY, McGraw-Hill, pp548–571.

Xing J, Andersen RA (2000) Memory activity of LIP neurons for sequential eye movements simulated with neural networks. *J Neurophysiol* 84(2), 651–665.

Yang T, Maunsell JH (2004) The effect of perceptual learning on neuronal responses in monkey visual area V4. *J Neurosci* 24, 1617–1626.

Part I

Retinal and Thalamic Plasticity

2

Synaptic Plasticity and Structural Remodeling of Rod and Cone Cells

ELLEN TOWNES-ANDERSON AND NAN ZHANG

Department of Neurology and Neurosciences, New Jersey Medical School,
University of Medicine and Dentistry of New Jersey, NJ, USA

Introduction

It is well documented that neural activity affects synaptic plasticity, a phenomenon known as activity-dependent plasticity. Synaptic plasticity in response to injury, known as injury-induced plasticity, is also well recognized. The vertebrate photoreceptor cell displays plasticity, in the form of structural synaptic change, in both these situations. Structural synaptic change in the photoreceptor refers to 1) change in the size and shape of the rod or cone terminal, 2) increase or decrease in the number of presynaptic active zones, or 3) new growth of neurites and/or development of new presynaptic varicosities. As the first synapse in the visual pathway, structural changes in the synapse between photoreceptors and second order neurons may influence all subsequent visual processing. Thus, an understanding of the mechanisms that initiate and promote plasticity in the outer synaptic layer of the retina where rod and cone cells interact with horizontal and bipolar neurons is critical to understanding the plasticity of the visual system as a whole.

Activity-dependent change of the photoreceptor synapse will be briefly discussed first. Photoreceptor synaptic plasticity in response to disease, is a more recently recognized phenomenon and will be reviewed in more detail. Our own current work has focused on the mechanisms that might be involved in injury-induced plasticity and an overview of this work will be presented. Finally, some future directions and outstanding questions will be discussed.

Types of structural plasticity in photoreceptors

Activity-dependent plasticity

Neurotransmission by rod and cone cells occurs by calcium-dependent vesicle exocytosis. However, the photoreceptor synapse differs from most chemical

R. Pinaud, L. A. Tremere, P. De Weerd(Eds.), Plasticity in the Visual System:
From Genes to Circuits, 13-31, ©2006 Springer Science + Business Media, Inc.

synapses because it is controlled by graded membrane potential changes, not action potentials. Its presynaptic active zone is distinguished by the presence of a flat sheet, known as the ribbon, surrounded by a halo of vesicles and attached to the plasma membrane by an arciform density. In addition, there are several molecular differences between photoreceptor synapses and conventional synapses. For instance, retinal ribbon synapses contain no synapsin (Mandell et al., 1990) and L-type calcium channels in rod cells and L-type and cGMP-gated channels in cone cells control exocytosis and neurotransmitter release (Rieke and Schwartz, 1994; Schmitz and Witkovsky, 1997; Taylor and Morgans, 1998; Nachman-Clewner et al., 1999; Morgans, 2001) whereas other calcium channel types are present in conventional synapses. Other molecular differences include the presence of syntaxin 3 in ribbon synapses, instead of syntaxin 1 for vesicle fusion (Morgans et al., 1996), and the protein RIBEYE, a unique component of the ribbon which may form the backbone of the ribbon structure (Schmitz et al., 2000).

In response to activity, both the size of the presynaptic terminal and the configuration of the ribbon change. In the dark, when the photoreceptor is depolarized, the size of the terminal increases whereas in the light, the terminal size decreases. This is presumably due to the balance of membrane exo- and endocytosis. Endocytosis dominates in the light but synaptic vesicle exocytosis proceeds continuously in the dark in the presence of endocytosis (Schaeffer and Raviola, 1978). Thus, in the light, plasma membrane is removed from the terminal whereas in the dark, more membrane is added than removed. For the ribbons, there are changes in shape, length, and location. Such changes occur in a diurnal pattern in many species (reviewed by Vollrath and Spiwoks-Becker, 1996). In rat rod cells, for instance, ribbons are flat and attached to the plasmalemma in the dark; in the light, spherical ribbons that are disconnected from the plasmalemma are more abundant (Adly et al., 1999). The change in ribbon shape to a spherical form and/or disassociation from the active zone is presumably due to inactivity and has also been observed in squirrel cone cells during hibernation (Remé and Young, 1977) and in cultured photoreceptors where the terminal is separated from its post-synaptic partners (Townes-Anderson, personal observations, Fig. 2.1). The molecular mechanism for the dissociation of the active zone in photoreceptors is unknown. However, mutant mice, lacking the protein bassoon in functional form, have aggregations of ribbons detached from the active zone (Dick et al., 2003) suggestive of what is seen with a reduction of activity (Abe and Yamamoto, 1984). It is possible therefore that bassoon plays some role in ribbon configuration and/or attachment to the membrane.

Injury-induced plasticity

Although there are changes in the shape of the photoreceptor presynaptic ending and its synaptic ribbon in the adult retina, the terminal normally stays within the outer synaptic layer of the retina where it is in contact with post-

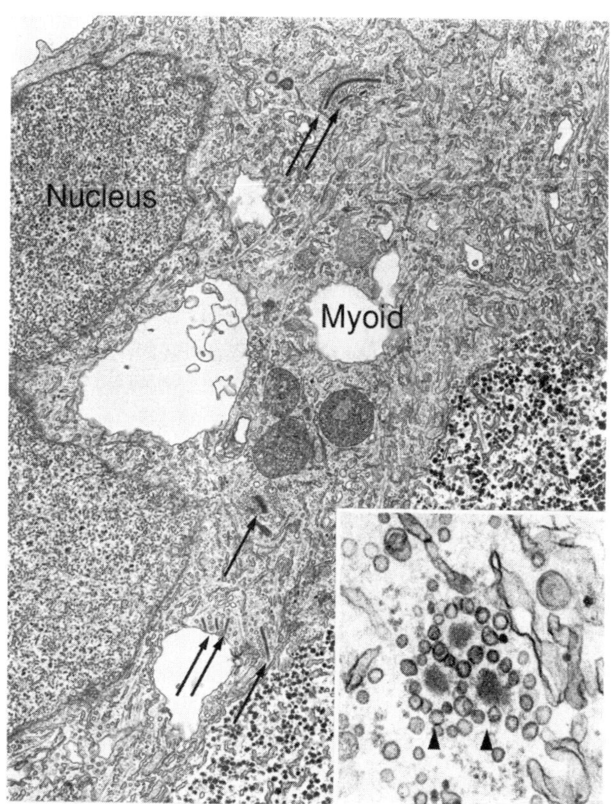

FIGURE 2.1. Movement and shape change are characteristics of ribbon plasticity. Electron micrographs from cultured cone cells isolated from the adult tiger salamander retina and maintained for 7 days in a defined medium. In vivo, cone terminals have multiple active zones and therefore ribbons. In cultured cone cells, ribbons are frequently observed in the myoid region (arrows), which normally contains only the cell organelles. The ribbons appear in linear form. Inset: Some ribbons, however, take on a spherical shape (arrowheads). Here three such ribbons share synaptic vesicles. It is highly likely that the spherical ribbons are detached from membrane. For linear ribbons, serial sectioning is necessary to determine whether or not they have remained associated with the plasma membrane.

synaptic bipolar and horizontal cell processes. In 1995, examination of several human retinas affected with retinitis pigmentosa (RP) revealed the presence of neurites coming from rod cells and extending into the inner retina (Li et al., 1995). These neurites could be observed because of the abnormally high levels of rod opsin in the plasma membrane of the cell. Some mutations of the opsin gene cause mislocalization of opsin because the C-terminus, which targets opsin to vesicles destined for the outer segment, is mutated.

In other cases, mislocalization of opsin occurs as the outer segment becomes disorganized and/or shortened. Opsin containing vesicles may then fuse with plasma membrane in the cell body by default. Finally, some conditions may lead to a fusion of the inner and outer segment membranes which allows a back flow of opsin from the outer to the inner segment and the rest of the cell (Spencer et al., 1988; Townes-Anderson, 1995). Regardless of the mechanism of mislocalization, the high density of the opsin protein in the plasmalemma of patients with RP demonstrated the extent of neurite growth and that the growth came from rod cell terminals. When stained for the presence of synaptic vesicles, the neurites were seen to have swellings along their length filled with synaptic vesicles. However, electron microscopy did not reveal differentiated synaptic contacts between these presynaptic varicosities and inner retinal neurons.

Retinitis pigmentosa is an inherited form of retinal degeneration that affects rod cells which degenerate and die leading to night blindness; cone cells die after rod cells have gone and the loss of cone cells results in total blindness. The mutations, which lead to blindness, occur in many different genes; most of them are for rod cell proteins (reviewed by Molday, 1998). The ectopic rod cell neurites were observed initially in retinas with autosomal dominant, X-linked, simplex and multiplex forms of the disease (Li et al., 1995; Milam et al., 1996). Since then, sprouting by rod cells has been observed in human retinas after laser photocoagulation, in age-related retinal degeneration, in retinal detachment, and in several animal models of retinal injury and disease (Table 2.1). In all cases, neurites were observed by immunolabeling for opsin; when synaptic immunolabels were applied, all presynaptic swellings were positive for synaptic vesicle proteins. Rodent models of retinal degeneration, however, do not show neuritic sprouting (Li et al., 1995). It has been suggested, for these models, that neurites are not formed due to the rapidity of the disease which does not provide time for neurite formation before cell death.

Rod axons have been reported to retract towards the cell body in response to injury as well. Initially observed as the early stages of rod cell degeneration several days after retinal detachment (Erickson et al., 1983), it is now apparent that axonal retraction occurs very soon after the detachment injury (Lewis et al., 1998) and that it does not necessarily lead to cell death: virtually all rod axons retract but not all rod cells die if reattachment is done in a timely fashion (Mervin et al., 1999). After reattachment of retina, rod cells have been shown, using anti-opsin immunolabeling, to extend neurites into the inner retina in a manner and with a morphology similar to what is seen in retinal disease (Table 2.1). Whether the same cell both retracts and extends neurites is likely and has been observed directly in cultured rod cells using video time-lapse microscopy (Nachmen-Clewner and Townes-Anderson, 1996).

Cone cells are affected by retinal detachment but appear not to dramatically retract their terminals. Instead the terminal shape changes by flattening so that the synaptic invagination is lost. This has been observed both in vivo and in vitro (Fisher et al., 2001; Khodair et al., 2003).

TABLE 2.1. Rod cell axonal plasticity in vivo.

Species	Response	Insult or disease	Reference
human	neurite sprouting/varicosity formation	retinitis pigmentosa (RP)	Li et al. 1995 Milam et al. 1996
human	neurite sprouting	laser irradiation in diabetic retina	Xiao et al. 1998
human	neurite sprouting/varicosity formation	reattachment after detachment	Lewis et al. 2002a
human	neurite sprouting	late-onset retinal degeneration	Gupta et al. 2003
human	neurite sprouting	age-related macular degeneration	Gupta et al. 2003
human	neurite sprouting/varicosity formation	detachment with proliferative retinopathy	Sethi et al. 2005
pig	filopodial growth	transgenic for RP	Li et al. 1998
cat	neurite sprouting	rod/cone dysplasia	Chong et al. 1999
cat	neurite sprouting/varicosity formation	reattachment after detachment	Lewis et al. 2002a
human	retraction of spherule	detachment after reattachment	Fisher and Lewis 2003
cat	retraction of spherule	3d after detachment	Erickson et al. 1983
cat	retraction of spherule	24hr after detachment	Lewis et al. 1998
mouse	increased synaptic contacts	rds mutation, homozygous	Jansen and Sanyal 1984 Jansen et al. 1997
mouse	increased synaptic contacts	rds mutation, heterozygous	Jansen and Sanyal 1992 Jansen at el. 1997
mouse	increased synaptic contacts	rd/wild type chimeras	Sanyal et al. 1992
mouse	increased synaptic contacts	constant light	Jansen and Sanyal 1987 Jansen at al. 1997
mouse	increased synaptic contacts	KO, cone cGMP channel & rod opsin	Claes et al. 2004

After retinal reattachment and in most diseases, cone cells do not form long neurites although smaller sprouts and filopodial extensions into the outer plexiform layer and sometimes into the inner retina have been described in retinitis pigmentosa (Table 2.2). These processes were observed with antibodies to cone transducin-α and synaptic proteins. It is possible that these markers are not as effective as rod opsin in highlighting cone synaptic change. In retinal detachment, for instance, the expression of many cone-specific proteins declines within a few days (Rex et al., 2002). Nonetheless, available evidence indicates that there is a significant difference in the response of rod and cone cell terminals to injury and disease.

TABLE 2.2. Cone cell axonal plasticity in vivo.

Species	Response	Insult or disease	Reference
human	axon elongation	retinitis pigmentosa (RP)	Li et al. 1995 Milam et al. 1996
cat	axon elongation	reattachment after detachment	Lewis et al. 2003
mouse	neurite sprouting/varicosity formation	rd1 mutation	Fei 2002
human	enlarged terminals	cone-rod dystrophy	Gregory-Evans et al. 1998
pig	increased synaptic contacts	transgenic for RP	Peng et al. 2000
mouse	increased synaptic contacts	rd1 mutation	Peng et al. 2000
rat	increased synaptic contacts	RCS strain	Peng et al. 2003
rat	increased synaptic contacts	transgenic for RP	Cuenca et al. 2004

At present, there is one instance where long neurites have been reported to come from cone cells (Fei, 2002). This is the mouse model rd1, which is an autosomal recessive type of retinal degeneration with a mutation in the beta subunit of rod phosphodiesterase. The cone processes were visualized using transgenic mice with GFP linked to a cone promoter. The presence of GFP delineated long neurites with varicosities going both into the inner retina and horizontally in the outer plexiform layer. These neurites arose from multiple locations, cone terminals, axons and cell bodies. Canine models of this type of RP (Aquirre et al., 1978) have not yet been examined for cone, or rod, cell synaptic plasticity.

Finally, photoreceptors have been reported to make new synapses in response to retinal degeneration (Tables 2.1 and 2.2). Within the outer plexiform layer, rod cells will increase the number of synapses they make with existing postsynaptic bipolar dendrites in the rd and rds mouse models of RP, after light damage, and in a rod opsin knockout mouse. It is possible that as adjacent rod cells die, remaining cells try to compensate by increasing their synaptic input to second order neurons. Cone cells also have been reported to form new synapses. In transgenic pigs with a mutation in the opsin gene, in the rd1 mouse, and in the RCS and P23H transgenic rat (Table 2.2), all models of RP, cone cells make synapses with rod bipolar cells that would not be present in the normal, healthy mammalian retina.

In summary, several forms of structural plasticity are present in the photoreceptors of injured (detachment, reattachment and excess light) and diseased (retinitis pigmentosa and age-related macular degeneration) retina: retraction of axons, sprouting of axons, neurite formation, development of presynaptic varicosities, and synaptogenesis (Fig. 2.2). These structural changes require special morphological techniques to be discerned and would not be easily visible in routine histopathology. They can occur in the absence of an outer segment and in the presence of mislocalized opsin but it is not known if these are requirements. They can be a prelude to cell death but not necessarily. Finally, it is not known if these changes are permanent or

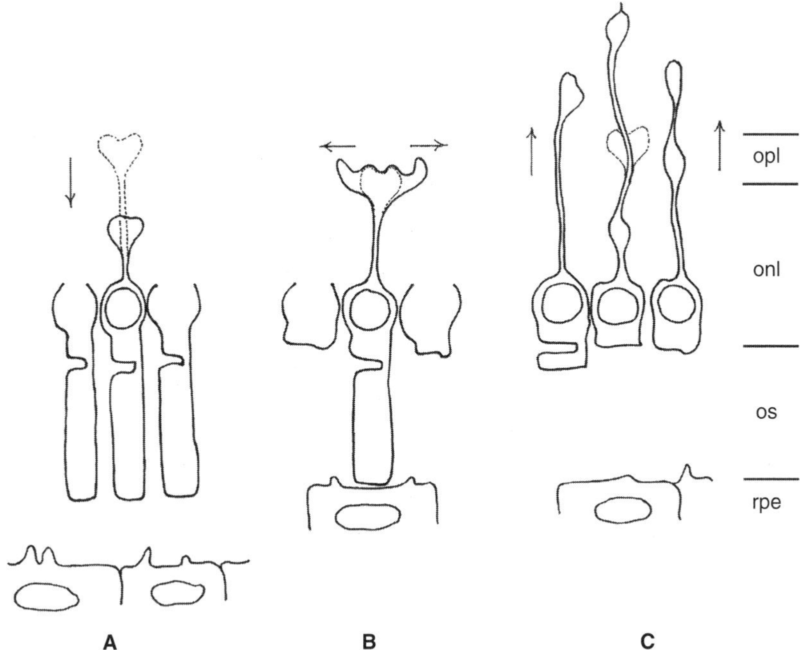

FIGURE 2.2. Summary of structural plasticity in adult photoreceptors. Rod cells show three general types of plasticity: A. retraction of the axon and terminal; B. enlargement of the terminal concomitant with new synaptic development; and C. neurite sprouting and varicosity development. Retraction occurs after retinal detachment, terminal enlargement occurs in retinal degeneration, and sprouting and varicosity formation have been seen in all types of degeneration and injury. Dotted lines indicate original synaptic outline before structural change. Arrows indicate overall direction of change. Cone cells (not shown) produce less plasticity. They do not exhibit retraction but do increase synaptic interaction in retinal disease. In one type of retinal degeneration (rd1) they produce varicosity-containing neurites similar to what is produced by rod cells. Opl, outer plexiform layer; onl, outer nuclear layer; os, outer segment layer; rpe, retinal pigmented epithelium.

transient or whether they can completely resolve or revert back to pre-injury morphology after therapeutic intervention (Lewis et al., 2002a, b, 2003).

Mechanisms producing structural plasticity in photoreceptors

Calcium

Because the plasticity described for rod and cone terminals is different, a key assumption that can be made about the mechanisms of structural plasticity is that they differ between the two sensory receptor types. A known difference between cone and rod terminals is the population of calcium channels that are present at the presynaptic active zone. In mammals, rod terminals have L-type channels with alpha-1F subunits whereas cone terminals have L-type channels with alpha-1D subunits (Taylor and Morgans, 1998; Morgans, 2001). Moreover, experiments in amphibians have shown that cone cells have cGMP-gated channels in their terminals which control exocytosis (Rieke and Schwartz, 1994; Savchenko et al., 1997).

Blockage of L-type and cGMP-gated channels affects plasticity. Retraction in rod axons can be prevented by blocking L-type channels (Nachman-Clewner et al., 1999) in cultured salamander rod cells. It is not known whether blocking L-type channels also affects the more subtle shape changes seen in cone cells. If blockage prevents retraction, it follows that influx of calcium must help initiate this activity. Retinal detachment, the injury associated with retraction, is accompanied by spreading depression, a depolarization that affects ion influxes and presumably would open L-type calcium channels. Thus, a possible scenario is that detachment causes an influx of calcium into the rod axon terminal and initiates cytoskeletal changes, which may include contraction of actinomyosin, that result in movement of the axon and terminal towards the cell body.

In cultures of salamander photoreceptors, about 24hrs after retraction of the axon, new filopodial growth is seen (Nachman-Clewner and Townes-Anderson, 1996). After 3 days in culture, thicker neuritic processes have formed and presynaptic varicosities start to develop. The outgrowth of rod processes is prevented by blocking L-type channels with nicardipine after 1 day in culture (Zhang and Townes-Anderson, 2002). Thus, the activity of L-type channels is also critical for development of neuritic structures in rod cells. In addition, blocking L-type channels prevents an increase in the production of the synaptic vesicle proteins SV2 and synaptophysin, which normally occurs by day 3 (Zhang and Townes-Anderson, 2002). This effect is specific for synaptic protein: blocking L-type channels has no effect on opsin synthesis. Other changes undoubtedly also occur with reduced calcium influx through L-type channels. Cone cell growth, however, is not noticeably affected. Instead, an antagonist to cGMP-gated channels, L-cis diltiazem, significantly reduces cone cell growth and the formation of varicosities (Zhang and Townes-Anderson, 2002). It appears, therefore, that both rod and cone cells require calcium influx for new neuritic development. The amount of required influx is probably either very small or highly localized since no effort was made to activate calcium channels in culture. Thus, the basal level of channel opening is adequate to sustain growth. Significantly, the channel *type* that primarily controls structural plasticity in rod and cone cells is different. With respect to disease and injury, one might speculate that because of the functional differences in these channels, one activated by voltage and the other by cGMP, the rod and cone photoreceptors will respond differently to the same insult depending on whether or not the function of L-type and/or cGMP-gated channel activity is affected.

cGMP

Since cGMP-gated channels appeared to be involved in cone cell axonal plasticity, it was reasonable to test whether increases in cGMP would promote cone cell growth. Application of an analog to cGMP, 8-bromo-cGMP caused an increase in varicosity formation in cone but not rod cells (Zhang and Townes-Anderson, 2002). More recently, a systematic investigation of the nitric oxide (NO)-cGMP signaling pathway has been made using agonists and antagonists to the components of the pathway (Zhang et al., 2005). NO is produced by nitric oxide synthase which can be stimulated by calcium; NO stimulates soluble guanylyl cyclase (sGC) to produce cGMP; cGMP in turn can activate 1) phosphodiesterase (PDE) to hydrolyze cGMP to 5'-GMP, 2) protein kinase G (PKG), and 3) cGMP-gated cationic channels. Most of the components of the NO-cGMP pathway are known to exist in cone cells. NOS is present in photoreceptors as well as adjacent Müller cells and other retinal neurons (Liepe et al., 1994; Habrecht et al., 1998). Soluble GC is also present in cone cells (Habebrect et al., 1998). Activating the pathway causes an increase in cone growth and inhibiting the pathway causes a decrease in growth (Zhang et al., 2005). The growth in cone cells presumably occurs because of calcium influx through cGMP-gated channels but also because of phosphorylation of protein by PKG. Inhibiting PKG using the cGMP analogue Rp-8-pCPT-cGMPs caused a decrease in the number of processes produced by cone cells. At present, we do not know what type of PKG is present in cone cells, nor what substrates it phosphorylates. It is possible that one of its targets is the cGMP-gated channel itself.

The effects of cGMP vary depending on the concentration of the reagent used. In cone cells, high concentrations of cGMP result in either no growth or inhibition of growth compared with lower levels of cGMP (Zhang et al., 2005). This suggests the presence of a feed-back loop which controls levels of cGMP in these photoreceptors. Such feedback could be provided by a cGMP-stimulated PDE. A similar feedback loop for cGMP exists in cerebellar neurons where it may play a role in controlling the form of synaptic plasticity known as long-term depression (Shimizu-Albergine et al., 2003).

Surprisingly and in contrast to cone cells, activation of the cGMP pathway in rod cells decreases growth, at all concentrations, whereas inhibiting the pathway increases rod cell growth. Thus, rod cells are also likely to contain the components of the NO-cGMP pathway although its effects on structural plasticity are different than those in cone cells. NOS is present in the inner segment of rod cells (Liepe et al., 1994). And we have recently been able to demonstrate the presence of sGC in cultured salamander rod cells (Zhang et al., 2005; Fig. 2.3). The enzyme is found not only in the cell body but in newly formed varicosities as well. Whether it is also present in the original axon terminal, before growth begins, is not known. It has been suggested that activation of the NO-cGMP pathway results in phosphorylation of L-type channels and their closure (Chik et al., 1995; Barnstable et al., 2004; Kourennyi et al., 2004).

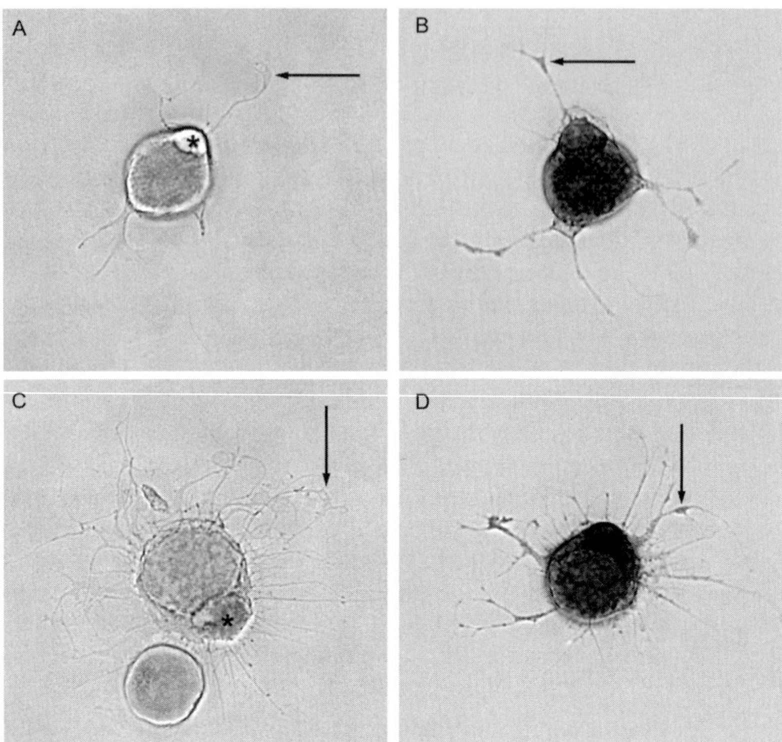

FIGURE 2.3. Both rod and cone cells contain soluble guanylyl cyclase (sGC) in the salamander. Bright field light micrographs from 3 day old cultures treated with an antibody against the alpha1 subunit of sGC (gift of Dr. Ari Sitaramayya) and ABC HRP immunocytochemistry. Photoreceptors have formed numerous new neuritic processes and created swellings along their length filled with synaptic vesicles, called presynaptic varicosities (arrows). *, the ellipsoid, a collection of mitochondria which is a characteristic of amphibian photoreceptors.

Cone cells: A. control without primary antibody, B. staining with anti-alpha1 antibody. In control cells, there was very little staining; with the antibody, staining was present in the cell soma as well as newly developed presynaptic varicosities.

Rod cells: C. control staining, D. staining with anti-alpha1 antibody. There was little staining in the control cells but dark staining in the cell soma and presynaptic varicosities when anti-alpha 1 was applied.

Reduced calcium influx would lead to reduced growth and account at least in part for the negative effects of cGMP on rod cell plasticity.

Rod and cone cells also contain the membrane form of guanylyl cyclase (pGC) in their synaptic terminals. Thus, pGC exists both in outer segments of photoreceptors where it functions in the phototransduction cascade, and in the axon terminal (reviewed by Duda and Koch, 2002). Cyclic GMP could therefore be increased both by NO activation of sGC and by pGC. Extracellular ligands that activate the pGC receptor are not known but calcium-binding proteins that control pGC activity are present in rod and cone terminals (Cuenca et al., 1998). How the membrane form of GC affects structural plasticity in photoreceptors remains to be determined.

The importance of the results regarding cGMP lies in their potential ability to explain the structural plasticity observed in rd1 mice. In this animal model of RP, the mutation in the beta subunit of rod PDE 6 results in very low PDE activity and high levels of cGMP in the photoreceptor layers (Lolley, 1994; Farber, 1995). Cone cells have long neuritic extensions (Fei, 2002) but rod cells have incomplete synaptic development (Blanks et al., 1974). This is precisely what increased levels of cGMP produced in cultured cells: increased cone cell growth and decreased rod cell growth. Since cone PDE is not affected in this form of retinal degeneration, cGMP must increase in cone cells by another route. It is possible that cGMP moves from rod to cone terminals via the gap junctions between these photoreceptors (Raviola and Gilula, 1973).

cAMP

Cyclic AMP also appears to play a role in photoreceptor plasticity. An analogue of cAMP, Sp-cAMP increases growth of processes and development of presynaptic varicosities in rod cells but does not significantly affect cone cells in salamander retinal cell cultures (Townes-Anderson et al., 2003). This suggests that cGMP-dependent processes stimulate cone cell growth whereas rod cells use cAMP-dependent processes. A different analogue of cAMP, pCPT-cAMP, has been applied to an in vitro model of photoreceptor transplantation. Sheets of photoreceptors that would be used as transplants are prepared by vibratomy from pig retina and maintained in culture (Khodair et al., 2003). In these sheets, axon retraction occurs as viewed by both synaptic vesicle and rod opsin antibodies (Khodair et al., 2003). Application of the cAMP analogue prevents this retraction; the effect is reversed if the analogue is removed (Khodair et al., 2005). In other cell types, neurite retraction is caused by cytoskeleton activity controlled by a small GTPase, RhoA (Jalink et al., 1994; Kozma et al., 1997; Lehmann et al., 1999). Cyclic AMP can inhibit RhoA and thereby inhibit retraction (Schoenwaelder and Burridge, 1999). The mechanism for retraction blockade in photoreceptors may therefore involve RhoA. RhoA has recently been immunolocalized in retina and is present in rod and cone terminals as well as the inner segments (Fontainhas et al., 2004). Cyclic GMP can also inhibit RhoA (Sauzeau et al., 2000) and high levels of cGMP in cone cells may allow cone cell growth in part because of the inhibition of RhoA.

Involvement of cAMP in rod cell plasticity in vitro is of interest because of the known increase in cAMP in some forms of retinal degeneration (Sanyal et al., 1984; Weiss et al., 1995; Nir et al., 2001; Traverso et al., 2002). If cAMP increases in the rod terminal, this may lead to the neurite extension seen in RP. Similar increases in cone cells, if they occur, would not result in similar growth if results from cultures are correct. And indeed there is little growth of cone cell terminals in RP. The reason for increased cAMP in disease is not known, however, it has been suggested that mislocalized opsin, which occurs in retinal disease and injury, when activated by light, can interact with signaling pathways in the inner segment that lead to stimulation of adenylyl cyclase (Alfinito and Townes-Anderson, 2002).

Extracellular Factors

Both different synaptic calcium channels and dependence on different cyclic nucleotide pathways could play a role in the differences observed in rod and cone cell structural plasticity. These might be considered intrinsic differences between the photoreceptor cell types. Environmental stimuli that activate or inhibit these pathways would then modify the expression of the plasticity. For instance, dopamine and adenosine released by the inner retina in response to light and dark affect the activity of photoreceptor adenylyl cyclase through their receptors. These neuromodulators may therefore affect the extent or time course of reactive neurite extension in rod cells. At present, however, the only transmitter that has been examined for possible effects on photoreceptor synaptic plasticity is GABA. In cultures where all retinal cell types are present, photoreceptors grow and contact GABAergic cells more often than other cell types (Sherry et al., 1996). In vivo, in retinas affected with RP, extensive sprouting by GABAergic amacrine cells has been observed and the rod neuritic sprouts appeared to grow towards and contact GABAergic amacrine cell somata preferentially (Fariss et al., 2000). The mechanism for this attraction is not known. Müller cells have also been suggested to be attractive to rod cell neuritic growth in RP retinas (Li et al., 1995). In one case of human autosomal dominant RP, synapse-like structures were observed between rod varicosities and Müller cells (Milam et al., 1996). The morphology of the new rod and cone growth strongly suggests that the inner retina is attractive: in all human and animal retinas where neuritic growth is observed, the processes extend into the inner retina with only one exception. In the exception, a retina with late-onset retinal degeneration, rod neurites filled the subretinal space (Gupta et al., 2003) and were not present in the inner retina. In salamander cultures, preliminary results indicate that the chemokine stromal-derived factor-1α (SDF-1α) promotes rod photoreceptor cell neuritic growth (Zhang and Townes-Anderson, unpublished results). These results are consistent with the concept that neuritic growth of rod cells requires activation of the cAMP signaling pathway. The receptor for SDF-1α is CXCR4 which is a G-protein-coupled receptor which can activate cAMP signaling (Chalasani et al., 2003). However, possible extrinsic factors that can influence plasticity of photoreceptors in either the inner or outer retina remain largely unexplored.

Although new rod and cone neuritic growth contains presynaptic varicosities, mature synapses are not observed between these varicosities and inner retinal neurons (Li et al., 1995). New synapses, however, are made between photoreceptors and bipolar cells in retinal disease: rod cells increase the number of synaptic contacts with bipolar cells with whom they are already in contact whereas cone cells have been shown to make synapses with new bipolar cells including neighboring rod bipolars denervated by rod cell death (Tables 2.1 & 2.2). So, synaptogenesis appears to be confined to the outer plexiform layer with cone cells innervating novel partners. This description of synaptogenesis in retinal disease is in sharp contrast to results from transplant studies where subretinal placement of photoreceptor cells does not yield synaptic interaction between the grafted photoreceptors and the host bipolar cells (reviewed by Lund et al., 2001). Most transplant paradigms use animals with rod-rich retinas. It is possible that cone cells will be more likely to make new synapses with host/novel bipolar cells. Recent work in culture suggests that both rod and cone cells will grow processes towards identified bipolar cells; moreover, cone cells seem to have a preference for bipolar over multipolar cells (Clarke et al., 2004). This is in contrast to the preference of rod cells for GABAergic multipolar cells (Sherry et al., 1996; Clarke et al., 2004).

Future Directions

The effects of structural plasticity at the rod and cone terminal on visual processing are not known. However, it is difficult to imagine that sprouting and differentiation of new neurites as well as retraction will not change retinal processing. Indeed, electroretinograms of human and animal degenerate retinas indicate that there are postreceptoral abnormalities (Milam et al., 1996; Banin et al., 1999), although the precise cause of these changes is not yet clear.

To preserve normal vision it may be necessary to prevent plasticity in diseases and after retinal injury. Knowing the intrinsic and external stimuli for synaptic change will make it possible to develop strategies to preserve stability. There has already been some interest in calcium channel blockers as a preventive of rod cell degeneration. Although these blockers have had mixed results regarding the delay of degeneration (Frasson et al., 1999; Pearce-Kelling et al., 2001), it would be of interest to know if they prevent synaptic plasticity since both rod and cone plasticity depends on calcium influx. Drugs which affect the NO-cGMP pathway may also, unintentionally, affect photoreceptor synapses. Viagra, for instance, which blocks PDE 5 and possibly also PDE 6 activity, increases cGMP levels and has been shown to have some effects on postreceptoral elements in the outer retina (Jägle et al., 2004). Are these drugs inducing synaptic plasticity in retinal cells? Thus both to protect normal vision and to reduce the deleterious effects of disease continued exploration of the mechanisms of plasticity is needed.

Calcium, cGMP, and cAMP have been demonstrated recently to play interconnected roles in axonal growth and guidance of cortical and spinal cord neurons in developing mammalian and amphibian CNS and in invertebrate nerve cells (Song et al., 1998; van Wagenen and Rehder, 1999; Polleux et al., 2000; van Wagenen and Rehder, 2001; Xiang et al., 2002). Increasing cyclic nucleotides can increase growth toward certain guidance molecules or make what was a repulsive guidance factor into an attractive one. The ratio of cAMP to cGMP may influence growth in developing neurons (Nishiyama et al., 2003). In addition, numerous external factors guide axonal and dendritic growth during development. These factors include molecules known as guidance factors but also chemokines and neurotransmitters (Xiang et al., 2002; Lipton and Kater, 1989). Many of the findings observed in developing nervous tissue could be easily tested in adult retina and with adult retinal neurons. Interestingly, in ferret retinal development, rod and cone cells extend long neurites, that look similar to processes extended in adulthood after injury and disease, into the inner retina (Johnson et al., 1999). As development proceeds and the outer synaptic layer forms, the processes are retracted. The ferret may provide an additional and somewhat unique model of photoreceptor plasticity. Although regeneration does not faithfully follow the details of ontogeny, it nonetheless appears that if reactive plasticity was viewed as a facet of developmental neurobiology, it would result in new and productive lines of research.

References

H. Abe and T. Y. Yamamoto, Diurnal changes in synaptic ribbons of rod cells of the turtle, *J Ultrastruct Res* **86**(3), 246–51 (1984).

M. A. Adly, I. Spiwoks-Becker and L. Vollrath, Ultrastructural changes of photoreceptor synaptic ribbons in relation to time of day and illumination, *Invest Ophthalmol Vis Sci* **40**(10), 2165–72 (1999).

P. D. Alfinito and E. Townes-Anderson, Activation of mislocalized opsin kills rod cells: a novel mechanism for rod cell death in retinal disease, *Proc Natl Acad Sci U S A* **99**(8), 5655–60 (2002).

G. Aquirre, D. Farber, R. Lolley, R. T. Fletcher and G. J. Chader, Rod-cone dysplasia in Irish setters: a defect in cyclic GMP metabolism in visual cells, *Science* **201**(4361), 1133–4 (1978).

E. Banin, A. V. Cideciyan, T. S. Aleman, R. M. Petters, F. Wong, A. H. Milam and S.G. Jacobson, Retinal rod photoreceptor-specific gene mutation perturbs cone pathway development, *Neuron* **23**(3), 549–57 (1999).

C. J. Barnstable, J. Y. Wei and M. H. Han, Modulation of synaptic function by cGMP and cGMP-gated cation channels, *Neurochem Int* **45**(6), 875–84 (2004).

J. C. Blanks, A. M. Adinolfi and R. N. Lolley, Photoreceptor degeneration and synaptogenesis in retinal-degenerative (rd) mice, *J Comp Neurol* **156**(1), 95–106 (1974).

S. H. Chalasani, K. A. Sabelko, M. J. Sunshine, D. R. Littman and J. A. Raper, A chemokine, SDF-1, reduces the effectiveness of multiple axonal repellents and is required for normal axon pathfinding, *J Neurosci* **23**(4), 1360–71 (2003).

C. L. Chik, Q. Y. Liu, B. Li, E. Karpinski and A. K. Ho, cGMP inhibits L-type Ca^{2+} channel currents through protein phosphorylation in rat pinealocytes, *J Neurosci* **15**(4), 3104–9 (1995).

N. H. Chong, R. A. Alexander, K. C. Barnett, A. C. Bird and P. J. Luthert, An immunohistochemical study of an autosomal dominant feline rod/cone dysplasia (Rdy cats), *Exp Eye Res* **68**(1), 51–7 (1999).

E. Claes, M. Seeliger, S. Michalakis, M. Biel, P. Humphries and S. Haverkamp, Morphological characterization of the retina of the CNGA3(-/-)Rho(-/-) mutant mouse lacking functional cones and rods, *Invest Ophthalmol Vis Sci* **45**(6), 2039–48 (2004).

R. Clarke, K. Hognason, R. Chawla and E. Townes-Anderson Photoreceptor interactions with second- and third-order retinal neurons after optical tweezer micromanipulation, ARVO: abstract#5368 (2004).

N. Cuenca, S. Lopez, K. Howes and H. Kolb, The localization of guanylyl cyclase-activating proteins in the mammalian retina, *Invest Ophthalmol Vis Sci* **39**(7), 1243–50 (1998).

N. Cuenca, I. Pinilla, Y. Sauve, B. Lu, S. Wang and R. D. Lund, Regressive and reactive changes in the connectivity patterns of rod and cone pathways of P23H transgenic rat retina, *Neuroscience* **127**(2), 301–17 (2004).

O. Dick, S. tom Dieck, W. D. Altrock, J. Ammermuller, R. Weiler, C. C. Garner, E. D. Gundelfinger and J. H. Brandstätter, The presynaptic active zone protein bassoon is essential for photoreceptor ribbon synapse formation in the retina, *Neuron* **37**(5), 775–86 (2003).

T. Duda and K. W. Koch, Calcium-modulated membrane guanylate cyclase in synaptic transmission?, *Mol Cell Biochem* **230**(1–2), 107–16 (2002).

P. A. Erickson, S. K. Fisher, D. H. Anderson, W. H. Stern and G. A. Borgula, Retinal detachment in the cat: the outer nuclear and outer plexiform layers, *Invest Ophthalmol Vis Sci* **24**(7), 927–42 (1983).

D. B. Farber, From mice to men: the cyclic GMP phosphodiesterase gene in vision and disease. The Proctor Lecture, *Invest Ophthalmol Vis Sci* **36**(2), 263–75(1995).

R. N. Fariss, Z. Y. Li and A. H. Milam, Abnormalities in rod photoreceptors, amacrine cells, and horizontal cells in human retinas with retinitis pigmentosa, *Am J Ophthalmol* **129**(2), 215–23 (2000).

Y. Fei, Cone neurite sprouting: an early onset abnormality of the cone photoreceptors in the retinal degeneration mouse, *Mol Vis* **8**, 306–14 (2002).

S. K. Fisher, J. Stone, T. S. Rex, K. A. Linberg and G. P. Lewis, Experimental retinal detachment: a paradigm for understanding the effects of induced photoreceptor degeneration, *Prog Brain Res* **131**, 679–98 (2001).

S. K. Fisher and G. P. Lewis, Müller cell and neuronal remodeling in retinal detachment and reattachment and their potential consequences for visual recovery: a review and reconsideration of recent data, *Vision Res* **43**(8), 887–97 (2003).

A. M. Fontainhas, N. Zhang, M. A. Khodair and E. Townes-Anderson Possible involvement of RhoA in synaptic plasticity of photoreceptors, ARVO: abstract#3646 (2004).

M. Frasson, J. A. Sahel, M. Fabre, M. Simonutti, H. Dreyfus and S. Picaud, Retinitis pigmentosa: rod photoreceptor rescue by a calcium-channel blocker in the rd mouse, *Nat Med* **5**(10), 1183–7 (1999).

K. Gregory-Evans, R. N. Fariss, D. E. Possin, C. Y. Gregory-Evans and A. H. Milam, Abnormal cone synapses in human cone-rod dystrophy, *Ophthalmology* **105**(12), 2306–12 (1998).

N. Gupta, K. E. Brown and A. H. Milam, Activated microglia in human retinitis pigmentosa, late-onset retinal degeneration, and age-related macular degeneration, *Exp Eye Res* **76**(4), 463–71 (2003).

M. F. Haberecht, H. H. Schmidt, S. L. Mills, S. C. Massey, M. Nakane and D. A. Redburn-Johnson, Localization of nitric oxide synthase, NADPH diaphorase and soluble guanylyl cyclase in adult rabbit retina, *Vis Neurosci* **15**(5), 881–90 (1998).

H. Jägle, C. Jägle, L. Sérey, A. Yu, A. Rilk, B. Sadowski, D. Besch, E. Zrenner and L. T. Sharpe, Visual short-term effects of Viagra: double-blind study in healthy young subjects, *Am J Ophthalmol* **137**(5), 842–9 (2004).

K. Jalink, E. J. van Corven, T. Hengeveld, N. Morii, S. Narumiya and W. H. Moolenaar, Inhibition of lysophosphatidate- and thrombin-induced neurite retraction and neuronal cell rounding by ADP ribosylation of the small GTP-binding protein Rho, *J Cell Biol* **126**(3), 801–10 (1994).

H. G. Jansen and S. Sanyal, Development and degeneration of retina in rds mutant mice: electron microscopy, *J Comp Neurol* **224**(1), 71–84 (1984).

H. G. Jansen and S. Sanyal, Synaptic changes in the terminals of rod photoreceptors of albino mice after partial visual cell loss induced by brief exposure to constant light, *Cell Tissue Res* **250**(1), 43–52 (1987).

H. G. Jansen and S. Sanyal, Synaptic plasticity in the rod terminals after partial photoreceptor cell loss in the heterozygous rds mutant mouse, *J Comp Neurol* **316**(1), 117–25 (1992).

H. G. Jansen, R. K. Hawkins and S. Sanyal, Synaptic growth in the rod terminals of mice after partial photoreceptor cell loss: a three-dimensional ultrastructural study, *Microsc Res Tech* **36**(2), 96–105 (1997).

P. T. Johnson, R. R. Williams, K. Cusato and B. E. Reese, Rods and cones project to the inner plexiform layer during development, *J Comp Neurol* **414**(1), 1–12 (1999).

M. A. Khodair, M. A. Zarbin and E. Townes-Anderson, Synaptic plasticity in mammalian photoreceptors prepared as sheets for retinal transplantation, *Invest Ophthalmol Vis Sci* **44**(11), 4976–88 (2003).

M. A. Khodair, M. A. Zarbin and E. Townes-Anderson, Cyclic AMP prevents retraction of axon terminals in photoreceptors prepared for transplantation: an in vitro study, *Invest Ophthalmol Vis Sci* **46**(3), 967–73 (2005).

D. E. Kourennyi, X. D. Liu, J. Hart, F. Mahmud, W. H. Baldridge and S. Barnes, Reciprocal modulation of calcium dynamics at rod and cone photoreceptor synapses by nitric oxide, *J Neurophysiol* **92**(1), 477–83 (2004).

R. Kozma, S. Sarner, S. Ahmed and L. Lim, Rho family GTPases and neuronal growth cone remodelling: relationship between increased complexity induced by Cdc42Hs, Rac1, and acetylcholine and collapse induced by RhoA and lysophosphatidic acid, *Mol Cell Biol* **17**(3), 1201–11 (1997).

M. Lehmann, A. Fournier, I. Selles-Navarro, P. Dergham, A. Sebok, N. Leclerc, G. Tigyi and L. McKerracher, Inactivation of Rho signaling pathway promotes CNS axon regeneration, *J Neurosci* **19**(17), 7537–47 (1999).

G. P. Lewis, K. A. Linberg and S. K. Fisher, Neurite outgrowth from bipolar and horizontal cells after experimental retinal detachment, *Invest Ophthalmol Vis Sci* **39**(2), 424–34 (1998).

G. P. Lewis, D. G. Charteris, C. S. Sethi and S. K. Fisher, Animal models of retinal detachment and reattachment: identifying cellular events that may affect visual recovery, *Eye* **16**(4), 375–87 (2002a).

G. P. Lewis, D. G. Charteris, C. S. Sethi, W. P. Leitner, K. A. Linberg and S. K. Fisher, The ability of rapid retinal reattachment to stop or reverse the cellular and molecular events initiated by detachment, *Invest Ophthalmol Vis Sci* **43**(7), 2412–20 (2002b).

G. P. Lewis, C. S. Sethi, K. A. Linberg, D. G. Charteris and S. K. Fisher, Experimental retinal reattachment: a new perspective, *Mol Neurobiol* **28**(2), 159–75 (2003).

Z. Y. Li, I. J. Kljavin and A. H. Milam, Rod photoreceptor neurite sprouting in retinitis pigmentosa, *J Neurosci* **15**(8), 5429–38 (1995).

Z. Y. Li, F. Wong, J. H. Chang, D. E. Possin, Y. Hao, R. M. Petters and A. H. Milam, Rhodopsin transgenic pigs as a model for human retinitis pigmentosa, *Invest Ophthalmol Vis Sci* **39**(5), 808–19 (1998).

B. A. Liepe, C. Stone, J. Koistinaho and D. R. Copenhagen, Nitric oxide synthase in Müller cells and neurons of salamander and fish retina, *J Neurosci* **14**(12), 7641–54(1994).

S. A. Lipton and S. B. Kater, Neurotransmitter regulation of neuronal outgrowth, plasticity and survival, *TINS* **12**(7), 265–70 (1989).

R. N. Lolley, The rd gene defect triggers programmed rod cell death. The Proctor Lecture, *Invest Ophthalmol Vis Sci* **35**(13), 4182–91 (1994).

R. D. Lund, A. S. Kwan, D. J. Keegan, Y. Sauve, P. J. Coffey and J. M. Lawrence, Cell transplantation as a treatment for retinal disease, *Prog Retin Eye Res* **20**(4), 415–49 (2001).

J. W. Mandell, E. Townes-Anderson, A. J. Czernik, R. Cameron, P. Greengard and P. De Camilli, Synapsins in the vertebrate retina: absence from ribbon synapses and heterogeneous distribution among conventional synapses, *Neuron* **5**(1), 19–33 (1990).

K. Mervin, K. Valter, J. Maslim, G. Lewis, S. Fisher and J. Stone, Limiting photoreceptor death and deconstruction during experimental retinal detachment: the value of oxygen supplementation, *Am J Ophthalmol* **128**(2), 155–64 (1999).

A. H. Milam, Z. Y. Li, A. V. Cideciyan and S. G. Jacobson, Clinicopathologic effects of the Q64ter rhodopsin mutation in retinitis pigmentosa, *Invest Ophthalmol Vis Sci* **37**(5), 753–65 (1996).

R. S. Molday, Photoreceptor membrane proteins, phototransduction, and retinal degenerative diseases. The Friedenwald Lecture, *Invest Ophthalmol Vis Sci* **39**(13), 2491–513 (1998).

C. W. Morgans, J. H. Brandstätter, J. Kellerman, H. Betz and H. Wässle, A SNARE complex containing syntaxin 3 is present in ribbon synapses of the retina, *J Neurosci* **16**(21), 6713–21 (1996).

C. W. Morgans, Localization of the alpha(1F) calcium channel subunit in the rat retina, *Invest Ophthalmol Vis Sci* **42**(10), 2414–8 (2001).

M. Nachman-Clewner and E. Townes-Anderson, Injury-induced remodelling and regeneration of the ribbon presynaptic terminal in vitro, *J Neurocytol* **25**(10), 597–613 (1996).

M. Nachman-Clewner, R. St Jules and E. Townes-Anderson, L-type calcium channels in the photoreceptor ribbon synapse: localization and role in plasticity, *J Comp Neurol* **415**(1), 1–16 (1999).

I. Nir, R. Haque and P. M. Iuvone, Regulation of cAMP by light and dopamine receptors is dysfunctional in photoreceptors of dystrophic retinal degeneration slow(rds) mice, *Exp Eye Res* **73**(2), 265–72 (2001).

M. Nishiyama, A. Hoshino, L. Tsai, J. R. Henley, Y. Goshima, M. Tessier-Lavigne, M. M. Poo and K. Hong, Cyclic AMP/GMP-dependent modulation of Ca^{2+} channels sets the polarity of nerve growth-cone turning, *Nature* **423**(6943), 990–5 (2003).

S. E. Pearce-Kelling, T. S. Aleman, A. Nickle, A. M. Laties, G. D. Aguirre, S. G. Jacobson and G. M. Acland, Calcium channel blocker D-cis-diltiazem does not slow retinal degeneration in the PDE6B mutant rcd1 canine model of retinitis pigmentosa, *Mol Vis* **7**, 42–7 (2001).

Y. W. Peng, Y. Hao, R. M. Petters and F. Wong, Ectopic synaptogenesis in the mammalian retina caused by rod photoreceptor-specific mutations, *Nat Neurosci* **3**(11), 1121–7 (2000).

Y. W. Peng, T. Senda, Y. Hao, K. Matsuno and F. Wong, Ectopic synaptogenesis during retinal degeneration in the royal college of surgeons rat, *Neuroscience* **119**(3), 813–20 (2003).

F. Polleux, T. Morrow and A. Ghosh, Semaphorin 3A is a chemoattractant for cortical apical dendrites, *Nature* **404**(6778), 567–73 (2000).

E. Raviola and N. B. Gilula, Gap junctions between photoreceptor cells in the vertebrate retina, *Proc Natl Acad Sci U S A* **70**(6), 1677–81 (1973).

C. E. Reme and R. W. Young, The effects of hibernation on cone visual cells in the ground squirrel, *Invest Ophthalmol Vis Sci* **16**(9), 815–40 (1977).

T. S. Rex, R. N. Fariss, G. P. Lewis, K. A. Linberg, I. Sokal and S. K. Fisher, A survey of molecular expression by photoreceptors after experimental retinal detachment, *Invest Ophthalmol Vis Sci* **43**(4), 1234–47 (2002).

F. Rieke and E. A. Schwartz, A cGMP-gated current can control exocytosis at cone synapses, *Neuron* **13**(4), 863–73 (1994).

S. Sanyal, R. Fletcher, Y. P. Liu, G. Aquirre and G. Chader, Cyclic nucleotide content and phosphodiesterase activity in the rds mouse (O2O/A) retina, *Exp Eye Res* **38**, 247–56 (1984).

S. Sanyal, R. K. Hawkins, H. G. Jansen and G. H. Zeilmaker, Compensatory synaptic growth in the rod terminals as a sequel to partial photoreceptor cell loss in the retina of chimaeric mice, *Development* **114**(3), 797–803 (1992).

V. Sauzeau, H. Le Jeune, C. Cario-Toumaniantz, A. Smolenski, S. M. Lohmann, J. Bertoglio, P. Chardin, P. Pacaud and G. Loirand, Cyclic GMP-dependent protein kinase signaling pathway inhibits RhoA-induced Ca^{2+} sensitization of contraction in vascular smooth muscle, *J Biol Chem* **275**(28), 21722–9 (2000).

A. Savchenko, S. Barnes and R. H. Kramer, Cyclic-nucleotide-gated channels mediate synaptic feedback by nitric oxide, *Nature* **390**(6661), 694–8 (1997).

S. F. Schaeffer and E. Raviola, Membrane recycling in the cone cell endings of the turtle retina, *J Cell Biol* **79**(3), 802–25 (1978).

Y. Schmitz and P. Witkovsky, Dependence of photoreceptor glutamate release on a dihydropyridine-sensitive calcium channel, *Neuroscience* **78**(4), 1209–16 (1997).

F. Schmitz, A. Königstorfer and T. C. Südhof, RIBEYE, a component of synaptic ribbons: a protein's journey through evolution provides insight into synaptic ribbon function, *Neuron* **28**(3), 857–72 (2000).

S. M. Schoenwaelder and K. Burridge, Bidirectional signaling between the cytoskeleton and integrins, *Curr Opin Cell Biol* **11**(2), 274–86 (1999).

C. S. Sethi, G. P. Lewis, S. K. Fisher, W. P. Leitner, D. L. Mann, P. J. Luthert and D. G. Charteris, Glial remodeling and neural plasticity in human retinal detachment with proliferative vitreoretinopathy, *Invest Ophthalmol Vis Sci* **46**(1), 329–42 (2005).

D. M. Sherry, R. S. St Jules and E. Townes-Anderson, Morphologic and neurochemical target selectivity of regenerating adult photoreceptors in vitro, *J Comp Neurol* **376**(3), 476–88 (1996).

M. Shimizu-Albergine, S. D. Rybalkin, I. G. Rybalkina, R. Feil, W. Wolfsgruber, F. Hofmann and J. A. Beavo, Individual cerebellar Purkinje cells express different cGMP phosphodiesterases (PDEs): in vivo phosphorylation of cGMP-specific PDE (PDE5) as an indicator of cGMP-dependent protein kinase (PKG) activation, *J Neurosci* **23**(16), 6452–9 (2003).

H. J. Song, G. L. Ming, Z. He, M. Lehmann, L. McKerracher, M. Tessier-Lavigne and M. M. Poo, Conversion of neuronal growth cone responses from repulsion to attraction by cyclic nucleotides, *Science* **281**(5382), 1515–8 (1998).

M. Spencer, P. B. Detwiler and A. H. Bunt-Milam, Distribution of membrane proteins in mechanically dissociated retinal rods, *Invest Ophthalmol Vis Sci* **29**(7), 1012–20 (1988).

W. R. Taylor and C. Morgans, Localization and properties of voltage-gated calcium channels in cone photoreceptors of Tupaia belangeri, *Vis Neurosci* **15**(3), 541–52 (1998).

E. Townes-Anderson, Intersegmental fusion in vertebrate rod photoreceptors. Rod cell structure revisited, *Invest Ophthalmol Vis Sci* **36**(9), 1918–33 (1995).

E. Townes-Anderson, R. Chawla and N. Zhang Differential roles of cAMP and cGMP in presynaptic plasticity of salamander cone and rod photoreceptors in vitro, ARVO: abstract#2847 (2003).

V. Traverso, R. A. Bush, P. A. Sieving and D. Deretic, Retinal cAMP levels during the progression of retinal degeneration in rhodopsin P23H and S334ter transgenic rats, *Invest Ophthalmol Vis Sci* **43**(5), 1655–61 (2002).

S. Van Wagenen and V. Rehder, Regulation of neuronal growth cone filopodia by nitric oxide, *J Neurobiol* **39**(2), 168–85 (1999).

S. Van Wagenen and V. Rehder, Regulation of neuronal growth cone filopodia by nitric oxide depends on soluble guanylyl cyclase, *J Neurobiol* **46**(3), 206–19 (2001).

L. Vollrath and I. Spiwoks-Becker, Plasticity of retinal ribbon synapses, *Microsc Res Tech* **35**(6), 472–87 (1996).

E. R. Weiss, Y. Hao, C. D. Dickerson, S. Osawa, W. Shi, L. Zhang and F. Wong, Altered cAMP levels in retinas from transgenic mice expressing a rhodopsin mutant, *Biochem Biophys Res Commun* **216**(3), 755–61 (1995).

Y. Xiang, Y. Li, Z. Zhang, K. Cui, S. Wang, X. B. Yuan, C. P. Wu, M. M. Poo and S. Duan, Nerve growth cone guidance mediated by G protein-coupled receptors, *Nat Neurosci* **5**(9), 843–8 (2002).

M. Xiao, S. M. Sastry, Z. Y. Li, D. E. Possin, J. H. Chang, I. B. Klock and A. H. Milam, Effects of retinal laser photocoagulation on photoreceptor basic fibroblast growth factor and survival, *Invest Ophthalmol Vis Sci* **39**(3), 618–30 (1998).

N. Zhang and E. Townes-Anderson, Regulation of structural plasticity by different channel types in rod and cone photoreceptors, *J Neurosci* **22**(16), 7065–79 (2002).

N. Zhang, A. Beuve and E. Townes-Anderson, The nitric oxide-cGMP signaling pathway differentially regulates presynaptic structural plasticity in cone and rod cells, *J Neurosci* **25**(10), 2761–70 (2005).

3

Retinal Remodeling: Circuitry Revisions Triggered by Photoreceptor Degeneration

ROBERT E. MARC, BRYAN W. JONES AND CARL B. WATT

University of Utah School of Medicine, John A. Moran Eye Center, Salt Lake City, UT, USA

1. Introduction

Structural, molecular and physiological responses of the *neural* retina to ablation of the *sensory* retina triggered by inherited or induced photoreceptor degeneration reveal a surprising degree of dynamic capacity in mature neurons. All classes of mature retinal neurons (bipolar, horizontal, amacrine and ganglion cells) show the capacity for five dynamic processes characteristic of developing neurons and stressed neurons displaying negative plasticity. These five dynamic processes – neurite formation, fascicle formation, synaptogenesis, self-signaling and migration – are unmasked when the neural retina is released from the control of the sensory retina. It is now clear that active mechanisms regulate and stabilize connectivity in the mature retina, contradicting a widely held presumption that the neural retina is a largely passive, stable and accessible substrate for cellular, bionic, and even molecular or genetic interventions to either restore visual capacity or retard visual loss in individuals suffering from inherited/acquired retinal degenerations. Put simply, the neural retina is not "Plug & Play."

This chapter will briefly review the concept of disease-related retinal remodeling and its three phases (Sections 2.1-2.3), the consequences of anomalous rewiring patterns (Section 2.4), culminating in a discussion of the clinical and biological implications of remodeling (Section 3). We will focus specifically on the forms and attributes of novel circuitries arising from synaptogenesis and the potential mechanisms associated with self-signaling. Analysis of retinal remodeling has a short history, but one key hypothesis emergent from analysis of numerous models of retinal degeneration is that mature retinal neurons appear to depend on a consistent excitatory input or trophic signal associated with functional synapses and that, following the loss of that signaling in retinal degenerations, neurites and synapse remodeling is initiated. More specifically, it is likely that neurite growth is inhibited or

R. Pinaud, L. A. Tremere, P. De Weerd(Eds.), Plasticity in the Visual System:
From Genes to Circuits, 33-54, ©2006 Springer Science + Business Media, Inc.

maintained in stasis by excitation/trophic signaling in the adult retina; that dynamic structural revisions and endogenous signaling are normal capacities of mature neurons held in check by systemic signaling.

2. Remodeling, Models, Rewiring and Self-signaling

Retinal remodeling overview

Retinal remodeling is phrase that refers to an array of molecular and cellular phenotype revisions of the *neural retina* in response to inherited or acquired degenerations of the *sensory retina*. Remodeling phenomena were first clearly defined for inherited retinal degenerations by Ann Milam and her colleagues (1995, Fariss et al., 2000, Milam et al., 1996, Milam et al., 1999) and for retinal detachments by Steven K. Fisher, Geoff Lewis and colleagues (reviewed in Fisher and Lewis, 2003). It has since become apparent that glial, vascular and neural remodeling are stereotyped responses to retinal disease stress (reviewed in Marc et al., 2003). Most of this review will deal with remodeling patterns associated with inherited retinal degenerations, where the development of natural genetic, crafted genetic and induced models have allowed us to screen a broad spectrum human disease homologues. In virtually all systems studied to date, retinal remodeling follows three well-defined phases. Phase 1 is the period of photoreceptor stress and phenotype deconstruction and is characterized by: photoreceptor neurite sprouting, photoreceptor phenotype deconstruction, bipolar cell dendrite and horizontal cell axon terminal pruning, horizontal cell axon sprouting. Phase 2 is the period of active photoreceptor cell death and sensory retina ablation and is characterized by: photoreceptor death and debris clearance, Müller cell distal process seal formation, bipolar cell phenotype revision. Finally, Phase 3 is the period of neuronal migration, rewiring and cell death, characterized by: Müller cell hypertrophy (perhaps hyperplasia?); bipolar, amacrine and ganglion cell neurite sprouting; mixed neurite fascicle formation; microneuroma formation; neuronal self-signaling; and neuronal cell death.

2.1. Remodeling Phase 1

In Phase 1 photoreceptors become stressed by direct inherited defects in photoreceptor protein expression, such as the absence of a functional PDEb6 subunit in rods of the classical rd1 mouse model (Bowes et al., 1990), dominant negative protein effects such as the P23H rhodopsin mutation (Illing et al., 2002), or by indirect mechanisms such as the *mertk*[-/-] RPE phagocytosis defect (D' Cruz et al., 2000). As many retinal degenerations are rod-based, such as retinitis pigmentosa (Daiger, 2004, Farrar et al., 2002), the secondary loss of cones falls into the indirect or "bystander" category, where cell death is activated and/or accelerated by a range of candidate processes such as microglial killing (Roque et al., 1999, Gupta et al., 2003, Zeiss and Johnson, 2004), gap junctions between dying rods and adjacent cones (Ripps, 2002), loss of survival factors produced by rods (Mohand-Said et al., 1998) or Müller cells (McGee Sanftner et al., 2001), or more complex effects such as enhanced generation of oxidizing species in the sensory retina as oxygen-consuming rods are lost (Valter et al., 1998, Stone et al., 1999).

The early phases of neuronal remodeling are manifest as changes in synaptic relations, as seen most dramatically in rod-based diseases. Rods appear to retract their synaptic terminals and then generate profuse bundles of rod neurites that bypass the outer plexiform layer and traverse the inner plexiform layer, often reaching the ganglion cell layer (Li et al., 1995, Fariss et al., 2000). Perhaps concurrently, rod bipolar cells retract their dendrites (Strettoi and Pignatelli, 2000, Strettoi et al., 2003, Strettoi et al., 2002) and apparently become unresponsive (Varela et al., 2003). This implies that the early loss of scotopic sensitivity characteristic of retinitis pigmentosa is likely a defect of synaptic signaling, and that photoreceptor death occurs much later.

Horizontal cells display a similar but more complex behavior. The rod-driven axon terminals of horizontal cells also retract their neurites, but the somas also generate new axons of more than one morphology, all of which appear to enter the inner plexiform layer (see review by Marc et al., 2003).

Cones also manifest sprouting behavior (Fei, 2002), likely just prior to cell death. But loss of cones can occur over different time scales and also display regional inhomogeneity. For example, secondary cone loss in the *rd1* mouse occurs on roughly the same time scale as rods and is complete by postnatal day (pnd) 90-100, except for a very small cohort of dorsal retina cones that seem to persist throughout life (Jiménez et al., 1996). Even more rapid loss occurs in the TG9N mouse (a transgenic mouse expressing a truncated RGS9 protein, likely in both rods and cones), where both classes are ablated by $\approx$ 60 days (Jones et al., 2003a). Conversely, cones in some transgenic retinitis pigmentosa models such as the P23H rat and P347L pig (expressing mutant rhodopsins on normal rhodopsin backgrounds) survive for many months past rod death and, in these models, the dendrites of rod bipolar cells actively switch to nearby cones (Peng et al., 2000, Cuenca et al., 2004). In the rodless *nrl*$^{-/-}$ mouse, both rod bipolar cells and rod-driven terminal fields of horizontal cells make contact with cones. While this switch is transient in retinal degeneration models, it appears to be permanent in the *nrl*$^{-/-}$ mouse. These phenomena presage the lack of specificity manifested by all retinal neurons in remodeling. Indeed, most rewiring in retinal degenerations may be opportunistic and unrelated to normal circuit assembly.

Phase 1 represents the initiation of a process whose details are yet unclear: phenotype deconstruction (Mervin et al., 1999, Lewis et al., 2002). Synapse disassembly argues that synapse-specific proteins are functionally compromised, either directly or through transcriptional modification. The transcription networks that maintain mature cell phenotypes could be become irreversibly corrupted or altered during photoreceptor stress as many photoreceptor-specific proteins are either misrouted or their expressions depressed (John et al., 2000, also discussed in Marc et al., 2003). Basic small molecule profiles, such as cyclic nucleotide levels, can also have dramatic effects on neurite stability in photoreceptors (Khodair et al., 2003, Zhang and Townes-Anderson, 2002), and since rho GTPases are critical signaling modules in neurite growth and stabilization (Li et al., 2000), disruptions in energy charge or cAMP/cGMP production could underlie synaptic disassembly.

2.2. Remodeling Phase 2

Phase 2 overlaps Phase 1 in that photoreceptor death likely begins while rod/cone phenotype deconstruction or revision enables ectopic sprouting. But the coherent loss of rods, triggering proliferation of microglia in the sensory retina (Gupta et al., 2003, Zeiss and Johnson, 2004) as well as bystander effects, likely activate more remodeling phenomena in the neural retina than have yet been found. At the very least, it is clear that bipolar cells mistarget their mGluR6 and Goα signaling proteins (Strettoi et al., 2003, Strettoi et al., 2002) and become unresponsive to glutamate (Varela et al., 2003). Whether they are capable of regenerating functional dendrites is not yet known. The only processes we are certain they can produce are additional copies of their axons and axon terminals (Jones et al., 2003a, Marc et al., 2003). In any case, loss of the mGluR6 pathway may be partially protective for remodeling rod bipolar cells, also known as depolarizing bipolar cells (DBCs) or ON bipolar cells. The mGluR6 pathway is net sign-inverting (reviewed in Marc, 2004) and the loss of glutamate signaling following photoreceptor ablation would theoretically lead to an intrinsic activation of the non-selective cation channels driving DBC signaling, and a constitutive loading with Ca^{+2}. As will be discussed below, this does not occur, consistent with the idea that key elements in the signaling pathway are deconstructed. Depending on the relative coherence of rod and cone death in Phase 2, cone bipolar cells are also likely to retract their dendrites (Strettoi et al., 2003).

Müller cell remodeling appears to accelerate in Phase 2. Since every photoreceptor inner segment is almost completely ensheathed in Müller cell distal processes that potentially influence every key metabolic pathway in rods and cones (Mata et al., 2002, Newman and Reichenbach, 1996, Pow and Robinson, 1994), the active ablation of photoreceptors must lead to both a structural compression of remnant distal processes and a change in the transport state of the distal Müller cell. This is manifest in part as a major increase in Müller cell glutamine content at the end of Phase 2 in all major retinal degeneration models (Jones et al., 2003a), including the albino rat light damage model (Jones et al., 2003b). Unlike retinal detachment, which leads to an increase Müller cell glutamate content due to down-regulation of glutamine synthetase activity (Marc et al., 1998), retinal degeneration leads to an inability of Müller cells to export glutamine or otherwise regulate intracellular levels. This implies a discrete revision in the steady-state phenotype of the

Müller cell cohort. Of course, the standard signature of Müller cell "stress" is the upregulation of GFAP expression, and this also occurs in retinal degenerations (Eisenfeld et al., 1984, Strettoi and Pignatelli, 2000, Wahlin et al., 2000). But, clearly, some pathways influencing Müller cell size and cell surface composition are important factors in triggering the neuronal cell migration and some of the fascicle extension that occurs in Phase 3. The definitive feature of the completion of Phase 2 is the formation of a dense seal of Müller cell processes at the distal margin of the neural retina, separating it from the retinal pigmented epithelium (RPE) over much of the retina. Remaining perforations in the seal seem to be areas of either active RPE cell invasion, vascular invasion or eruption of large neuropil tufts reaching the remnant RPE (Jones et al., 2003a).

2.3. *Remodeling Phase 3*

Phase 3 remodeling begins after the large-scale entombment of the neural retina beneath the glial seal. Though the most aggressive remodeling occurs in this period, it has been studied least because it does not emerge until after the cone photoreceptors have been ablated. The time course of remodeling is not well-known yet and varies across models (Jones et al., 2003a, Marc et al., 2003), but phase 3 events do not appear until $\approx$ pnd 160 in the fastest models (e.g. the TG9N mouse) and more commonly take a year to emerge. As discussed in Marc et al. (2003), the same time scale seems to operate in humans: after cone death, neurite extension, fascicle formation, microneuroma formation, neuronal migration and neuronal death all emerge. The earliest sign of phase 3 remodeling is the elaboration of small microneuromas in the distal retina beneath the glial seal (Jones et al., 2003a). Microneuromas are collections of new synapses that form largely in the distal retina within the remnant inner plexiform layer, representing the terminations of extensive bundles of neurite fascicles. The ultrastructure of microneuromas is indistinguishable from the inner plexiform layer in general, but when explored in detail by serial section reconstruction, their functional circuitry is clearly anomalous (Watt et al., 2004). This demonstrates that mature retinal neurons are perfectly capable of synaptogenesis but are apparently capricious in their selection of synaptic partners. Indeed, it is very likely that an individual neuron cannot "know" anything about the larger performance attributes of circuit in which it is embedded. The implications of this will be addressed in section 2.4.

The emerge of new processes from bipolar cells that have previously ablated their dendrites (Pignatelli et al., 2004, Strettoi and Pignatelli, 2000, Strettoi et al., 2003, Strettoi et al., 2002) does not appear to be dendritogensis, but rather production of supernumerary axons, as microneuromas display numerous profiles containing synaptic ribbons. While bipolar cells appear to able to generate new, mistargeted dendrites in retinal detachments (Lewis et al., 1998), this occurs in cells that have not lost their dendrites. The slow, persistent phenotype deconstruction of bipolar cells in retinal degenerations is clearly a different process, and the loss of bipolar cell dendrites seems irreversible. This is consistent with the observation by Watt et al. (2004) that a given bipolar cell sends a single unbranched process into a given microneuroma, where it makes ribbon contacts with targets and receives mostly conventional synaptic inputs from amacrine cells. Mapping of neuronal signatures by computational molecular phenotyping (CMP), allows complete identification of the phenotypes of every cell type in a retinal sample (Jones et al., 2003a, Kalloniatis et al., 1996, Marc and Cameron, 2002, Marc and Jones, 2002, Marc et al., 1995), and reveals that every type of retinal neuron can generate processes anew. The new processes from GABAergic and glycinergic amacrine cells are indistinguishable from the dendrodendritic synaptic forms they take in the normal inner plexiform layer. Though we have only documented that new ganglion cell processes assume their classic postsynaptic positions in microneuromas, we have not excluded the possibility that they can form new excitatory presynaptic assemblies via generation of intraretinal axons. The emergence of new, myelinated axonal profiles in inherited (Jones et al., 2003a, Pow and Sullivan, 2002) and induced retinal degenerations (Sullivan et al., 2003, Jones et al., 2003b) is circumstantial evidence that ganglion cells produce intraretinal axons, and some of the new, surprising forms of conventional synapses found in microneuromas are reminiscent of excitatory synapses in the CNS (Jones et al., 2003a). Finally, though horizontal cells are the sparsest retinal neuron class, it appears that every surviving horizontal cell sends one or more processes, some axonal and some dendritic in form, into the inner plexiform layer (Park et al., 2001, Strettoi et al., 2002). One balance, there is no evidence for any cell type in the retina that cannot generate new processes.

Fasciculation of mixed processes, as described by Jones et al. (2003a) is a remarkable finding, because retinal neurons in general have only short range and precisely laminar projections, with very restricted fascicle formation. Ganglion cell axons are the only processes known to fasciculate extensively in native retina, but it appears that every class of retinal neuron is fully capable of recognizing other processes and following them through permissive regions. The molecular mechanisms of fasciculation and pathfinding of new fascicles within the retina are unknown and may be as random as the synaptic connectivity of microneuromas, since mixed fascicles are unusual in the vertebrate CNS.

As rewiring progresses, the early cell death described by Strettoi et al. (Strettoi et al., 2002) accelerates and, depending on the coherence and speed of the specific degeneration, may result in a large depletion of neurons through cell death, though the exact mechanism of killing remains unknown (Jones et al., 2003a, Jones et al., 2003b, Marc et al., 2003, Marc et al., 2001). Combined with the apparent hypertrophy of Müller cells and perhaps invasion of blood vessels from both the vitreal and choroidal margins, this excavation of the neural retina appears to facilitate the final major transformation of the retina: neuronal migration. It has long been presumed that post-mitotic, mature retinal cells do not move long distances, though they may engage in fine mosaic adjustments via tangential dispersion in late development (Reese and Galli-Resta, 2002). However, bidirectional migration between the remnant inner nuclear layer and the ganglion cell layer occur in large streams of cells mixed within and around large columns of Müller cells (Jones et al., 2003a). *Eversions* of ganglion cells to the outer retina appear less frequently than eversions of amacrine cells to the distal margin of the retina and *inversions* of amacrine and bipolar cells into the inner plexiform layer and the ganglion cell layer. A more dramatic version of migration occurs in the light damage model of retinal degeneration in albino rats, where severe remodeling of the distal retina (including RPE and choriocapillaris ablation) is apparently permissive for both glial and neuronal migration through the remnant Brüch's membrane into the choroid, depleting the retina of survivor neurons (Jones et al., 2003b, Sullivan et al., 2003). The stimulus that triggers the emigration of retinal neurons is simply unknown.

2.4. Rewiring and its consequences

The mammalian retina contains 55–80 classes of retinal neurons connected in highly stereotyped modes (Marc and Jones, 2002, Masland, 2001, Wässle, 2004), with specific neurochemical rules for signaling (Marc, 2004). Most retinal neurons are associated with the major parallel pathways: the ON center cone pathway formed by bipolar cells expressing the class C GPCR group III metabotropic mGluR6 receptor and the OFF center cone pathway formed by bipolar cells expressing either AMPA or kainate (KA) receptors (reviewed in Marc, 2004). This normal circuitry (Fig. 3.1) is heavily transformed by remodeling, especially during phase 3. In brief, new neurites from every known superclass (bipolar, horizontal, amacrine and ganglion cells) are elaborated, collect in new fascicles that can traverse hundreds of micrometers of retina, and terminate in synaptic aggregates known as microneuromas (Figs. 3.1, 3.2). Microneuromas range from tiny foci less than 10 μm in diameter to large eruptions of processes from the inner plexiform layer forming zones over 50 μm in length and containing hundreds of new synapses. In mid-phase 3, microneuromas are abundant and can exceed 10,000 in a single mouse retina (Jones et al., 2003a). As GABAergic amacrine cells dominate all vertebrate retinas (Kalloniatis et al., 1996, Marc, 2004, Marc and Cameron, 2002, Marc et al., 1995), microneuroma formation is most easily visualized with quantitative GABA immunocytochemistry. Figure 3.2 is a thin section of phase 3 remodeling in the human rhodopsin-GFP fusion protein knock-in mouse generated by Chan et al. (2004), showing profuse microneuromas formation by GABAergic amacrine cells, including invasion of the RPE.

Microneuromas are easily visualized and have been demonstrated to be zones of abundant synaptogenesis. This raises four key questions. Is the connectivity generated by survivor neurons typical of the native retina? Is the deafferented retina active? Does the inner plexiform layer retina retain its normal wiring? Why does the retina rewire at all? We can answer the first two

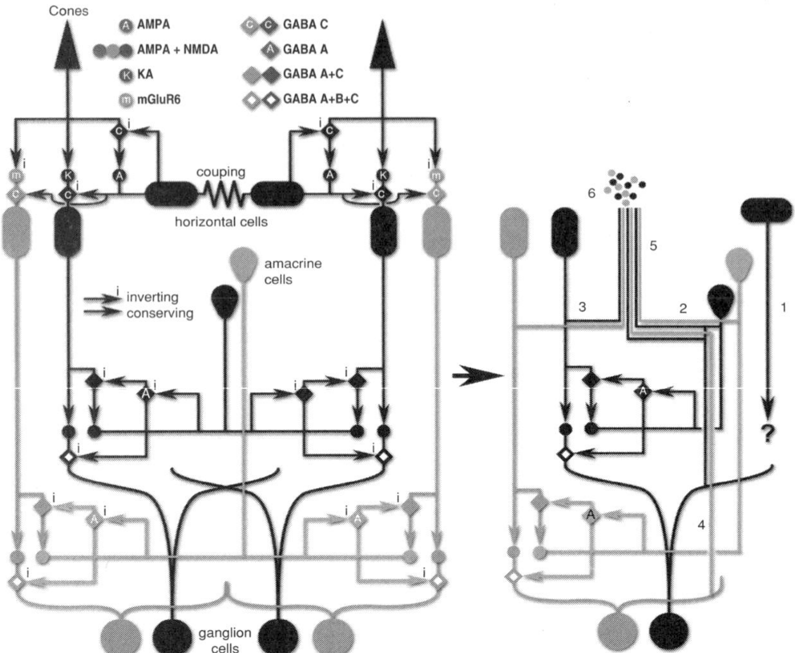

FIGURE 3.1. Native (left) and remodeled (right) photopic circuitry of the mammalian retina with the basic glutamatergic (AMPA, AMPA+NMDA, KA and mGluR6) and GABAergic (GABAA, GABAB and/or GABAC) transfers indicated. Black cells hyperpolarize in response to light and contribute to the OFF pathway. Grey cells depolarize in response to light and are part of the ON pathway. The primary, stable sign-conserving and sign-inverting transfers that construct the receptive field surrounds of the ON and OFF pathways are indicated for amacrine cell systems as nested feedback elements (see text). In phase 3 remodeling, the photoreceptors and the dendritic inputs to bipolar and horizontal cells are lost and survivor neurons begin to generate new processes. Horizontal cells all generate new axon-like processes that enter the inner plexiform layer (1). Amacrine cells appear to generate additional dendrite-like processes (2) that make both pre- and postsynaptic contacts. Bipolar cells appear to regenerate axon-like (3) but not dendritic processes, while ganglion cells appear to produce both (4). All types of new processes can be found in mixed fascicles of neurites (5) surrounded by Müller cell process, and these terminate in dense aggregates of synaptic contacts (6) termed microneuromas.

concretely by considering how local circuits operate and exploring their behaviors in remodeling retinas.

The retina has three major synaptic transfer modes of relevance to remodeling:

- High gain, sign conserving >> synapses where signals are decoded by ionotropic AMPA (+/-NMDA) or KA receptors.

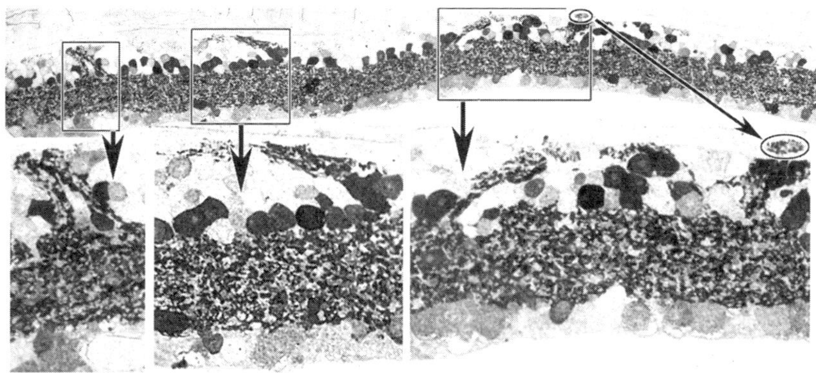

FIGURE 3.2. GABA immunoreactivity visualized by silver-immunogold detection in a 200 nm section of a retina from the hrhoG knock-in mouse model (Chan et al., 2004): 13 month old albino female. Top panel image width = 0.75 mm. Profuse microneuromas erupt from the inner plexiform layer and everted amacrine cells across the entire width of the retina. Three zones are enlarged below the panel. Left lower panel: Clusters of GABAergic dendrites arise directly from the inner plexiform layer. Middle lower panel: Microneuromas are formed from neurites arising in parted from everted amacrine cells. Right lower panel: Numerous focal microneuromas are formed, including one (ellipse) that is actually located in the RPE layer.

- High gain, sign inverting $>>\circ$ synapses where signals are decoded by metabotropic mGluR6 receptors
- Low gain, sign inverting $>\circ$ synapses where signals are predominantly decoded by ionotropic GABA and glycine receptors

These transfers mediate the assembly of formal circuits resulting in the 15–20 ganglion cell pathways that inform the CNS of key parameters of the visual world. The basic transfers between cones [C], horizontal cells [HC], hyperpolarizing or OFF bipolar cells [HBC], depolarizing or bipolar cells [DBC], amacrine cells [AC], and ganglion cells are: C >> HBC or HC, C >>∘ DBC, BC >> any target cell, and AC >∘ any target cell. There are important exceptions to these rules (e.g. cholinergic cells), but their consideration changes no conclusions regarding the anomalous circuit effects of remodeling.

The assembly of these transfer operations into long, branched and re-entrant chains results in the generation of micronetworks (Marc and Liu, 2000) that form the basic elements from which ganglion cell receptive fields are built. The key characteristic of these micronetworks is their fundamental stability, like well-designed operational amplifiers, and the key biological mechanism underlying this stability is a highly patterned set of GABAergic feedback events that operate locally to ensure 'good' bipolar to ganglion cell transfers. Figure 3.3 summarizes how these feedback events operate by mathematically simulating the signal seen by a ganglion cell dendrite driven by a bipolar cell. If a periodic stimulus is injected into the bipolar cell and its frequency swept from low

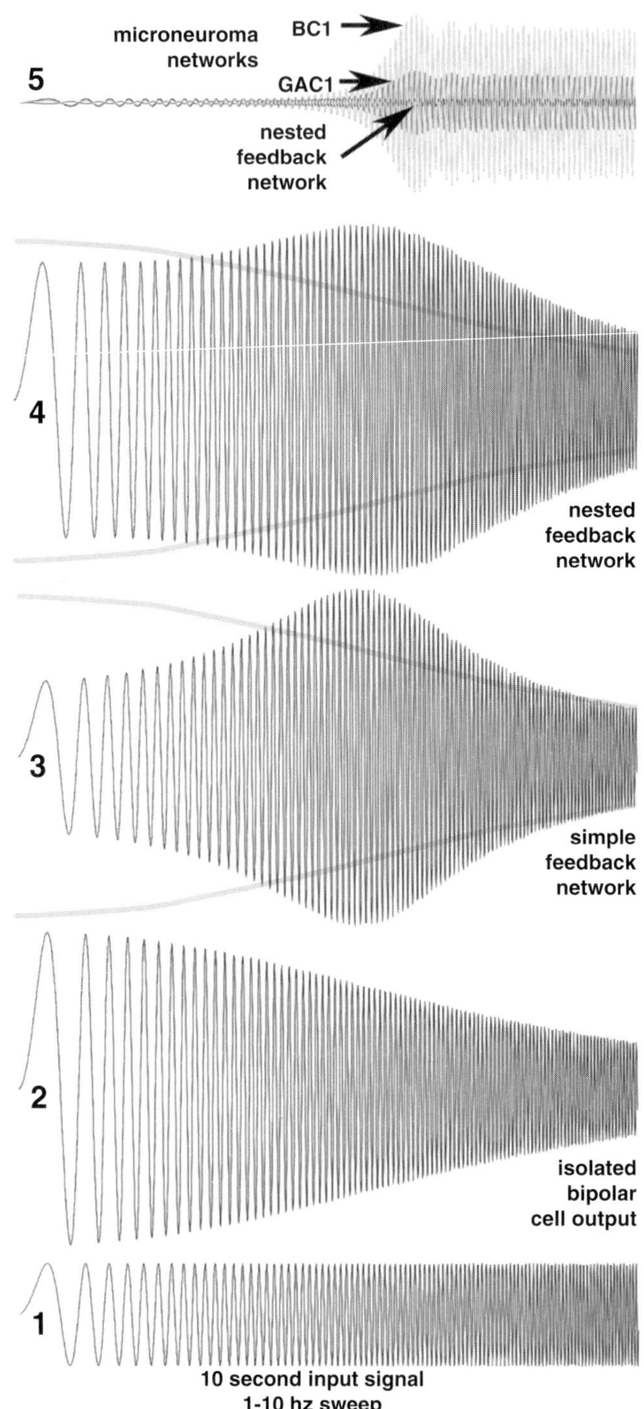
microneuroma
networks
BC1
5
GAC1
nested
feedback
network
4
nested
feedback
network
3
simple
feedback
network
2
isolated
bipolar
cell output
1
10 second input signal
1-10 hz sweep

to high, the envelope of voltage responses of the transfer gives us a rough idea of the fidelity of the system. Thus a signal injected into a presumed "isolated" bipolar cell gives a characteristic response where amplification is good at low frequencies, but "rolls off" at higher frequencies (Fig. 3.3 trace 2). The classical solution to the problem of high frequency roll-off is low-gain negative feedback, such as that provided by reciprocal and lateral AC >∘ BC GABAergic transfers via both GABAA and GABAC receptors (Marc, 1992, Marc, 2004, Marc and Liu, 2000): BC1 >> AC >∘ BC1. This improves middle-frequency response, at the price of low frequency attenuation, producing a band-pass envelope (Fig. 3.3 trace 3). One basic engineering solution to such non-ideal effects is the judicious use of nested feedback (Xie et al., 1999), as is evident in the abundant AC >. AC signaling of the vertebrate retina (Marc and Liu, 2000). The synaptic implementation is equivalent to providing another level of feedback within the network: (BC1 >> AC1 >∘ BC1) + (BC1 >> AC1 >∘ AC2 >∘ BC1). This second low-gain positive feedback channel has a bias for low frequencies and preferentially enhances signaling in the range where simple feedback does the most attenuation (Fig. 3.3 trace 4), resulting in a more uniform response function over a wide frequency range. The important features about all of these transfers are that vertebrate retinas implement anatomical nested feedback and are highly stable under all stimulus conditions. This strongly constrains the performance parameters of models.

Given that we cannot yet record from neurons that send their processes into microneuromas, our knowledge of the stability of retinal circuits is a powerful reference: we can compare simulations of native micronetworks using normal model parameters with simulations of micronetworks derived from reconstructions of microneuromas (Watt et al., 2004). We have several samples of rewired neuropil in remodeling retinas and it is clear that microneuromas contain circuits that are not compatible with visual processing. The RCS rat *mertk$^{-/-}$* system displays well over 10^4 microneuromas, formed *de novo* in phase 3. A microneuroma containing 36 separate processes was mapped by a fusion of

FIGURE 3.3. Five simulation traces generated from a formal model of retinal nested feedback circuitry (for details, see Marc and Liu, 2000). To assess the basic frequency response characteristics of native and remodeled circuits, a simple sinusoidal input was frequency modulated linearly from 1–10 Hz (Trace 1) and applied to the input of a given cell model (Extend, v6, ImagineThat Inc, San Jose, CA). Trace 2: A simple, isolated bipolar cell with no feedback generates a smooth response envelope with a characteristic high-frequency rolloff. Trace 3: Addition of a single stage of feedback improves higher frequency responses at the expense of low frequency gain. The envelope of the simple response is shown to scale. Trace 4: Nested feedback recovers the low frequency gain and generates a flatter response as a function of frequency. Trace 5: Superimposed traces from elements in a microneuroma from the RCS rat. Trace four is labeled "nested feedback network" and scaled down 33-fold. The anomalous, resonant outputs from driving at glycinergic amacrine cell (GAC1) and a bipolar cell (BC1) in the microneuroma are indicated in grey and light grey respectively.

electron microscopy and CMP, identifying circuits centered around bipolar cell BC1, which sends a single process into a microneuroma and therein drives four cells, including another bipolar cell, a GABAergic amacrine cell, a glycinergic amacrine cell and a ganglion cell neurite (GC1), which we can use as a reference to model the output of this novel network (Fig. 3.4). With the same 1–10 Hz linear frequency sweep stimulus used in Fig. 3.3, we can drive any element in the network using the transfer rules and parameters derived for native retinal circuits. Two of the most well-behaved are shown in Fig. 3.3, trace 5 superimposed on the properly scaled response of a standard nested feedback network (from Fig. 3.3 trace 4). If the sweep is injected into cell GAC1, the system responds at about 50% gain at very low frequencies, but once the input reaches about 5 Hz, the network quickly increases its gain to many-fold the correct amplitude. Driving BC1 directly activates this anomalous gain at about 4 Hz and quickly saturates at well over 10x the correct gain and driving AC1 creates an even worse saturation. All of these responses would likely drive a ganglion cell into depolarization block or fatigue, but also mean that this network cannot faithfully encode intensity, form or velocity. Whatever GC1 is receiving at this dendrite is unlike any signal developmentally constructed to serve sight. Less obvious at first glance is the fact that these responses are resonant and their impulse responses ring at around 5 Hz when activated, regardless of the frequency of the input (Watt et al., 2004). Put simply, the networks in microneuromas are corruptive and cannot be used as input for the visual system. Two things are clear from all of our analyses. (1) Retinal neurons do not intrinsically "know" how to make proper sensory signal processing networks. (2) The networks that they make in abundance are incompatible with visual processing, so connecting with them via transplantation (Klassen et al., 2004, Young et al., 2000) or bionic implant (Margalit et al., 2002, Zrenner, 2002) strategies is unlikely to restore adequate form or motion vision. We have little data to determine the extent of remodeling in the inner plexiform layer proper, but subtle changes in synaptic morphology in the *rd1* mouse (Strettoi et al., 2003, Strettoi et al., 2002) and RCS rat (Jones et al., 2003a) suggest that it would be imprudent to assume that circuitry in the inner plexiform layer was resistant to remodeling.

2.5 Self-signaling

Of course degenerated retinas no longer have sensory afferents to activate circuits, so what does such modeling really mean? It is almost certain that remodeling retinas have some means of self-activation, at least while a significant number of neurons remain alive, and the resonant behavior of these networks means that they have somehow been stabilized as satisfactory elements. Three pieces of evidence support the notion of self-signaling. First, retinitis pigmentosa patients often complain of photopsias, which are erratic, scintillating illusions resembling showers of stars (Delbeke et al., 2001, Heckenlively et al., 1988, Weleber and Gregory-Evans, 2001). The fact that

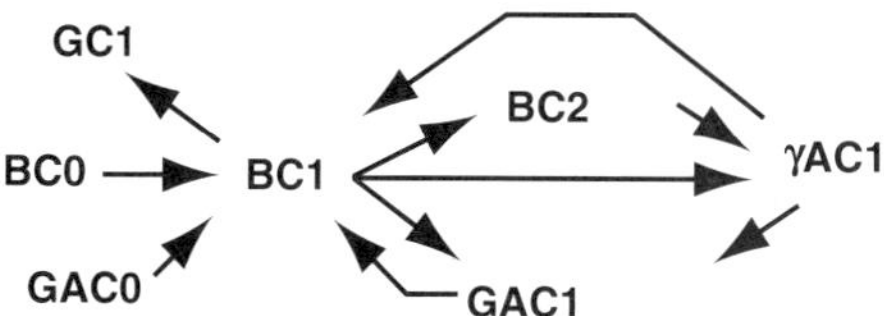

FIGURE 3.4. The fundamental circuitry of bipolar cell 1 (BC1) in a microneuroma from the retina of a phase 3 RCS rat. The basal microneuroma circuit includes three bipolar cells (BC0, BC1, BC2), two glycinergic amacrine cells (GAC0, GAC1), one GABA AC (γAC1) and one ganglion cell (GC1), which is used as the sensor for the output of modeling. There are, in fact, several other target ACs that provide no input to the network, but which clearly make contacts in other retinal loci.

photopsias can be re-activated by ocular stimulation long after vision has been lost argues strongly that they arise from stable aberrant retinal circuits (Delbeke et al., 2001). Second, single unit recordings from ganglion cells in the *rd1* mouse retina show that they experience wave-like but erratic excitation long after all photoreceptor input has been ablated (Stasheff, 2004). Finally, *in vivo* excitation mapping using the organic cation 1-amino-4-guanidobutane or AGB (Marc, 1999b, Marc, 1999a) shows extensive ionotropic glutamate receptor activation in many amacrine and ganglion cells, but relatively few bipolar cells (Marc et al., 2004, Marc et al., 2003) in the *rdcl* and *hrhoG* mouse retinal degeneration models. All of these data indicate that intrinsic mechanisms are capable of initiating signal activity that propagates through remodeling networks. There are a number of candidates for these generators including unstable bipolar cell-driven networks (Marc et al., 2004, Marc et al., 2003), reactivation of developmental periodic excitatory waves from cholinergic starburst amacrine cells (Feller, 1999, Feller, 2002), or spontaneous glutamate release from non-bipolar sources. These could be cryptic glutamate-releasing amacrine cells such as the vGlut3 glycinergic amacrine cell (Haverkamp and Wassle, 2004), type 1 dopaminergic amacrine cells (Jones et al., 2004), or even retinal ganglion cells that generate new intraretinal axons (Jones et al., 2003a, Pow and Sullivan, 2002, Sullivan et al., 2003). However, there are arguments that make any one of these candidates unlikely to be the major source of signaling. The most compelling feature of endogenous glutamate signaling from bipolar cells as the major process supporting self-signaling is the fact that AGB is permeant at glutamate-gated but not acetylcholine-gated channels (Marc, 1999b) and that a large fraction of all amacrine and ganglion cells are driven by some endogenous glutamate source. None of the alternative glutamate sources have a wide distribution of outputs and would be unlikely to drive the inner plexiform layer uniformly. But since bipolar cells themselves are largely inactive, the driver must be elsewhere.

The most likely endogenous oscillators are metabolic clocks intrinsic to neurons themselves that lead membrane depolarization: specifically a cAMP oscillator that mediates coherent pulse-like excitation events lasting some

1–10 sec in the developing ferret retina with a period of $\approx$ 45–60 sec (Stellwagen et al., 1999). This signaling appears to arise from a subset of neurons in which the cAMP oscillator functions autonomously, even in amacrine cells in culture (Firth and Feller, 2004). In principle, cycling of activated PKA levels could then modulate neuronal K^+ channels, changing the resting potential of the isolated cells, the degree activation of voltage-sensitive Ca^{+2} channels and the extent of transmitter release. Returning to Fig. 3.3, we see that oscillations in membrane potential of any amacrine cell can also drive the entire network. But as classical amacrine cells that drive feedback are GABAergic, mediating AC >∘ BC signaling through activation of Cl-channel permeation, AGB-permeant ionotropic glutamate receptor-gated channels will not be activated and most bipolar cells will remain AGB-(Marc et al., 2004, Marc et al., 2003). These bipolar cells will nevertheless drive target amacrine cell and ganglion cells via AMPA and NMDA receptors, consistent with the high level of endogenous AGB permeation in these neuronal classes *in vivo*. Thus the genesis of self-signaling could simply be the unmasking of endogenous oscillators after loss of light-driven signaling.

Finally, some bipolar cells are clearly AGB^+ and this implies intrinsic activation of ionotropic glutamate receptors or re-expression of a functional mGluR6 transduction sequence that permits constitutive non-selective cation channel opening. In the former case, this is consistent with observations of possible BC >> BC connections (Marc et al., 2004, Marc et al., 2003) that are forbidden in native retina. This also implies that some dendritic properties emerge in the new axon-like processes of bipolar cells targeting microneuromas. Clearly, the structures, potential phenotype revisions and intrinsic activities of neurons involved phase 3 remodeling require much more analysis.

3. Discussion

The existence of retinal remodeling impacts every modality of clinical intercession yet proposed. Gene therapies have been explored for early intercession to repair loss-of-function defects before cell death occurs (Dejneka et al., 2003). We now see that remodeling in phase 1 can be extensive and overlaps critical developmental periods. Trophic signaling between photoreceptors and bipolar cells is clearly critical for the preservation of functional synapses, and it is likely that therapeutic time windows may be much more restricted than previously believed. When does glial and neuronal phenotype revision become irreversible? What are the earliest signatures of phenotype revision? Fletcher (1996, 2000) noted that glutamine levels are elevated in Müller cells of the RCS rat at pnd 16, long before any significant degeneration begins in the sensory retina. Müller cell glutamine levels never regularize once this major increase occurs (Fletcher, 2000, Jones et al., 2003a, Jones et al., 2003b, Marc et al., 2003) and this implies that the expressions of some key components of glutamate recycling in Müller cells are permanently modulated by environmental signals such as trophic RPE signaling. While much emphasis has been placed on Müller cells as intermediaries in neurotrophin signaling (Wahlin et al., 2000), it is clear that they are also early disease response elements. The qualitative preservation of visual performance achieved in the Briard dog model of Leber Congenital Amaurosis (Acland et al., 2001) indicates that some of these effects may be reversible or that their consequences are more difficult to detect. In any case, early genetic intervention may the strategy least impacted by remodeling of the neural retina.

In dominant disease, stress triggered by transduction or protein processing defects (Hao et al., 2002, Illing et al., 2002) can potentially be ameliorated with molecular tools that promote photoreceptor survival (McGee Sanftner et al., 2001) or knock down expression of mutant genes (LaVail et al., 2000, Lewin et al., 1998). As these strategies would likely be implemented in phases 1 or 2, remodeling will be well underway, including retraction of bipolar cell dendrites. This raises the possibility that not all phenotype revisions may reverse (Marc et al., 2003), even in photoreceptors themselves (Marc, 2002). Since most analysis of survival factors have concentrated on rods expressing mutant genes, we know little of the effects of neurotrophins on remodeling in the neural retina. Endogenous BDNF secretion in response to ischemia or other neuronal traumas appears to be a leading candidate for the activation of ectopic axonal growth characteristic of temporal lobe epilepsy (Koyama et al., 2004) and the effects of any exogenous neurotrophin signaling in remodeling is clearly a major target for future investigations.

Cellular therapies focusing on the transplantation of fetal photoreceptors (Aramant and Seiler, 2002), cultured mature cells (Lund et al., 2000, Lund et al., 2001a, Lund et al., 2001b, Lund et al., 1999) or neural progenitor cells (Klassen et al., 2004, Young et al., 2000) are impacted in several ways by the existence of remodeling. Transplanted photoreceptors or neuroprogenitor cells intended to restore functional photoreceptor signaling will have to cope with the fact that bipolar cells will already have lost their neurites in most instances and the questions is simply, will they grow new ones? The dense Müller cell distal seal that encases survivor neurons will somehow have to be breached and there is no evidence that this is possible short of explicit trauma. If remodeling has already reached phase 3, with significant microneuroma formation, neuronal death and survivor migration, it remains doubtful that any transplantation of neural cells can "restore" a retina to functionality. As remodeled networks can drive anomalous central percepts, we know that some pathways remain functionally connected to the CNS, but that does not imply proper signaling. Any new synapses formed via transplantation may be corruptive, fictive or silent.

Finally, bionic therapy via implants of prosthetic subretinal and epiretinal semiconductor arrays will have to cope with the propensity of Müller cells to form seals and remodel the ganglion cell layer via rewiring, cell migration and neuronal death in the second. The ability of neurons from degenerating retinas to send processes through apertures in surface-modified artificial films placed in the subretinal space suggests that explicit control of synaptogenesis and ectopic electrode contact patterns is possible (Palanker et al., 2004). It remains to be seen whether any strategy, cellular or bionic, changes the course of remodeling, but 'bionic' tools may allow exploration of the processes that trigger neurite extension in remodeling neurons. We hypothesize that all neurons require a basal level of Ca^{+2} permeation to maintain activation of Ca^{+2}–gated gene expression, especially that associated with dendritic growth or maintenance (Redmond et al., 2002). If it is possible to use bionic schemes as basal stimulators that maintain Ca^{+2} signaling by field activation of voltage-gated Ca^{+2} channels, either the loss of dendrites or control of repatterning may be achieved.

There are far too many biological implications of remodeling to summarize here, but the ability of adult retinal neurons to migrate without losing their mature neurotransmitter phenotypes; the ability of adult retinal neurons to engage in abundant new process formation and synaptogenesis (even if wiring is misdirected); and the ability of retinal neurons to activate self-signaling after deafferentation from sensory drive implies a range of capacities in native mammalian retina that may mimic the dynamic plasticities associated with many non-mammalians. Moreover, the recent discoveries of critical periods for synaptic maturation in the mammalian retina (Tian, 2003, Tian and Copenhagen, 2001, Tian and Copenhagen, 2003, Vistamehr and Tian, 2004, Xu and Tian, 2004) and remarkable influences of visual complexity on retinal gene expression (Pinaud et al., 2003, Pinaud et al., 2002) imply that the wiring of the mature retina simply may not be as static as once thought.

Acknowledgements and funding: The work described in this review was supported by National Eye Institute grants EY002576 and EY015128, and Research to Prevent Blindness Inc.

References

G. M. Acland, G. D. Aguirre, J. Ray, Q. Zhang, T. S. Aleman, A. V. Cideciyan, S. E. Pearce-Kelling, V. Anand, Y. Zeng, A. M. Maguire, S. G. Jacobson, W. W. Hauswirth and J. Bennett (2001) Gene therapy restores vision in a canine model of childhood blindness. Nature Genetics 28, 92–5.

R. B. Aramant and M. J. Seiler (2002) Retinal transplantation - advantages of intact fetal sheets. Progress in Retinal and Eye Research 21, 57–73.

C. Bowes, T. Li, M. Danciger, L. C. Baxter, M. L. Applebury and D. B. Farber (1990) Retinal degeneration in the rd mouse is caused by a defect in the beta subunit of rod cGMP-phosphodiesterase. Nature 347, 677–80.

F. Chan, A. Bradley, T. G. Wensel and J. H. Wilson (2004) Knock-in human rhodopsin-GFP fusions as mouse models for human disease and targets for gene therapy. Proc. Natl. Acad. Sci. 101, 9109–14.

N. Cuenca, I. Pinilla, Y. Sauvé, B. Lu, S. Wang and R. Lund (2004) Regressive and reactive changes in the connectivity patterns of rod and cone pathways of P23H transgenic rat retina. Neurosience 127, 301–17.

P. M. D' Cruz, D. Yasumura, J. Weir, M. T. Matthes, H. Abderrahim, M. M. LaVail and D. Vollrath (2000) Mutation of the receptor tyrosine kinase gene Mertk in the retinal dystrophic RCS rat. Human Molecular Genetics 9, 645–51.

S. P. Daiger (2004) Identifying retinal disease genes: how far have we come, how far do we have to go? Novartis Foundation Symposium 255, 17–27.

N. S. Dejneka, T. S. Rex and J. Bennett (2003) Gene therapy and animal models for retinal disease. Dev Ophthalmol 37, 188–98.

J. Delbeke, D. Pins, G. Michaux, M.-C. Wanet-Defalque, S. Parrini and C. Veraart (2001) Electrical stimulation of anterior visual pathways in retinitis pigmentosa. Invest Ophthalmol Vis Sci 42, 291–7.

A. Eisenfeld, A. H. Bunt-Milam and P. V. Sarthy (1984) Muller cell expression of glial fibrillary acidic protein after genetic and experimental photoreceptor degeneration in the rat retina. Invest Ophthalmol Vis Sci 25, 1321–8.

R. N. Fariss, Z. Y. Li and A. H. Milam (2000) Abnormalities in rod photoreceptors, amacrine cells, and horizontal cells in human retinas with retinitis pigmentosa. American Journal of Ophthalmology 129, 215–23.

G. J. Farrar, P. F. Kenna and P. Humphries (2002) On the genetics of retinitis pigmentosa and on mutation-independent approaches tro therapeutic intervention. EMBO Journal 21, 857–64.

Y. Fei (2002) Cone neurite sprouting: An early onset abnormality of the cone photoreceptors in the retinal degeneration mouse. Mol Vis 8, 306–14.

M. B. Feller (1999) Spontaneous correlated activity in developing neural circuits. Neuron 22, 653–6.

M. B. Feller (2002) The role of nAChR-mediated spontaneous retinal activity in visual system development. Journal of Neurobiology 53, 556–67.

S. I. Firth and M. B. Feller (2004) Cellular basis of periodic bursts of spontaneous activity in cultured retinal neurons. Invest Ophthalmol Vis Sci 45, E-Abstract 5311.

S. K. Fisher and G. P. Lewis (2003) Muller cell and neuronal remodeling in retinal detachment and reattachment and their potential consequences for visual recovery: a review and reconsideration of recent data. Vision Research 43, 887–97.

E. L. Fletcher (2000) Alterations in neurochemistry during retinal degeneration. Microsc. Res. Tech. 50, 89–102.

E. L. Fletcher and M. Kalloniatis (1996) Neurochemical architecture of the normal and degenerating rat retina. Journal of Comparative Neurology 376, 343–60.

N. Gupta, K. E. Brown and A. H. Milam (2003) Activated microglia in human retinitis pigmentosa, late-onset retinal degeneration, and age-related macular degeneration. Exp Eye Res 76, 463–71.

W. Hao, A. Wenzel, M. S. Obin, C. K. Chen, E. Brill, N. V. Krasnoperova, P. Eversole-Cire, Y. Kleyner, A. Taylor, M. I. Simon, C. Grimm, C. E. Remé and J. Lem (2002) Evidence for two apoptotic pathways in light-induced retina degeneration. Nature Genetics 32, 254–60.

S. Haverkamp and H. Wassle (2004) Characterization of an amacrine cell type of the mammalian retina immunoreactive for vesicular glutamate transporter 3. Journal of Comparative Neurology 468, 251–63.

J. R. Heckenlively, S. L. Yoser, L. H. Friedman and J. J. Oversier (1988) Clinical findings and common symptoms in retinitis pigmentosa. Am J Ophthalmol 105, 504–11.

M. E. Illing, R. S. Rajan, N. F. Bence and R. R. Kopito (2002) A rhodopsin mutant linked to autosomal dominant retinitis pigmentosa is prone to aggregate and interacts with the ubiquitin proteasome system. Journal of Biological Chemistry 37, 34150–60.

A. J. Jiménez, J.-M. García-Fernández, B. González and R. Foster (1996) The spatio-temporal pattern of photoreceptor degeneration in the aged rd/rd mouse retina. Cell and Tissue Research 284, 193–202.

S. K. John, J. E. Smith, G. D. Aguirre and A. H. Milam (2000) Loss of cone molecular markers in rhodopsin-mutant human retinas with retinitis pigmentosa. Molecular Vision 6, 204–15.

B. W. Jones, R. E. Marc and C. B. Watt (2004) Dopaminergic amacrine and inter-plexiform cells exhibit glutamatergic signatures. Invest Ophthalmol Vis Sci 45, E Abstract 5435.

B. W. Jones, C. B. Watt, J. M. Frederick, W. Baehr, C. K. Chen, E. M. Levine, A. H. Milam, M. M. LaVail and R. E. Marc (2003a) Retinal remodeling triggered by photoreceptor degenerations. Journal of Comparative Neurology 464, 1–16.

B. W. Jones, C. B. Watt, D. K. Vaughan, D. Organisciak and R. E. Marc (2003b) Retinal remodeling triggered by light damage in the albino rat. invest Ophthalmol Vis Sci Suppl 44, E-Abstract 5124.

M. Kalloniatis, R. E. Marc and R. F. Murry (1996) Amino acid signatures in the primate retina. Journal of Neuroscience 16, 6807–29.

M. A. Khodair, M. Zarbin and E. Townes-Anderson (2003) Synaptic plasticity in mammalian photoreceptors prepared as sheets for retinal transplantation. invest Ophthalmol Vis Sci 44, 4976–88.

H. J. Klassen, T. F. Ng, Y. Kurimoto, I. Kirov, M. Shatos, P. Coffey and M. J. Young (2004) Multipotent retinal progenitors express developmental markers, differentiate into retinal neurons, and preserve light-mediated behavior. Invest Ophthalmol Vis Sci 45, 4167–73.

R. Koyama, M. K. Yamada, S. Fujisawa, R. Katoh-Semba, N. Matsuki and Y. Ikegaya (2004) Brain-derived neurotrophic factor Induces hyperexcitable reentrant circuits in the dentate gyrus. Journal of Neuroscience 24, 7215–24.

M. M. LaVail, D. Yasumura, M. T. Matthes, K. A. Drenser, J. G. Flannery, A. S. Lewin and W. W. Hauswirth (2000) Ribozyme rescue of photoreceptor cells in P23H transgenic rats: long-term survival and late-stage therapy. Proc. Natl. Acad. Sci. 97, 11488–93.

A. S. Lewin, K. A. Drenser, W. W. Hauswirth, S. Nishikawa, D. Yasumura, J. G. Flannery and M. M. LaVail (1998) Ribozyme rescue of photoreceptor cells in a transgenic rat model of autosomal dominant retinitis pigmentosa. Nature Medicine 4, 967–71.

G. P. Lewis, D. G. Charteris, C. S. Sethi, W. P. Leitner, K. A. Linberg and S. K. Fisher (2002) The ability of rapid retinal reattachment to stop or reverse the cellular and molecular events initiated by detachment. Investigative Ophthalmology & Visual Science 43, 2412–20.

G. P. Lewis, K. A. Linberg and S. K. Fisher (1998) Neurite outgrowth from bipolar and horizontal cells after experimental retinal detachment. Investigative Ophthalmology & Visual Science 39, 424–34.

Z. Li, L. Van Aelst and H. T. Cline (2000) Rho GTPases regulate distinct aspects of dendritic arbor growth in Xenopus central neurons in vivo. Nature Neuroscience 3, 217–25.

Z. Y. Li, I. J. Kljavin and A. H. Milam (1995) Rod photoreceptor neurite sprouting in retinitis pigmentosa. Journal of Neuroscience 15, 5429–38.

R. Lund, Y. Sauve, D. J. Keegan, H. Winton, P. Coffey and S. Wang (2000) Immortalised human RPE cells retain phenotypical characteristics and limit deterioration of vision in the RCS rat. Investigative Ophthalmology & Visual Science 41, S857–S.

R. D. Lund, P. Adamson, Y. Sauve, D. J. Keegan, S. V. Girman, S. M. Wang, H. Winton, N. Kanuga, A. S. L. Kwan, L. Beauchene, A. Zerbib, L. Hetherington, P. O. Couraud, P. Coffey and J. Greenwood (2001a) Subretinal transplantation of genetically modified human cell lines attenuates loss of visual function in dystrophic rats. Proceedings of the National Academy of Sciences of the United States of America 98, 9942–7.

R. D. Lund, A. S. L. Kwan, D. J. Keegan, Y. Sauve, P. J. Coffey and J. M. Lawrence (2001b) Cell transplantation as a treatment for retinal disease. Progress in Retinal and Eye Research 20, 415–49.

R. D. Lund, S. J. Whiteley, Y. Sauve, T. Pheby and J. M. Lawrence (1999) Schwann cell grafting into the retina of the dystrophic RCS rat prolongs photoreceptor survival. Investigative Ophthalmology & Visual Science 40, S728–S.

R. E. Marc (1992) The structure of GABAergic circuits in ectotherm retinas. In: GABA in the Retina and Central Visual System (edited by R. Mize, R. E. Marc and A. Sillito), pp. 61–92. Amsterdam: Elsevier.

R. E. Marc (1999a) Kainate activation of horizontal, bipolar, amacrine, and ganglion cells in the rabbit retina. Journal of Comparative Neurology 407, 65–76.

R. E. Marc (1999b) Mapping glutamatergic drive in the vertebrate retina with a channel-permeant organic cation. Journal of Comparative Neurology 407, 47–64.

R. E. Marc (2002) A review of: Retinal Cell Rescue – The Sixth Annual Vision Research Conference. Investigational Drugs: Weekly Highlights June, 53–60.

R. E. Marc (2004) Retinal Neurotransmitters. In: The Visual Neurosciences (edited by E. Chalupa and Werner), pp. 315–30: MIT Press.

R. E. Marc and D. A. Cameron (2002) A molecular phenotype atlas of the zebrafish retina. Journal of Neurocytology 30, 593–654.

R. E. Marc and B. W. Jones (2002) Molecular phenotyping of retinal ganglion cells. Journal of Neuroscience 22, 412–27.

R. E. Marc, B. W. Jones and R. J. Lucas (2004) Excitatory self–signaling In retinal remodeling. Invest Ophthalmol Vis Sci 45, E-Abstract 833.

R. E. Marc, B. W. Jones, C. B. Watt and E. Strettoi (2003) Neural Remodeling in Retinal Degeneration. Prog Ret Eye Res 22, 607–55.

R. E. Marc, B. W. Jones, J. H. Yang, M. V. Shaw and A. H. Milam (2001) Molecular phenotyping of the neural retina in retinitis pigmentosa. Invest Ophthalmol Vis Sci 42, S117.

R. E. Marc and W. Liu (2000) Fundamental GABAergic amacrine cell circuitries in the retina: nested feedback, concatenated inhibition, and axosomatic synapses. Journal of Comparative Neurology 425, 560–82.

R. E. Marc, R. F. Murry and S. F. Basinger (1995) Pattern recognition of amino acid signatures in retinal neurons. Journal of Neuroscience 15, 5106–29.

R. E. Marc, R. F. Murry, S. K. Fisher, K. A. Linberg and G. P. Lewis (1998) Amino acid signatures in the detached cat retina. Investigative Ophthalmology & Visual Science 39, 1694–702.

E. Margalit, M. Maia, J. D. Weiland, R. J. Greenberg, G. Y. Fujii, G. Torres, D. V. Piyathaisere, T. M. O'Hearn, W. Liu, G. Lazzi, G. Dagnelie, D. A. Scribner, E. J. de

Juan and M. S. Humayun (2002) Retinal prosthesis for the blind. Survey of Ophthalmology 47, 335–56.

R. H. Masland (2001) Neuronal diversity in the retina. Current Opinion in Neurobiology 11, 431–6.

N. L. Mata, R. A. Radu, R. S. Clemmons and G. H. Travis (2002) Isomerization and oxidation of Vitamin A in cone-dominant retinas: A novel pathway for visual-pigment regeneration in daylight. Neuron 36, 69–80.

L. H. McGee Sanftner, H. Abel, W. W. Hauswirth and J. G. Flannery (2001) Glial cell line derived neurotrophic factor delays photoreceptor degeneration in a transgenic rat model of retinitis pigmentosa. Mol Therapy 4, 622–9.

K. Mervin, K. Valter, J. Maslim, G. Lewis, S. Fisher and J. Stone (1999) Limiting photoreceptor death and deconstruction during experimental retinal detachment: The value of oxygen supplementation. American Journal of Ophthalmology 128, 155–64.

A. H. Milam, Z. Y. Li, A. V. Cideciyan and S. G. Jacobson (1996) Clinicopathologic effects of the Q64ter rhodopsin mutation in retinitis pigmentosa. Invest Ophthalmol Vis Sci 37, 753–65.

A. H. Milam, Z. Y. Li and R. N. Fariss (1999) Histopathology of the human retina in retinitis pigmentosa. Prog Ret Eye Res 18,175–205.

S. Mohand-Said, A. Deudon-Combe, D. Hicks, M. Simonutti, V. Forster, A.-C. Fintz, T. Léveillard, H. Dreyfus and J.-A. Sahel (1998) Normal retina releases a diffusible factor stimulating cone survival in the retinal degeneration mouse. Proc. Natl. Acad. Sci. 95, 8357–62.

E. Newman and A. Reichenbach (1996) The Muller cell: a functional element of the retina. Trends in Neuroscience 19, 307–12.

D. Palanker, P. Huie, A. Vankov, R. B. Aramant, M. J. Seiler, H. Fishman, M. Marmor and M. Blumenkranz (2004) Migration of retinal cells through a perforated membrane: implications for a high-resolution prosthesis. Invest Ophthalmol Vis Sci 45, 3266–70.

S. J. Park, I. B. Kim, K. R. Choi, J. I. Moon, S. J. Oh, J. W. Chung and M. H. Chun (2001) Reorganization of horizontal cell processes in the developing FVB/N mouse retina. Cell and Tissue Research 306, 341–6.

Y.-W. Peng, Y. Hao, R. M. Petters and F. Wong (2000) Ectopic synaptogenesis in the mammalian retina caused by rod photoreceptor-specific mutations. Nature Neuroscience 3, 1121–7.

V. Pignatelli, C. L. Cepko and E. Strettoi (2004) Inner retinal abnormalities in a mouse model of Leber's congenital amaurosis. Journal of Comparative Neurology 469, 351–9.

R. Pinaud, P. De Weerd, R. W. Currie, M. Fiorani, F. F. Hess and L. A. Tremere (2003) NGFI-A immunoreactivity in the primate retina: Implications for genetic regulation of plasticity. International Journal of Neuroscience 113, 1275–85.

R. Pinaud, L. A. Tremere, M. R. Penner, F. F. Hess, S. Barnes, H. A. Robertson and R. W. Currie (2002) Plasticity-driven gene expression in the rat retina. Molecular Brain Research 98, 93–101.

D. V. Pow and S. R. Robinson (1994) Glutamate in Some Retinal Neurons Is Derived Solely from Glia. Neuroscience 60, 355–66.

D. V. Pow and R. Sullivan (2002) Abnormal myelination and migration of adult neurones along glial scaffolds in degenerating adult rat retinas. Invest Ophthalmol Vis Sci 43, E-Abstract 2720.

L. Redmond, A. H. Kashani and A. Ghosh (2002) Calcium regulation of dendritic growth via CaM Kinase IV and CREB-mediated transcription. Neuron 34, 999–1010.

B. E. Reese and L. Galli-Resta (2002) The role of tangential dispersion in retinal mosaic formation. Prog Ret Eye Res 21, 153–68.

H. Ripps (2002) Cell death in retinitis pigmentosa: gap junctions and the "bystander" effect. Exp Eye Res 74, 327–36.

R. S. Roque, A. A. Rosales, L. Jingjing, N. Agarwal and M. R. Al-Ubaidi (1999) Retina-derived microglial cells induce photoreceptor cell death in vitro. Brain Research 836, 110–9.

S. F. Stasheff (2004) Increased spontaneous activity and high frequency rhythmic firing among retinal ganglion cells accompanies outer retinal degeneration in the rd1 mouse. Invest Ophthalmol Vis Sci 45, E-Abstract 5070.

D. Stellwagen, C. J. Shatz and M. B. Feller (1999) Dynamics of retinal waves are controlled by cyclic AMP. Neuron 24, 673–85.

J. Stone, J. Maslim, K. Valter-Kocsi, K. Mervin, F. Bowers, Y. Chu, N. Barnett, J. Provis, G. Lewis, S. K. Fisher, S. Bisti, C. Gargini, L. Cervetto, S. Merin and J. Pe'er (1999) Mechanisms of photoreceptor death and survival in mammalian retina. Progress in Retinal and Eye Research 18, 689–735.

E. Strettoi and V. Pignatelli (2000) Modifications of retinal neurons in a mouse model of retinitis pigmentosa. Proc. Natl. Acad. Sci. 97, 11020–5.

E. Strettoi, V. Pignatelli, C. Rossi, V. Porciatti and B. Falsini (2003) Remodeling of second-order neurons in the retina of rd/rd mutant mice. Vision Research 43, 867–77.

E. Strettoi, V. Porciatti, B. Falsini, V. Pignatelli and C. Rossi (2002) Morphological and functional abnormalities in the inner retina of the rd/rd mouse. Journal of Neuroscience 22, 5492–504.

R. Sullivan, P. Penfold and D. V. Pow (2003) Neuronal migration and glial remodeling in degenerating retinas of aged rats and in nonneovascular AMD. Invest Ophthalmol Vis Sci 44, 856–65.

N. Tian (2003) Light deprivation blocks the developmental changes of ganglion cell dendritic ramification in mouse retina. Investigative Ophthalmology & Visual Science 44, E-Abstract 3237.

N. Tian and D. R. Copenhagen (2001) Visual deprivation alters development of synaptic function in inner retina after eye opening. Neuron 32, 439–49.

N. Tian and D. R. Copenhagen (2003) Visual stimulation is required for refinement of ON and OFF pathways in postnatal retina. Neuron 39, 85–96.

K. Valter, J. Maslim, F. Bowers and J. Stone (1998) Photoreceptor dystrophy in the RCS rat: roles of oxygen, debris, and bFGF. Invest Ophthalmol Vis Sci 39, 2427–42.

C. Varela, I. Igartua, E. J. De la Rosa and P. De la Villa (2003) Functional modifications in rod bipolar cells in a mouse model of retinitis pigmentosa. Vision Research 43, 879–85.

S. Vistamehr and N. Tian (2004) Light deprivation suppresses the light response of inner retina in both young and adult mouse. Visual Neuroscience 21, 23–37.

K. J. Wahlin, P. Campochiaro, D. J. Zack and R. Adler (2000) Neurotrophic factors cause activation of intracellular signaling pathways in Müller cells and other cells of the inner retina, but not photoreceptors. Invest Ophthalmol Vis Sci 41, 927–36.

C. B. Watt, B. W. Jones, J. H. Yang, R. E. Marc and M. M. LaVail (2004) Complex rewiring In retinal remodeling. Invest Ophthalmol Vis Sci 45, E-Abstract 777.

R. Weleber and K. Gregory-Evans (2001) Retinitis pigmentosa and allied disorders. In: Retina: Basic science, inherited disease and tumors (edited by T. E. Ogden and D. R. Hinton), pp. 362–460. St Louis: Mosby.

H. Wässle (2004) Parallel processing in the mammalian retina. Nature Reviews Neuroscience 5, 747–57.

X. Xie, M. C. Schnieder, E. Sa'nchez-Sinencio and S. H. K. Embabi (1999) Sound design of low power nested transconductance-capacitance compensation amplifiers. IEEE Electronics Letters 35, 956–8.

H. P. Xu and N. Tian (2004) Pathway-specific maturation, visual deprivation, and development of retinal pathway. Neuroscientist 10, 337–46.

M. J. Young, J. Ray, S. J. O. Whiteley, H. Klassen and F. H. Gage (2000) Neuronal differentiation and morphological integration of hippocampal progenitor cells transplanted to the retina of immature and mature dystrophic rats. Molecular and Cellular Neuroscience 16, 197–205.

C. J. Zeiss and E. A. Johnson (2004) Proliferation of microglia, but not photoreceptors, in the outer nuclear layer of the rd-1 mouse. Invest Ophthalmol Vis Sci 45, 971–6.

N. Zhang and E. Townes-Anderson (2002) Regulation of structural plasticity by different channel types in rod and cone photoreceptors. Journal of Neuroscience 22, 7065–7079.

E. Zrenner (2002) Will Retinal Implants Restore Vision? Science 295, 1022–5.

4

Retinal Plasticity and Interactive Cellular Remodeling in Retinal Detachment and Reattachment

GEOFFREY P. LEWIS AND STEVEN K. FISHER

Neuroscience Research Institute, University of California Santa Barbara,
Santa Barbara, CA, USA

Introduction

During development, the CNS demonstrates a remarkable ability to regenerate damaged neuronal processes. New processes are able to grow through lesioned areas and even re-establish connections to their targets (Saunders et al., 1995). This ability has traditionally been considered lost in the adult mammalian CNS. Indeed, molecules have been identified on some CNS glia that specifically inhibit axonal growth in the adult (Chen et al., 2000). Recent evidence, however, suggests that the adult CNS is in fact capable of a greater repertoire of responses to various forms of insult than previously thought, with many of these mediated through the up-regulation of growth associated proteins (see Emery et al., 2003 for review). Furthermore, there is considerable evidence that new neurons and glia are generated in specific regions of the adult brain as part of a normal renewal process (Gage, 2000). These responses are not limited to the brain and spinal cord. The adult mammalian retina also exhibits responses indicative of plasticity or remodeling as demonstrated by its ability to respond to enriched sensory environments (Pinaud et al., 2002) or injury (Coblentz et al., 2003) in up-regulating growth associated proteins, as well as generating new cells through cell division after injury (Fisher et al., 1991). Moreover, retinal neurons appear to have the ability to initiate the growth of neurites that can extend great lengths throughout all layers of the tissue. These responses appear to be relatively common to different species as well to different types of insult. Animal and human inherited degenerations, as well as various forms of trauma, all elicit some form of neuronal and glial remodeling in the adult retina (see Marc et al., 2003 for review).

In this chapter, retinal detachment and reattachment in the feline retina will be used as the primary form of injury that initiates dramatic remodeling

R.Pinaud, L.A. Tremere, P. De Weerd(Eds.), Plasticity in the Visual System:
From Genes to Circuits, 55-78, ©2006 Springer Science + Business Media, Inc.

of both neurons and glia thus illustrating the plasticity retained in the adult retina. A benefit to this model is that the timing of both the injury and the recovery can be precisely controlled and the sequence of changes followed. Using this model we have observed that the growth of neuronal processes initiated by detachment occurs rapidly and in many cases does not appear to be a random process. Rather, much of the remodeling occurs as a result of morphological changes of other cell types. This "interactive cellular remodeling" begins with photoreceptors and continues to second and third order neurons as the detachment interval proceeds. We have proposed that the initiating factor of this remodeling is hypoxia of the outer retina (Lewis et al., 1999b). This results in programmed cell death of some photoreceptors (Cook et al., 1995) and programmed deconstruction of surviving photoreceptors (Mervin et al., 1999) which then sets in motion a specific sequence of remodeling events including both neurons and glia. Following retinal reattachment and a reperfusion of retinal oxygen levels, many of the glial and neuronal changes that begin as a result of detachment are halted and some are reversed. However, reattachment also elicits its own set plastic responses (Lewis et al., 2003) many of which are dependent on other cell types.

Retinal detachment is defined as the separation of the neural retina from the underlying retinal pigment epithelial (RPE) layer. At a cellular level, this results in a withdrawal of the rod and cone outer segments from the ensheathing microvillous processes of the RPE (Fig. 4.1, blue cells). Initially, this causes only minor structural damage to the outer segments but an immediate decline in vision. Moreover, this begins a cascade of cellular and molecular events in the retina that will continue as long as the retina remains detached. The first and most obvious structural change that can be observed is the degeneration of the photoreceptor outer segments. This portion of the cell becomes progressively truncated with time until there is eventually only an inner segment with a short connecting cilium present. In perhaps a remarkable example of neuronal regeneration, the highly degenerate outer segments can regenerate following retinal reattachment and most of them eventually appear morphologically normal. This, however, leads to an interesting dichotomy. On the one hand, the retina appears to be able recover from the loss of outer segments. On the other hand, visual deficits are not uncommon following successful reattachment surgery. Indeed it has been shown that only 20–60% of successful macular reattachments achieve a visual acuity of 20/50 or better (Burton et al., 1980; Tani et al., 1980; 1981; Ross, 2002). Moreover, it appears that the process of visual recovery is slow, in some cases continuing to change for months or years following reattachment (Liem et al., 1994; Ross, 2002). If the outer segments can recover relatively quickly (over a few weeks), what is it that accounts for the continued change in vision over time? We would like to propose here that changes in photoreceptors initiate downstream changes in many retinal cell types, resulting in a significant "rewiring" of the retina and ultimately leading to changes in vision. Recovery

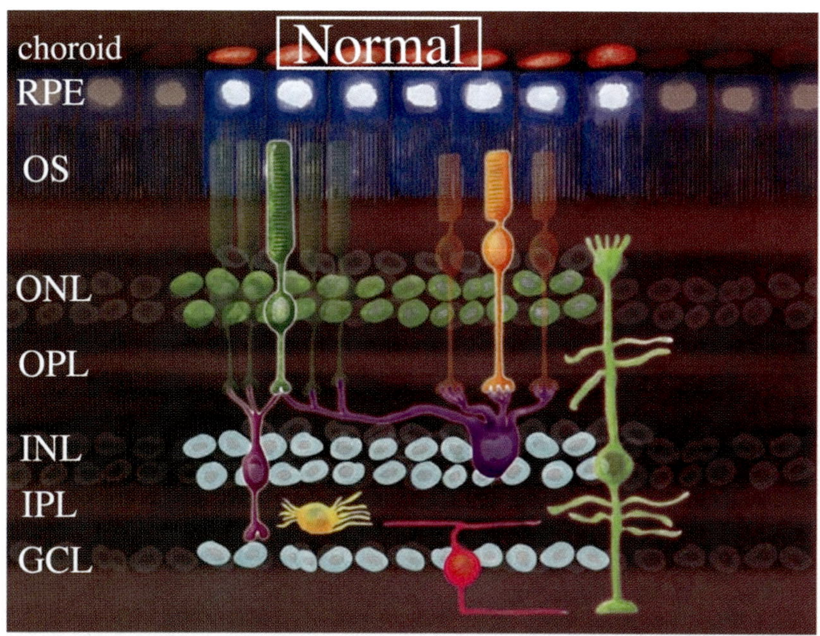

choroid
Normal
RPE
OS
ONL
OPL
INL
IPL
GCL

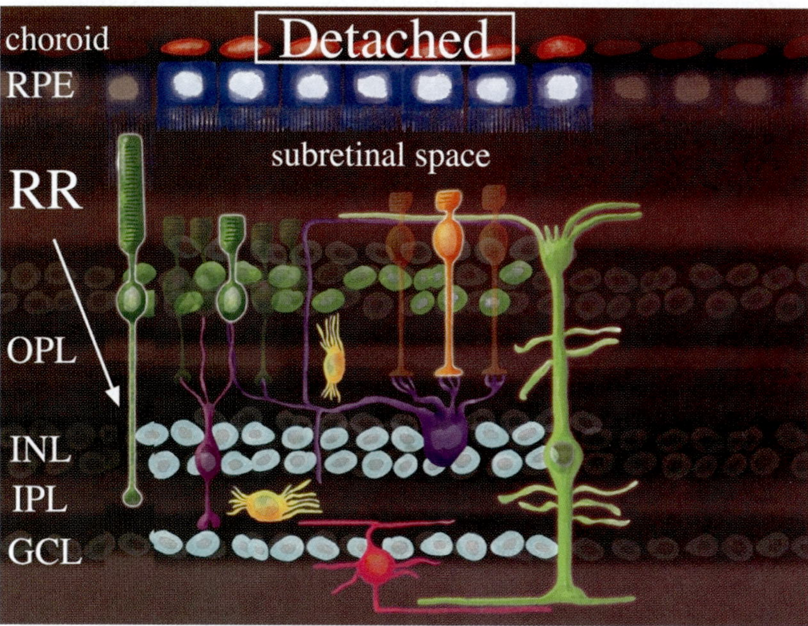

choroid
Detached
RPE
subretinal space
RR
OPL
INL
IPL
GCL

itself would also require either physical rewiring or some other form of compensation for recovery of normal vision.

Surgical Procedures

Retinal detachments in our laboratory are created by first removing the lens and vitreous and then infusing a solution of sodium hyaluronate (Healon; 0.25%), a natural component of the vitreous and interphotoreceptor matrix, in a balanced salt solution between the retina and RPE via a glass micropipette (Lewis et al., 1999a). The Healon is used to prevent the retina from spontaneously reattaching. The pipette, which is secured to a micromanipulator, is attached to tubing that connects to an infusion pump. This allows precise control of the position and size of the detachment. The pipette, which creates a small hole as it passes through the retina, produces what we believe models rhegmatogenous detachments. In humans, a rhegmatogenous detachment refers to the fact that a hole or tear is present in the retina. As a result, fluid passes through the break in the neural retina to the subretinal space, thus creating the detachment. This is an important distinction since retinal detachments that occur without the presence of a hole, instead developing as a result of fluid accumulating under the retina (in what is termed a "serous" detachment), often show very few signs of photoreceptor degeneration or visual loss. The reason for the differences in the responses is unclear, however, it has been suggested that the composition of the subretinal fluid remains relatively normal because it does not mix with fluid from the vitreous, perhaps retaining factors in the subretinal space that the preserve the retina. Although in this animal model there is no large tear in the retina as there often is in human rhegmatogenous detachments, the retinal changes documented so far in

FIGURE 4.1. Diagram of the morphology of retinal cell types in the normal and detached retina, illustrating the remodeling that occurs following detachment. Retinal detachment occurs between the RPE (blue) and the photoreceptor OS. This initiates the degeneration and shortening of rod (dark green) and cone (yellow) OS as well as a retraction of rod axon terminals to their cell body into the ONL. Rod bipolar cell dendrites (light purple) and neurites from the axon terminals of horizontal cells (dark purple) extend adjacent to the retracted rod terminals. Horizontal cell processes continue to grow wildly into the subretinal space. Ganglion cell bodies (red) in the GCL sprout neurites. Microglia (yellow) in the IPL become activated and migrate towards the degenerating photoreceptors. Müller cells (light green), which span the width of the retina, hypertrophy, extend processes into the subretinal space, and act as substrates for the growth of horizontal cells and ganglion cells. Retinal reattachment (RR) stimulates the growth of rod axons into the inner retina (arrow) as well as the growth of Müller cells onto the vitreal surface of the retina that then serves as a substrate for the growth of ganglion cell neurites. (GCL, ganglion cell layer; IPL, inner plexiform layer; INL, inner nuclear layer; ONL, outer nuclear layer; OS, outer segments; RPE, retinal pigment epithelium.)

human detachments are essentially identical to those observed in the feline retinas (Chang et al., 1996; Sethi et al., 2005).

Experimentally, retinal reattachments are created by pneumatic retinopexy. That is, the fluid is removed from the vitreous chamber, allowing fluid to flow out of the subretinal space through the hole and the retina to settle back down on the RPE. The eye is then flushed with 20% sulfur hexafluoride (mixed with filtered room air) that acts as a tamponade to keep the retina adjacent to the RPE until the 2 layers become physiologically attached.

Retinal Detachment and Reattachment

The retina is a highly organized tissue consisting of 3 nuclear layers and 2 synaptic layers (Fig. 4.1A). Because it is so highly structured, any significant change from its normal architecture is easily noted by conventional histological procedures. As early as 1968 investigators used light and electron microscopy to reveal that detachment induced a number of changes in photoreceptors including outer segment degeneration, the loss of mitochondria, and the disorganization of organelles involved in protein synthesis and trafficking (Machemer, 1968; Kroll and Machemer, 1969a, 1969b; Anderson et al., 1983). It wasn't until the introduction of immunocytochemistry, however, that it became apparent that cells within the adult retina do not simply degenerate over time but are actually capable of significant cellular remodeling. While evidence suggests that the photoreceptors are the first neuronal cell type to respond to detachment, Müller cells also respond rapidly to detachment and since they appear to be intimately involved in all levels of the neuronal remodeling, they will be introduced first.

Müller cells

Müller cells are radial glial cells that span the entire width of the retina (Fig. 4.1A, light green cell). The inner most portion of the cell lies adjacent to the vitreous and possesses a club shaped "endfoot" while the cell body lies in the central part of the retina, in the inner nuclear layer (INL). The outer portion of the cell consists of numerous branches that weave around photoreceptors and ends with a tuft of microvilli that interdigitate between the photoreceptor inner segments. Because of this they are uniquely positioned to detect changes in photoreceptors or the interphotoreceptor matrix that occurs as a result of detachment. These polarized cells contribute to the 2 boundaries of the retina. Their endfeet and basement membrane (the "inner limiting membrane", ILM) form the vitreal border of the neural retina while at the photoreceptor border, adherens junctions between the Müller cells and photoreceptors make up the "outer limiting membrane" (OLM). For many years the exact function of Müller cells was enigmatic and they were generally described as playing only a supportive role. It is now becoming increasingly evident that they are intimately involved in many essential functions including glutamate recycling, potassium regulation and pH balance in the retina (see Sarthy and Ripps, 2001 for review).

Following retinal detachment the highly polarized morphology of Müller cells is rapidly altered as they undergo a number of biochemical and structural changes (see Fisher and Lewis, 2005 for review). These changes can be traced to events that occur within minutes of the retina becoming detached. Indeed, it has been shown that the extracellular signal–regulated kinase (ERK) is phosphorylated within 15 minutes of creating the detachment, and immunoreactivity for the immediate-early response genes, cFos and cJun, increases within 2 hours in these cells (Geller et al., 2001). Moreover, the FGF receptor, previously identified in Müller cells (Yamamoto et al., 1996), is phosophorylated with 15 minutes of detachment and dephosphorylated by 2 hours (Geller et al., 2001). Perhaps as a result of these early events Müller cells, normally quiescent, soon begin to proliferate. While the peak number of cells dividing occurs 3 days after detachment the Müller cells continue divide as long as the retina remains detached (Fisher et al., 1991; Geller et al., 1995). The details of the mitotic events in these cells are unknown, however, the presence of large rounded mitotic figures near the outer retinal border suggests that they must lose their complex morphology as they dedifferentiate and re-differentiate as part of the cell cycle. Morphological remodeling of Müller cells can be observed as early as one day after detachment (Lewis et al., 1994; 1995). At this time, they begin to hypertrophy with their main trunk and lateral branches expanding in size. Also at this time they greatly alter their expression for many proteins with glutamine synthetase, carbonic anhydrase II, and cellular retinaldehyde binding protein decreasing in expression (Lewis et al., 1989) and two intermediate filament proteins, glial fibrillary acidic protein (GFAP) and vimentin dramatically increasing. Expression of the latter are particularly useful to mark these cells as they remodel. In the normal feline retina these proteins are mainly restricted to the endfoot of the Müller cell (Fig. 4.2A). Following detachment these proteins rapidly increase their expression so that within a few days they fill the entire cytoplasm of the cell (Fig. 4.2B). Even though both GFAP and vimentin are upregulated, the balance of the two shifts over the course of the detachment time (Lewis and Fisher, 2003). In the normal retina, vimentin is the predominant protein in the endfoot with GFAP being less abundant. Following detachment there is a shift to more GFAP expression as the endfeet rapidly take on a more elongated and branched appearance compared to that in normal retina. This is readily seen in whole mount preparations where the endfeet are transformed from their clubbed appearance to long processes extending laterally along the inner border of the retina (Figs. 4.2D, 4.2E). The remodeling of the endfeet then progresses to a significant expansion of the entire cell with GFAP becoming the predominant intermediate filament protein. Interestingly, a sub-population of Müller cells express predominantly vimentin in the outer retina and it are these cells that extend beyond their normal boundary at the OLM into the subretinal space (Fig. 4.2B, arrow).

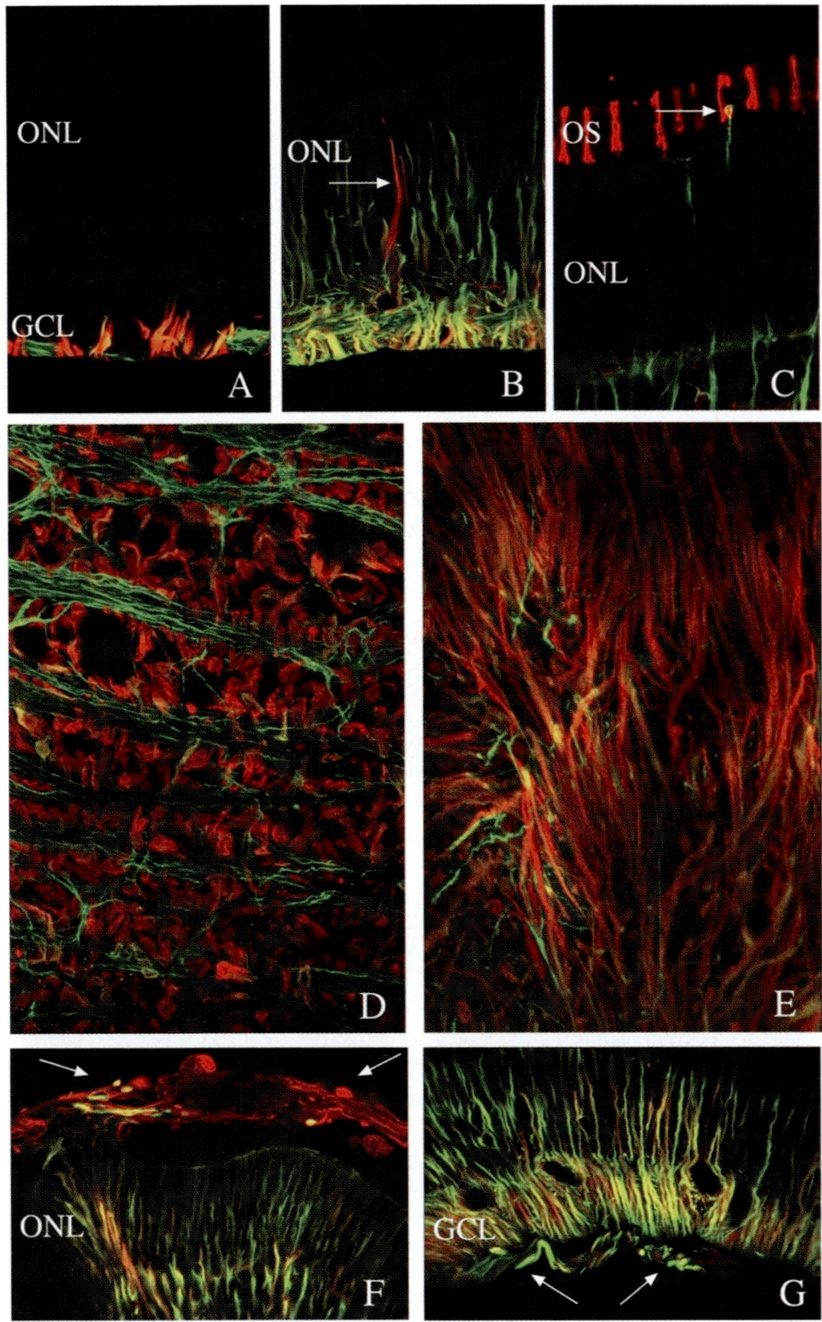
ONL
GCL
A
ONL
B
OS
ONL
C
D
E
ONL
F
GCL
G

The growth of these cells into the subretinal space, however, is not random. Indeed, those cells that initially extend past the OLM invariably do so adjacent to cones (Fig. 4.2C, arrow) (Lewis et al., 2000). Once this growth is initiated the Müller cells continue to expand over the surface of photoreceptors creating what is termed a glial scar (Fig. 4.2F, arrows). The consequences of the presence of the glial scar are several-fold. First, photoreceptor outer segments cannot regenerate and reform a connection with the RPE if this scar is present (Anderson et al., 1983). Second, photoreceptor cell bodies appear to be carried into this scar as Müller cells hypertrophy, thereby depleting the retina of photoreceptors. Third, the scar appears to act as a substrate for the growth of aberrant neuronal processes originating from horizontal and ganglion cells (discussed below; Fisher and Lewis, 2003).

Retinal reattachment appears to greatly slow much of the intraretinal hypertrophy of the Müller cells as well as their growth into the subretinal space (Lewis et al., 2002; 2003). It does not, however, completely halt their reactivity since cell division can still be observed a month after retinal reattachment, albeit at low levels. In addition, reattachment appears to be the stimulus in redirecting the growth of Müller cells from the subretinal space towards the vitreal surface of the retina; Müller cell growth on the vitreal surface is never observed in the detached retinas. As was the case with the growth of Müller cells into the subretinal space, the growth onto the vitreal surface also can have devastating consequences for vision. These cells can

FIGURE 4.2. Laser scanning confocal images demonstrating the remodeling of Müller cells after detachment and reattachment. (**A,B**) Sections are labeled with antibodies to the intermediate filament proteins vimentin (red) and glial fibrillary acidic protein (GFAP, green). (**A**) In the normal retina vimentin is the predominant intermediate filament in the Müller cell endfeet in the GCL while GFAP is less prevalent. (Anti-GFAP labeling also occurs in the astrocytes in the GCL.) (**B**) Following a 7-day detachment both proteins are upregulated although GFAP is expressed more widespread. A few cells express predominantly vimentin (arrow) and it are these cells that will grow into the subretinal space. (**C**) The initial growth of Müller cells (GFAP, green) into the subretinal space (arrow) occurs adjacent cone photoreceptor OS (labeled with peanut agglutinin, red). (**D,E**) Retinal flat-mounts focused at the level of the GCL and labeled with anti-vimentin (red) and anti-GFAP illustrating the remodeling of glia. (**D**) In normal retina the endfeet (mostly red) appear club-like while astrocyte processes (green) are organized in parallel arrays. (**E**) Following a 7-day detachment the endfeet hypertrophy and grow laterally across the retina while the astrocyte processes become highly disorganized. (**F,G**) Sections of detached retina illustrating the differential expression of intermediate filaments in sub-(**F**) and epiretinal (**G**) membranes. (**F**) The Müller cells that grow into the subretinal space following detachment express predominantly vimentin (red, arrows). (**G**) Those that grow onto the vitreal surface of the retina following reattachment express predominantly GFAP (green, arrows). (GCL, ganglion cell layer; ONL, outer nuclear layer; OS, outer segments).

form significant membranes, termed "epiretinal membranes" that contract and cause a re-detachment of the retina. In addition, we have observed, both in feline and human samples, that epiretinal membranes can act as substrates for the growth of ganglion cell neurites (discussed below). The mechanism that redirects the growth from the subretinal to the vitreal surface is not known, however, intermediate filaments once again appear to be involved. The Müller cells that first penetrate the inner limiting membrane and enter the vitreous cavity express predominantly GFAP over vimentin; just the opposite to the expression profile for those growing into the subretinal space (Fig. 4.2G, arrows). Even as the membranes mature and become quite large, GFAP predominates on the leading edge (Lewis and Fisher, 2003). From these examples it is clear that Müller cells are highly involved in the interactive cellular remodeling that occurs as a result of detachment; they are interacting with cone photoreceptors in the formation of subretinal glial membranes and they are interacting with second and third ordering neurons in creating a substrate for neurite sprouting and growth.

Astrocytes

Astrocytes, the only other glial cell type in the retina, also undergo cellular remodeling after detachment. In the normal retina they are located within the ganglion cell layer, their processes intimately associated with the axons of the ganglion cells (Figs. 4.2A, 4.2D). Following detachment they follow the same proliferation time scale as Müller cells, peaking at day 3 (Fisher et al., 1991; Geller et al., 1995) but cell division continues at low levels even after retinal reattachment (Lewis et al., 2003). It is difficult to examine the change in intermediate filament levels on sections because the large increase in intermediate filaments in Müller cells overwhelms that of the astrocytes. When viewed in a wholemount preparation, however, it is clear that these cells dramatically change their morphology in response to detachment. In normal retina astrocyte processes label more heavily with anti-GFAP than the Müller cell endfeet, which express predominantly vimentin (Fig. 4.2D). These GFAP labeled processes can be seen extending in organized parallel arrays as they follow bundles of ganglion cell axons. Following a detachment interval of just 3 days these processes begin to appear disorganized and by day 7, they extend in what appears to be a random orientation (Fig. 4.2E). In addition, following retinal reattachment, astrocytes migrate out of the retina, perhaps using Müller cells as a substrate, and contribute to the composition of epiretinal membranes (data not shown).

Photoreceptors

The photoreceptor cell is highly polarized with an outer segment at the distal end (which contains the molecular machinery for phototransduction) an inner segment, cell body, axon and synaptic terminal (Fig. 4.1A, dark green = rod cell & yellow = cone cell). These cells, considered the "first order neurons" of the retina since the visual transduction signal begins here and is transferred via 2nd and 3rd order neurons to the brain, are the first neuronal cell type in the retina to respond to detachment, with virtually every region of the cell undergoing change. Both rods and cones show evidence of plasticity but with striking differences in their response; rods very actively remodel while cones appear to become almost quiescent (Rex et al., 2002). This difference is evident structurally but even more so when examining the expression of their respective opsins, the photopigments responsible for capturing light. In the normal retina these proteins are restricted to the outer segment (Fig. 4.3A, rod opsin) but beginning within a few days after detachment, concomitant with outer segment degeneration, immunolabeling shows that the entire rod cell plasma membrane contains relatively high levels of rod opsin (Fig. 4.3B). Cones show a similar response initially but within a few days of detachment most stop expressing their opsins altogether. This pattern of protein expression continues as long as the retina remains detached. Following retinal reattachment, as the outer segments regenerate, the expression of cone opsin returns and rod opsin once again is restricted to the outer segments (Fig. 4.3C) (Lewis et al., 2003).

Antibodies to synaptic proteins can be used to visualize the synaptic terminal morphology of both rods and cones. The results indicate that both appear to change but in different ways. Early in detachment, the synaptic terminals of many rods begin to retract from the outer plexiform layer where they connect to second order neurons (the rod bipolar and horizontal cells) (Erickson et al., 1983; Lewis et al., 1998). The end result is that rod terminals, often with shortened synaptic ribbons, become located directly adjacent to their nucleus, appearing much like they do in the developing retina (Fig. 4.3D). The cone terminals, however, do not retract. They do change their morphology, with their multiple invaginations having a more shallow appearance or being lost completely, when compared to normal retina (Lewis et al., 2003).

Another remarkable display of cellular plasticity is observed in rods following retinal reattachment (Lewis et al., 2003). In retinas detached for 3 days and reattached for 28, the distribution of rod opsin, that at one time outlined the entire cell, usually appears relatively normal and is restricted to the regenerated outer segment (Fig. 4.3C). In some focal regions across the retina, however, the recovery appears to be incomplete; outer segments are shorter and opsin levels remain elevated in the plasma membrane. In these focal regions, rod axons often can be observed extending well beyond their normal target in the outer plexiform layer deep into the inner retina (Fig. 4.3F, arrows). The retraction and growth of rod axons appears to be a recapitulation of development; the retracted rod terminals appear structurally as if they were from an early period of development and the overgrowth of some rod axons appear similar to those observed deep in the inner layers of the normal developing ferret retina (Johnson et al., 1999). As of now there is no evidence that these terminals form synaptic connections in either case.

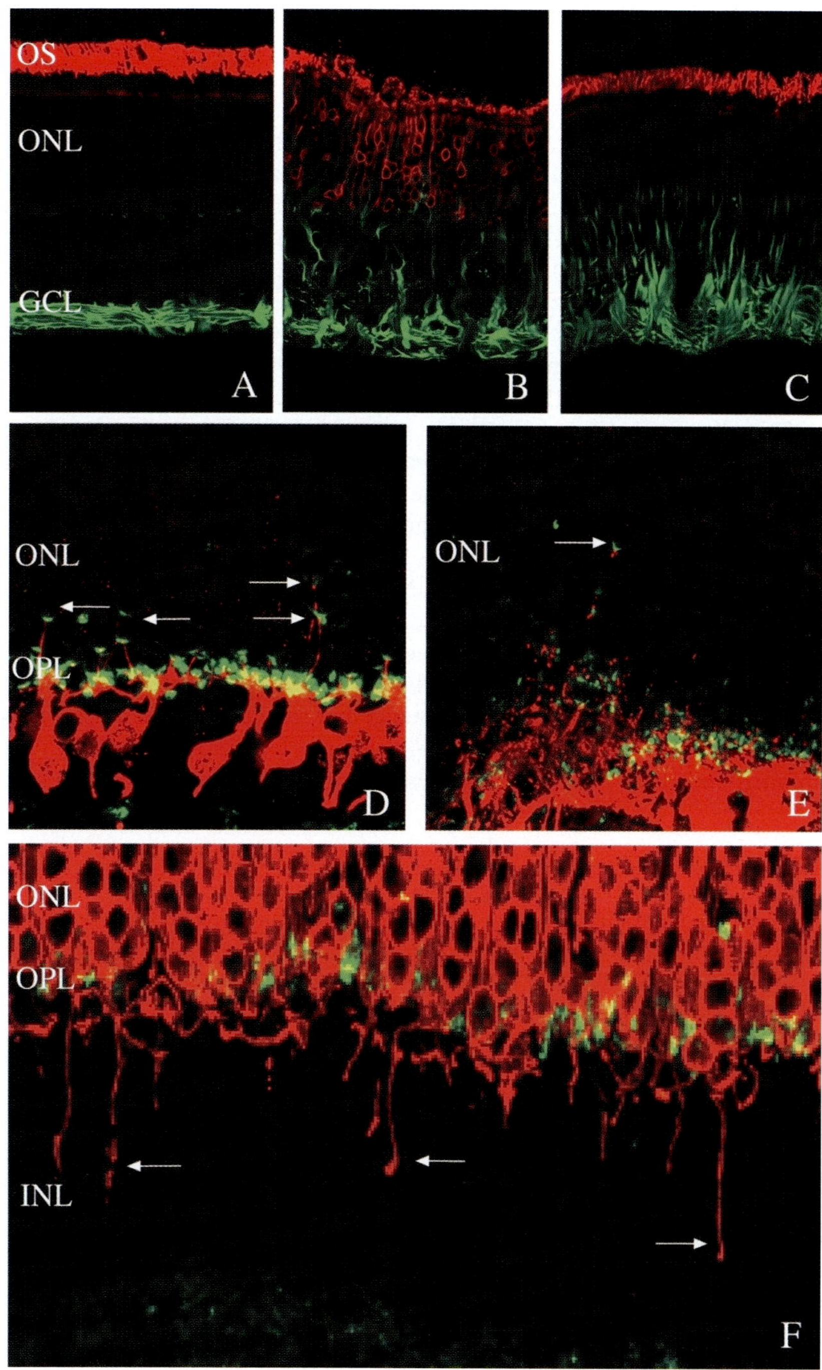

We also find no evidence of cone axon growth and this may be related to the fact that these terminals do not retract following detachment.

The factor(s) that initiates the re-growth of rod axons is not known but may be related to a physical interaction of photoreceptors with the RPE apical surface since we do not observe it in the detached retina. As mentioned above, over much of the reattached retina the outer segments and their interface with the RPE are indistinguishable from normal retina. However, in the abrupt patchy regions where rod and cone outer segment regeneration is incomplete, the cone-ensheathing apical RPE apical processes are either short or missing indicating that a truly normal physical interaction of RPE and outer segments had not been re-established (Lewis et al., 2003). Fortuitously, in these patchy regions, there is continued redistribution of rod opsin in the cell bodies and axons allowing for visualization of the aberrant rod axon growth beyond their normal targets. Without the elevated opsin levels the rod extensions would not have been observed. Thus we can't rule out the possibility that axon extension occurs in these other areas. Rod axon extensions have also been found in the human retinal degenerations, retinitis pigmentosa (RP) and age related macular degeneration (AMD) (Li et al., 1995; Fariss et al., 2000; Gupta et al., 2003), and in human detached retinas with "complex detachments" (i.e. the retinas had been reattached at least once before the tissue samples were obtained) (Sethi et al., 2005). In all of these examples the interaction between the RPE and outer segments may have been compromised. The molecular signals that stimulate the growth of rod axons have not been extensively studied but recent evidence indicates that regulation of Ca2+ influx may be involved since L-type channel antagonists inhibited rod axon outgrowth from salamander photoreceptors in culture

FIGURE 4.3. (**A,B,C**) Laser scanning confocal images of sections labeled with anti-rod opsin (red) and anti-GFAP (green) illustrating the degeneration and regeneration of rods as well as the hypertrophy of Müller cells. In normal retina rod opsin is restricted to the OS and GFAP is found only in Müller cell endfeet and astrocyte processes. (**B**) Following a 3-day detachment the outer segments are truncated, rod opsin is redistributed to the rod cell bodies in the ONL, and GFAP increases in Müller cells. (**C**) Following 3 days of detachment and 28 days of reattachment, rod outer segments regenerate and have a near normal morphology, rod opsin is no longer present in the rod cell bodies and the increase in GFAP is halted. (**D,E**) Confocal images of 7-day detachments demonstrating that neurite outgrowth from rod bipolar cells (**D**, red, anti-protein kinase C) and horizontal cells (**E**, red, anti-calretinin) is initially associated with the retraction of rod synaptic terminals (green, anti-synaptophysin, arrows). (**F**) Confocal image showing rod axon growth into the inner retina observed following retinal reattachment (red, anti-rod opsin, arrows). Anti-synaptophysin labeling (green) is observed in the terminals of rods and cones, both in their normal location in the OPL and on occasion, in the terminals of the axon extensions. (GCL, ganglion cell layer; INL, inner nuclear layer; OPL, outer plexiform layer; ONL, outer nuclear layer; OS, outer segments).

(Zhang and Townes-Anderson, 2002). The only correlation we have observed that may be relevant is an apparent association of these rod processes with activated microglia.

Microglia

Microglial cells are in fact not true glia but derived from bone marrow precursor cells that enter the retina at early stages of development from the optic nerve and lie "dormant", usually in the inner and outer plexiform layers of the adult retina. Following trauma and disease however, they are stimulated to assume macrophage-like functions, migrating into the damaged area and phagocytosing debris. Microglia have previously been shown to become activated in various forms of retinal degeneration including the RCS rat (Thanos et al, 1992; Roque et al., 1996), light damaged mice (Ng et al., 2001; Harada et al., 2002), glaucoma (Naskar et al., 2002; Lam et al., 2003), laser photocoagulation (Humphrey et al., 1996), and human RP (Gupta et al., 2003). It has been proposed that stressed or dying photoreceptors attract microglia which then eventually kill the photoreceptors since inhibition of their activation quantitatively reduces photoreceptor degeneration (Thanos et al., 1995). Indeed, microglia are known to secrete molecules that kill neurons such as pro-inflammatory cytokines, proteases, tumor necrosis factor, and nitric oxide (Lee et al., 2002). It is not known if microglia kill photoreceptor cells following retinal detachment, but they do become activated and migrate throughout the retina, eventually becoming prominent among the degenerating photoreceptors (Figs. 4.4A, 4.4B) (Lewis et al. 2005). They begin their migration towards the outer retina within a day of detachment and continue to be present at higher levels as long as the retina remains detached. When the retina is reattached after 3 days of detachment (a time when significant numbers of microglia are present in the outer retina) and examined a month later the microglial cell response appears, for the most part, to be reversed (Fig. 4.4C, left half of image). However, in focal regions of the retina where photoreceptor regeneration appears to lag behind, the number of microglia remains high, at a level similar to that observed during detachment (Fig. 4.4C, right half of image). At this time it is uncertain if the microglia in the disrupted areas are simply phagocytosing debris from dying photoreceptors or if they are actually contributing to photoreceptor cell death. Data from Harada et al. (2002) using light to induce photoreceptor degeneration, suggest that microglia derived neurotrophic factors can also modulate trophic factor expression in Müller cells. The result of such a microglial-Müller cell interaction could be beneficial or harmful at the level of individual photoreceptors depending on the type and balance of factors released. Because microglia may also control, at some level, the responses of Müller cells, they may also be key to modifying the glial responses that occur after detachment. Controlling this balance of survival and death factors may turn out

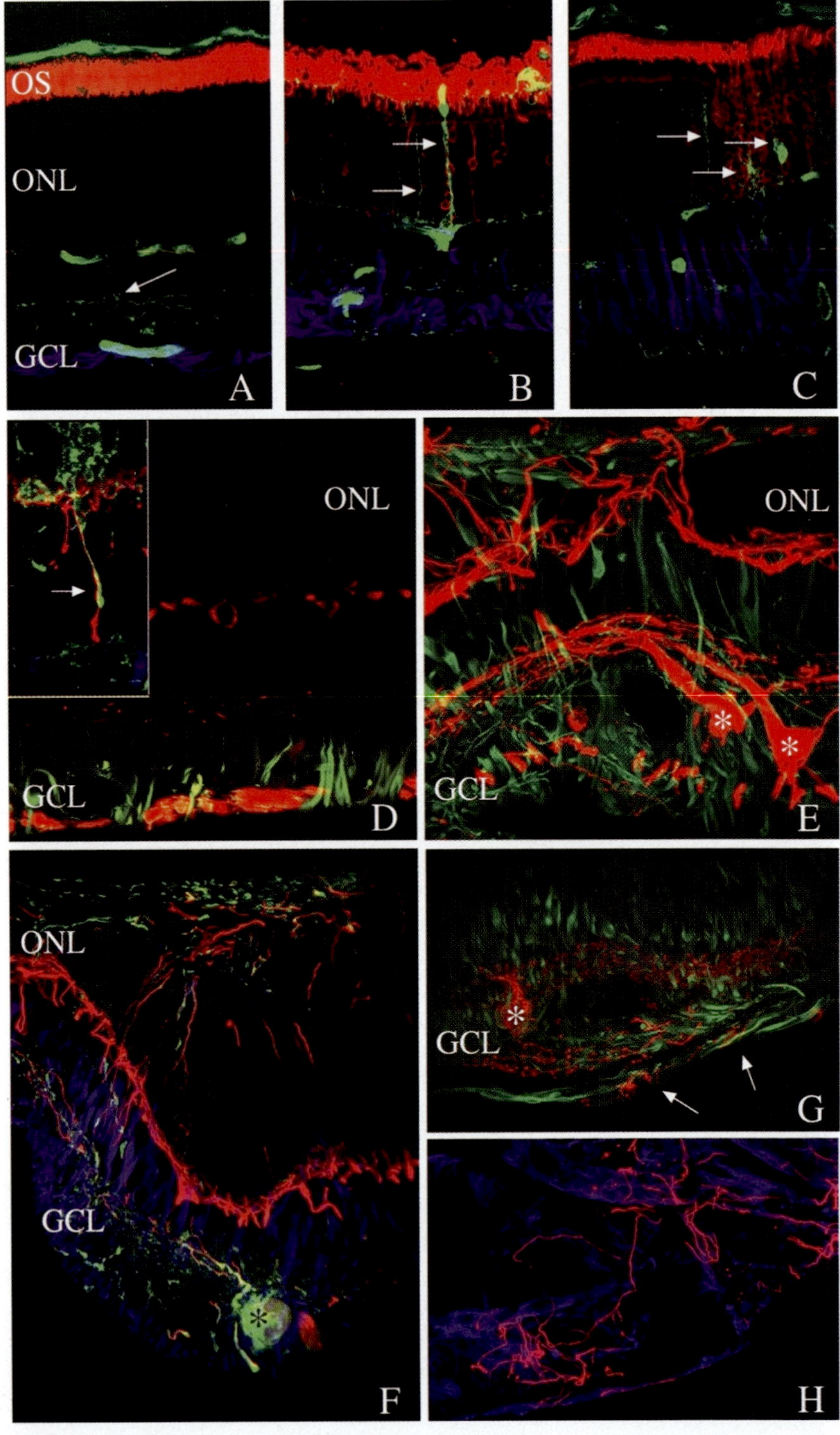

to be an important step in preventing photoreceptor degeneration. If, however, microglia are releasing some type of "cell killing" factor after detachment, its effects are highly specific in the feline retina because only photoreceptors appear to die in this model.

We have observed a structural interaction between microglia and the remodeling of rod axons as they extend deep into the inner retina following retinal reattachment. Fine microglia cell processes can often be observed directly adjacent to the elongated rod axons (Fig. 4.4D, inset). The fact that microglia become activated upon detachment prior to the outgrowth of rod axons might suggest that its their presence that stimulates rod axon outgrowth, if they are involved at all. This hypothesis may be testable by using agents that will inhibit the microglia response (Thanos et al., 1995).

FIGURE 4.4. (**A,B,C**) Laser scanning confocal images illustrating the activation of microglia (green; isolectin B4) following detachment. (Sections are also labeled with anti-rod opsin, red, and anti-GFAP, blue, to follow the responses of photoreceptors and Müller cells, respectively.) (**A**) In the normal retina microglia are located in the IPL (arrow). (Note: blood vessels are also labeled with the lectin used to label microglia and appear green.) (**B**) Following a 3-day detachment microglia migrate into the ONL (arrows). (**C**, left half of image) When the retina is reattached for 28 days following a 3-day detachment the activation of microglia is for the most part reversed. (**C**, right half of the image) In focal regions where there is poor photoreceptor recovery as illustrated by increased opsin redistribution (red) in the ONL, there is continued activation of microglia (arrows). (**D, inset**) At higher magnification in regions of poor photoreceptor recovery in the reattached retinas, microglia (green) are often observed in association with rod axons (red) that extend into the inner retina. (**D,E**) Confocal sections labeled with anti-neurofilament protein (red) and anti-GFAP (green) illustrating the upregulation of neurofilament protein that occurs after detachment. (**D**) In normal retina anti-neurofilament protein (red) labels ganglion cell axons in the GCL and horizontal cells; anti-GFAP (green) labels only the endfeet of Müller cells. (**E**) Following 28 days of detachment ganglion cell bodies upregulate neurofilament protein (asterisks) and horizontal cells sprout neurites that grow along the processes of reactive Müller cells into the subretinal space. (**F,G,H**) Confocal images illustrating the growth of neurites from ganglion cells. (**F**) Following 28 days of detachment ganglion cell bodies upregulate GAP 43 (green, asterisk) and sprout processes that grow throughout the retina and into the subretinal space adjacent to reactive Müller cells (blue, anti-GFAP). These processes are distinct from the horizontal cell processes labeled with anti-neurofilament (red) also present adjacent to Müller cells in the subretinal space. (**G**) Anti-GAP 43 labeled ganglion cell neurites (red) can also be found in epiretinal membranes (green, anti-GFAP, arrows) in retinas reattached for 28 days following a 3-day detachment. Note the ganglion cell body also labeled with anti-GAP 43 (asterisk). (**H**) Ganglion cell neurites (red, anti-neurofilament) are also observed growing in epiretinal membranes removed from human patients. These membranes consist largely of Müller cells (blue, anti-GFAP). (GCL, ganglion cell layer; ONL, outer nuclear layer; OS, outer segments).

Bipolar and Horizontal cells

Second order neurons (Fig. 4.1, purple cells) begin to remodel in apparent response to the changes initiated by the photoreceptors in a classic example of interactive cellular remodeling. As the rod axon terminals retract towards their cell body, the rod bipolar dendrites and neurites from what appear to be the horizontal cell axon terminals (Lewis et al., 1998; Linberg et al., 2004) grow into the outer nuclear layer, remaining adjacent to these retracted terminals (Figs. 4.3D, 4.3E, arrows). It is not known if the processes of the second order neurons are actually physically connected to the rod terminals as the remodeling occurs or if they first become disconnected and then reconnect to their "lost" terminal at some later time. Evidence from culture systems indicates that neurite growth can be initiated by applying mechanical tension to the cell (Lamoureux et al., 2002). This suggests that if rod terminals remain connected to the processes of second order neurons, physical retraction may exert sufficient mechanical tension on the second order neurons to stimulate neurite growth. Interestingly, rod bipolar dendrites that have grown into the outer nuclear layer are almost always found adjacent to a retracted rod terminal even in very degenerate retinas. This is unlike the response of the horizontal cells that seem to have an additional growth phase. Once the horizontal cell outgrowth starts after detachment, some processes can grow wildly throughout the retina and even into the subretinal space (Figs. 4.4D, 4.4E, 4.4F). The structural evidence points to a tantalizing interaction between the Müller cells and these aberrant horizontal cell processes. These longer processes are invariably found adjacent to reactive Müller cells (based on increased intermediate filament labeling) (Figs. 4.4E, 4.4F). As the Müller cells hypertrophy throughout the retina and into the subretinal space, horizontal cells appear to follow, using these processes as a scaffold or substrate for growth. Indeed, horizontal cell neurites can be observed extending great distances in the subretinal space in association with a Müller cell "scar". While the molecular signals that attract the horizontal cells are unknown, interestingly the bipolar cells do not appear to use these cues in their course of remodeling. This type of interactive remodeling, where neurons appear to grow adjacent to reactive Müller cells has also been observed in other forms of retinal degeneration (see Marc et al., 2003 for review) indicating that this phenomenon is not unique to retinal detachment.

Ganglion cells

Ganglion cells (Fig. 4.1, red cell), the third order neuron and the retinal cell type responsible for relaying the visual information to higher centers in the brain, also undergo significant structural and molecular remodeling as a result of detachment. The first evidence that ganglion cells respond to detachment was the observation that GAP 43 is upregulated in a subpopulation of these cells (Figs. 4.4F, 4.4G; Coblentz et al., 2003). During development, as new processes are being elaborated in the retina, GAP 43 is expressed at high levels in ganglion cells. As the animal matures this protein is lost almost entirely from the cell, with the only anti-GAP 43 labeling occurring in 2 bands in the inner plexiform layer. Upregulation of the protein is first detected by immunocytochemistry in some ganglion cell bodies at 3 days of detachment with the number of labeled cells increasing at later time-points. As with GAP 43, the 70 and 200 kDa polypeptide forms of neurofilament protein are not detectable in the cell bodies of ganglion cells until the retina is detached at least 3 days (Fig. 4.4E). The cells showing an increase in GAP 43 and neurofilament protein expression also undergo significant structural remodeling by elaborating fine neurites from their cell body (Coblentz et al., 2003). Since these antibodies do not label cell bodies and branches from their main dendritic trunks in non-detached retina, the morphology of these cells cannot be directly compared to ganglion cells in the normal retina. However, when compared to data from studies using Golgi impregnation or dye injection, it becomes apparent that these cells have a morphology unlike any described previously for feline ganglion cells. The most compelling morphological data for this remodeling comes from the many angular neurites on the basal surface of the cell body and the very long processes that extend into the outer retina, the subretinal space or into the vitreous (discussed below).

At this point, the mechanisms underlying ganglion cell remodeling is unknown. There is no evidence that the cells directly connecting to ganglion cells (the cone bipolar cells and amacrine cells) actively remodel following detachment. Amacrine cells, however, have been shown to sprout new processes in the human genetic retinal degeneration, RP (Fariss et al., 2000), so it is possible that a similar response is occurring after detachment and it has yet to be observed. Although there is no evidence that Müller cells induce neurite sprouting from ganglion cells, there is evidence for an interaction between Müller cells and neurites growing from ganglion cells. Ganglion cell processes can be observed extending adjacent to reactive Müller cells through the entire width of the retina and into the subretinal space as well as in association with Müller cells that have grown into the vitreous cavity (Figs. 4.4F, 4.4G). This is apparently not an uncommon phenomenon since neurites from ganglion cells can be found in both sub- and epiretinal membranes removed from human patients at the time of reattachment surgery (Fig. 4.4H, epiretinal membrane).

Mechanisms of Retinal Remodeling

Retinal detachment results in remodeling of at least 6 cell types in the feline retina and a diagram of these events is shown in Figure 4.1B. The remodeling illustrated throughout this chapter may be interpreted as a cascade of cellular interactions that are initiated by the separation of neural retina and RPE. The specific mechanisms that begin the process, however, are unknown and most likely involve multiple factors acting in concert. For example, detachment results in the disruption of the interphotoreceptor matrix and the release of endogenous growth factors which may initiate some of the events that follow (Hageman et al., 1991). One of these components is basic fibroblast growth factor (bFGF; FGF II). When injected into a normal eye bFGF can induce similar remodeling events in Müller cells to those observed after retinal detachment (Lewis et al., 1992) and neurite outgrowth is greatly accentuated by intraocular administration of bFGF when the retina is detached (G. Lewis, unpublished data). In addition, bFGF has been observed in the subretinal fluid taken from human patients at the time of reattachment surgery (La Heij et al., 2001) and the bFGF receptor becomes phosphory-lated in the retina within minutes after production of a detachment (Geller et al., 2001). These data suggest a significant role for bFGF in the remodel-ing of retinal neurons and glia that occurs after detachment. Finally, bFGF (and other neurotrophins) can also protect photoreceptors from degeneration (La Vail et al. 1992; Lewis et al., 1999a) thus increasing a likely role for fac-tors of this type in retinal remodeling. Intraocular delivery of factors that modulate the expression or action of bFGF could be used to further eluci-date its role in retinal detachment.

The evidence that many of the remodeling events are "downstream" from photoreceptor degeneration stems from experiments using hyperoxia to treat detachment. When experimental animals were placed immediately into a hyperoxic environment (70%) following the production of a retinal detach-ment and examined on day 3 (the peak of photoreceptor cell death and non-neuronal cell proliferation) there was a significant preservation of photoreceptor structure, a reduction of photoreceptor cell death, and less Müller cell reactivity (Mervin et al., 1999; Lewis et al., 1999b). Hyperoxia also maintained the overall structure of the interphotoreceptor matrix and apparently prevented the dispersal of bFGF from its stores in that location, thus potentially explaining the lack of Müller cell proliferation and hyper-trophy and giving further evidence for the role of bFGF in retinal detach-ment. In experiments where hyperoxia was delayed until a day after the production of a detachment and retinas examined 6 days later, much of the remodeling of neurons that would normally occur at this time (rod terminal retraction, neurite sprouting from 2nd and 3rd order neurons) was prevented (Lewis et al., 2004). This again supports the hypothesis that preservation of photoreceptors prevents downstream changes in other neuronal cell types.

Retinal remodeling has now been observed in many forms of induced and inherited degenerations and this topic was extensively reviewed in a recent paper by Marc et al. (2003). As pointed out in that paper remodeling is observed in RP, AMD, rd and Crx -/- mice, RCS rats, Abyssinian cats, light damage, glaucoma, ischemia, drug toxicity, and even aged animals. What becomes apparent from this list is that remodeling can occur quickly or occur over the course of decades. A common denominator, however, appears to be the compromising of photoreceptors. Following trauma such as detachment, many of the photoreceptor cell changes (terminal retraction, outer segment deconstruction, death) occur quickly, within the first few days and the remodeling occurs rapidly as well. In the inherited degenerations, photoreceptor changes often occur over the course of years and likewise the neuronal remodeling also appears to occur over years seemingly without subsiding. Another common factor is that in almost all cases, whether it is a result of induced or inherited degeneration, the neuronal remodeling is accompanied by glial remodeling, and based on immunocytochemical observations, there is often an interaction between these two cell types. It is fairly easy to envision remodeling as a cascade of events starting with photoreceptors and proceeding to 2nd and 3rd order neurons. As indicated by Marc et al. (2003) remodeling is most extreme when photoreceptor loss is essentially complete, resulting in "deafferentation" of the inner retina. Müller cells and other non-neuronal cells may react more independently from photoreceptor cell death since Müller cell events are among the earliest identified after detachment (Geller et al., 2001).

Conclusion

Understanding retinal detachment has assumed broader relevance in recent years because short-term retinal detachment occurs as part of experimental therapies in use or proposed for retinal degenerative diseases (e.g., RP, AMD) including retina–RPE transplantation (Bok 1993; Del Cerro et al., 1997), macular translocation (de Juan et al., 1998; Eckardt et al., 1999), or the introduction of trophic factors or vectors for transfection of retinal cells (Lewin et al., 1998). Thus, it is essential that we gain an understanding of the consequences of inducing a detachment if we are going to optimize the regenerative capacity of the retina. A critical theme that continues to return since first introduced in 1986 (Anderson et al.), is that reattachment does not simply return the retina to its pre-detachment state. Many of the changes induced by detachment such as photoreceptor cell death are permanent and reattachment will have no effect on these cells. Other changes, however, such as outer segment growth, protein distribution, rod terminal growth and glial hypertrophy *are* affected by reattachment and indicate that the retina has the ability to re-remodel during its recovery phase. This remodeling, however, does not always result in normal retinal morphology; the RPE/photoreceptor interface often appears structurally abnormal, rod axons over-grow, missing their normal targets, and Müller cells re-direct their growth from the subretinal space into the vitreous cavity often forming epiretinal membranes. Outer segment regeneration, therefore, appears to be only part of the story in regards to recovery of vision; re-establishing the RPE/photoreceptor interface, synaptic circuitry and normal expression of Müller cell proteins will most likely all be significant components of the recovery process.

The functional implication of the remodeling that occurs in retinal detachment and other retinal degenerations is not known. It's quite possible that the extent of remodeling observed after an event such as short-term detachment does not dramatically affect vision. On the other hand, continued long-term neuronal remodeling as part of the recovery process could explain the continued change in visual acuity that has been reported to occur for years after successful reattachment surgery. Strategies to improve visual recovery after detachment may be to provide a retinal environment that optimizes photoreceptor survival and prevents Müller cell (and possibly microglial cell) activation. The strategies that have shown some success experimentally include delivering drugs intraocularly that accelerate the physical process of reattachment (Nour et al., 2003), administering trophic factors (Lewis et al., 1999a) or hyperoxia (Lewis et al., 1999b; 2004) to prevent photoreceptor cell death, or delivering pharmacological agents to prevent the activation of microglia (Thanos et al., 1995). If proven to be effective at improving recovery of vision, strategies such as these could become an important adjunct to retinal detachment surgery.

Acknowledgements: The authors would like to thank Drs. Charanjit Sethi, David Charteris and Robert Avery for providing human retinal specimens as well as for their scientific and clinical input. We would also like to thank Ken Linberg, Kevin Talaga, and Katrina Carter for their technical assistance and informative discussions.

References

D. H. Anderson, W. H. Stern, S. K. Fisher, P. A. Erickson, and G. A. Borgula, Retinal detachment in the cat: the pigment epithelial-photoreceptor interface, *Invest. Ophthalmol. Vis. Sci.* **24**, 906–926 (1983).

D. Bok, Retinal transplantation and gene therapy: Present realities and future possibilities, *Invest. Ophthalmol. Vis. Sci.* **34**, 473–6 (1993).

T.C. Burton, Recovery of visual acuity after retinal detachment involving the macula, *Trans. Am. Ophthalmol. Soc.* **80**, 475–97 (1982).

M. S. Chen, A. B. Huber, M. E. van der Haar, M. Frank, L. Schnell, A. A. Spillmann, F. Christ, and M. E. Schwab, Nogo-A is a myelin-associated neurite outgrowth inhibitor and an antigen for monoclonal antibody IN-1. *Nature* **403**, 434–439 (2000).

Y. Chu, M. F. Humphrey, and I. J. Constable, Horizontal cells of the normal and dystrophic rat retina: a wholemount study using immunolabeling for the 28 kDa calcium binding protein, *Exp. Eye Res.* **57**, 141–148 (1993).

F. E. Coblentz, M. J. Radeke, G. P. Lewis, and S. K. Fisher, Evidence that ganglion cells react to retinal detachment, *Exp. Eye Res.* **76**, 333–342 (2003).

B. Cook, G. P. Lewis, S. K. Fisher, and R. Adler, Apoptotic photoreceptor degeneration in experimental retinal detachment. *Invest. Ophthalmol. Vis. Sci.* **36**, 990–996 (1995).

E. de Juan, A. Loewenstein, N. M. Bressler, and J. Alexander, Translocation of the retina for management of subfoveal choroidal neovascularisation II: a preliminary report in humans, *Am. J. Ophthalmol.* **125**, 635–646 (1998).

M. Del Cerro, E. S. Lazar, and D. Diloreto, The first decade of continuous progress in retinal transplantation, *Micros. Res. Tech.* **36**, 130–141 (1997).

C. J. Chang, W. W. Lai, D. P. Edward, and M. O. Tso, Apoptotic photoreceptor cell death after traumatic retinal detachment in humans, *Arch Ophthalmol.* **113**, 880–886 (1995).

D. L. Emery, N. C. Royo, I. Fischer, K. E. Saatman, and T. K. Mcintosh, Plasticity following injury to the adult central nervous system: is recapitulation of a developmental state worth promoting? J. *Neurotrauma* **20**, 1271–1292 (2003).

P. Erickson, S. Fisher, D. Anderson, W. Stern, and G. Borgula, Retinal detachment in the cat: the outer nuclear and outer plexiform layers, *Invest. Ophthalmol. Vis. Sci.* **24**, 927–942 (1983).

C. Eckardt, U. Eckardt, and H. D. Conrad, Macular rotation with and without counter rotation of the globe in patients with age-related macular degeneration, *Graefe's Arch. Clin. Exp. Ophthalmol.* **237**, 313–25 (1999).

R. N. Fariss, Z. Y. Li, and A. H. Milam, Abnormalities in rod photoreceptors, amacrine cells, and horizontal cells in humans with retinitis pigmentosa, *Am. J. Ophthalmol.* **129**, 215–223 (2000).

S. K. Fisher, P. A. Erickson, G. P. Lewis, and D. H. Anderson, Intraretinal proliferation induced by retinal detachment. *Invest. Ophthalmol. Vis. Sci.* **32**, 1739–1748 (1991).

S. K. Fisher, and G. P. Lewis, Müller cell and neuronal remodeling in retinal detachment and reattachment and their potential consequences for visual recovery, *Vis. Res.* **43**, 887–897(2003).

S. K. Fisher and G. P. Lewis, Cellular effects of detachment on the neural retina and the retinal pigment epithelium, in *Retina*, edited by S. J. Ryan and C. P. Wilkinson (Mosby, St. Louis, 2004) Vol. 3. "Surgical Retina" (2005, in press).

F. H. Gage, Mammalian neural stem cells, *Science* **287**, 1433–1438 (2000).

S. F. Geller, G. P. Lewis, D. H. Anderson, and S. K. Fisher, S.K. Use of the MIB-1 antibody for detecting proliferating cells in the retina, *Invest. Ophthalmol. Vis. Sci.* **36**, 737–744 (1995).

S. F. Geller, G. P. Lewis, and S. K. Fisher, FGFR1, signaling, and AP-1 expression following retinal detachment: reactive Müller and RPE cells, *Invest. Ophthalmol. Vis. Sci.* **42**, 1363–1369 (2001).

N. Gupta, K. E. Brown, and A. H. Milam, Activated microglia in human retinitis pigmentosa, late-onset retinal degeneration, and age-related macular degeneration, Exp. Eye Res. 76, 463–471 (2003).

G. S. Hageman, M. A. Kirchoff-Rempe, G. P. Lewis, S. K. Fisher, and D. H. Anderson, Sequestration of basic fibroblast growth factor in the primate retinal interphotoreceptor matrix. *Proc. Nat. Acad. Sci.* **88**, 6706–6710 (1991).

T. Harada, C. Harada, S. Kohsaka, E. Wada, K. Yoshida, S. Ohno, H. Mamada, K. Tanaka, L. F. Parada, and K. Wada, Microglia-Müller glia cell interactions control neurotrophic factor production during light-induced retinal degeneration. *J. Neurosci.* **22**, 9228–9236 (2002).

E. C. La Heij, M. P. H. Van de Waarenburg, H. G. T. Blaauwgeers, A. G. H. Kessels, J. de Vente, A. T. A. Liem, H. Steinbusch, and F. Hendrikse, Levels of basic fibroblast growth factor, glutamine synthetase, and interleukin-6 in subretinal fluid from patients with retinal detachment, *Am. J. Ophthalmol.* **132**, 544–550 (2001).

M. F. Humphrey and S. R. Moore, Microglial responses to focal lesions of the rabbit retina: correlation with neural and macroglial reactions, *Glia* **16**, 325–341 (1996).

P. T. Johnson, R. R. Williams, K. Cusato, and B. E. Reese, Rods and cones project to the inner plexiform layer during development, J. *Comp. Neurol.* **414**, 1–12 (1999).

A. J. Kroll, and R. Machemer, Experimental retinal detachment in the owl monkey. V. Electron microscopy of reattached retina, *Am. J. Ophthalmol.* **67**, 117–130 (1969a).

A. J. Kroll, and R. Machemer, Experimental retinal detachment and reattachment in the rhesus monkey, *Am. J. Ophthalmol.* **68**, 58–77 (1969b).

S. Kusaka, A. Toshino, Y. Ohashi, and E. Sakaue, Long-term visual recovery after scleral buckling for macula-off retinal detachments, *Jap. J. Ophthalmol.* **42**, 218–222 (1998).

T. T. Lam, J. M. Kwong, and M. O. Tso, Early glial responses after acute elevated intraocular pressure in rats, *Invest. Ophthalmol. Vis. Sci.* **44**, 638–645 (2003).

P. Lamoureux, G. Ruthel, R. E. Buxbaum, and S. R. Heidemann, Mechanical tension can specify axonal fate in hippocampal neurons, *J. Cell Biol.* **159**, 499–508 (2002).

M. M. LaVail, K. Unoki, D. Yasumura, M. T. Matthes, G. D. Yancopoulos, and R. H. Steinberg, Multiple factors, cytokines, and neurotrophins rescue photoreceptors from the damaging effects of constant light, *Proc. Natl. Acad. Sci. USA* **89**, 11249–11253 (1992).

Y. B. Lee, A. Nagai, A., and S. U. Kim, Cyotkines, chemokines and cytokine receptors in human microglia, *J. Neurosci. Res.* **69**, 94–103 (2002).

S. Lewin, K. A. Drenser, W. W. Hauswirth, S. Nishikawa, D. Yasamura, J. G. Flannery, and M. M. LaVail, Ribozyme rescue of photoreceptor cells in a transgenic rat model of autosomal dominant reinitis pigmentosa. *Nature Medicine* **4**, 967–971 (1998).

G. P. Lewis, P. A. Erickson, C. J. Guerin, D. H. Anderson, and S. K. Fisher, Changes in the expression of specific Müller cell proteins during long-term retinal detachment, *Exp. Eye Res.* **49**, 93–111 (1989).

G. P. Lewis, P. A. Erickson, C. J. Guerin, D. H. Anderson, and S. K. Fisher, Basic fibroblast growth factor: A potential regulator of proliferation and intermediate filament expression in the retina, *J. Neurosci.* **12**, 3968–3978 (1992).

G. P. Lewis, C. J. Guerin, D. H. Anderson, B. Matsumoto, and S. K. Fisher, Rapid changes in the expression of glial cell proteins caused by experimental retinal detachment, *Am. J. Ophthalmol.* **118**, 368–376 (1994).

G. P. Lewis, B. Matsumoto, and S. K. Fisher, Changes in the organization of cytoskeletal proteins during retinal degeneration induced by retinal detachment *Invest. Ophthalmol. Vis. Sci.* **36**, 2404–2416, (1995).

G. P. Lewis, K. A. Linberg, and S. K. Fisher, Neurite outgrowth from bipolar and horizontal cells following experimental retinal detachment, *Invest. Ophthalmol. Vis. Sci.* **39**, 424–434 (1998).

G. P. Lewis, K. A. Linberg, S. F. Geller, C. J. Guerin, and S. K. Fisher, Effects of the neurotrophin BDNF in an experimental model of retinal detachment. *Invest. Ophthalmol. Vis. Sci.* **40**, 1530–1544 (1999a).

G. P. Lewis, K. Mervin, K. Valter, J. Maslim, P. J. Kappel, J. Stone, and S. K. Fisher, Limiting the proliferation and reactivity of retinal Müller cells during detachment: the value of oxygen supplementation, *Am. J. Ophthalmol.* **128**, 165–172 (1999b).

G. P. Lewis and S. K. Fisher, Müller cell outgrowth following experimental detachment: association with cone photoreceptors, *Invest. Ophthalmol. Vis. Sci.* **41**, 1542–1545 (2000).

G. P. Lewis, C. S. Sethi, D. G. Charteris, W. P. Leitner, K. A. Linberg, and S. K. Fisher, The ability of rapid retinal reattachment to stop or reverse the cellular and molecular events initiated by detachment, *Invest Ophthalmol. Vis. Sci.* **43**, 2412–2420 (2002).

G. P. Lewis, C. S. Sethi, K. A. Linberg, D. G. Charteris, and S. K. Fisher, Experimental retinal reattachment: A new perspective, *Molec. Neurobiol.* **28**, 159–175 (2003).

G. P. Lewis and S. K. Fisher, Upregulation of GFAP in response to retinal injury: Its potential role in glial remodeling and a comparison to vimentin expression, in *Int. Rev. Cytol. A Survey of Cell Biology,* edited by K. W. Jeon (Elsevier, Academic Press, San Diego CA, 2003), Vol. 230, pp. 263–290.

G. P. Lewis, C. S. Sethi, K. M. Carter, D. G. Charteris, and S. K. Fisher, Microglial cell activation following retinal detachment (RD): A comparison between species, *Molecular Vision,* 11,491–500 (2005).

G. P. Lewis, K. C. Talaga, K. A. Linberg, R. L. Avery, and S. K. Fisher, The efficacy of delayed oxygen therapy in the treatment of experimental retinal detachment. *Am. J. Ophthalmol.* 137,1085–1095 (2004)..

Z. Y. Li, I. J. Kljavin, and A. H. Milam, Rod photoreceptor neurite sprouting in retinitis pigmentosa, J. *Neurosci.* 15, 5429–5438 (1995).

A. T. A. Liem, J. E. E. Keunen, G. J. Meel, and D. Norrern, Serial foveal densitometry and visual function after retinal detachment surgery with macular involvement. *Ophthalmol.* 101, 1945–1952 (1994).

K. A. Linberg, G. P. Lewis, K. M. Carter, and S. K. Fisher, Immunocytochemical evidence that B-type horizontal cell axon terminals remodel in response to retinal detachment in the cat, *Invest. Ophthalmol. Vis. Sci.* 45, ARVO E-Abstract 4604 (2004).

R. Machemer, Experimental retinal detachment in the owl monkey: IV, The reattached retina, *Am. J. Ophthalmol.* 66, 1075–1091 (1968).

K. Mervin, K. Valter, J. Maslim, G. P. Lewis, S. K. Fisher, and J. Stone, J., Limiting the death and deconstruction during retinal detachment: the value of oxygen supplementation, *Am. J. Ophthalmol.* 128, 155–164 (1999).

R. E. Marc, B. W. Jones, C. B. Watt, and E. Strettoi, Neural remodeling in retinal degeneration, in *Progress in Retinal and Eye Research,* (Pergamon Press, 2003), Vol. 22, pp. 607–655.

R. Naskar, M. Wissing, and S. Thanos, Detection of early neuron degeneration and accompanying microglial responses in the retina of a rat model of glaucoma, 43, 2962–2968 (2002).

T. F. Ng and J. W. Streilein, Light induced migration of retinal microglia into the subretinal space, *Invest. Ophthalmol. Vis. Sci.* 42, 3301–3310 (2001).

M. Nour, A. B. Quiambao, W. M. Peterson, M. R. Al-Ubaidi, and M. I. Naash, P2Y(2) receptor agonist INS37217 enhances functional recovery after detachment caused by subretinal injection in normal and rds mice, *Invest. Ophthalmol. Vis. Sc.* 44, 4505–4514 (2003).

S. J. Park, I. B. Kim, K. R. Choi, J. I. Moon, S. J. Oh, J. W. Chung, and M. H. Chun, Reorganization of horizontal cell processes in the developing FVB/N mouse retina, *Cell Tissue Res.* 306, 341–346 (2001).

T. S. Rex, R. N. Fariss, G. P. Lewis, K. A. Linberg, I. Sokal, and S. K. Fisher, A survey of molecular expression by photoreceptors after experimental retinal detachment, *Invest. Ophthalmol. Vis. Sci.* 43, 1234–1247 (2002).

R. S. Roque, C. J. Imperial, and R. B. Caldwell, Microglial cells invade the outer retina as photoreceptors degenerate in Royal College of Surgeons Rats, *Invest. Ophthalmol. Vis. Sci.* 37, 196–203 (1996).

W. H. Ross, Visual recovery after macula-off retinal detachment, *Eye* 16, 440–446 (2002).

V. Sarthy and H. Ripps, The Retinal Müller cell, in *Perspectives in Vision Research,* edited by C. Blakemore (Plenum Press, New York, 2001).

N. R. Saunders, A. Deal, G. W. Knott, Z. M. Varga, and J. G. Nicholls, Repair and recovery following spinal cord injury in a neonatal marsupial (*Monodelphis domestica*). *Clin. Exp. Pharmacol. Physiol.* **22**, 518–526 (1995).

C. S. Sethi, G. P. Lewis, S. K. Fisher, W. P. Leitner, D. L. Mann, P. J. Luthert, and D. G. Charteris, Glial remodeling and neuronal plasticity in human retinal detachment with proliferative vitreoretinopathy, *Invest. Ophthalmol. Vis. Sci.* **46**, 329–342 (2005).

E. Strettoi, V. Pignatelli, C. Rossi, V. Porciatti, and B. Falsini, Remodeling of second-order neurons in the retina of rd/rd mutant mice, *Vis. Res.* **43**, 867–877 (2003).

P. Tani, D. M. Robertson, and A. Langworthy, Rhegmatogenous retinal detachment without macular involvement treated with scleral buckling. *Am. J. Ophthalmol.* **90**, 503–508 (1980).

P. Tani, D. M. Robertson, and A. Langworthy, Prognosis for central vision and anatomic reattachment in rhegmatogenous retinal detachment with macula detached. *Am. J. Ophthalmol.* **92**, 611–20 (1981).

S. Thanos, Sick photoreceptors attract activated microglia from the ganglion cell layer: a model to study the inflammatory cascades in rats with inherited retinal dystrophy, *Brain Res.* **14**, 21–28 (1992).

S. Thanos, W. Richter, and H. J. Thiel, The protective role of microglia-inhibiting factor in the retina suffering of hereditary photoreceptor degeneration, in *New strategies in prevention and therapy. Cell and tissue protection in ophthalmology,* edited by U. Pleyer, K. Schmidt, and H. J. Thiel (1995), pp. 195–203.

C. Yamamoto, N. Ogata, M. Matsushima et al., Gene expressions of basic fibroblast growth factor and its receptor in healing of rat retina after laser photocoagulation, *Jpn. J. Ophthalmol.* **40**, 480–490 (1996).

N. Zhang and E. Townes-Anderson, Regulation of structural plasticity by different channel types, *J. Neurosci.* **22**, 7065–7079 (2002).

5

Experience-Dependent Rewiring of Retinal Circuitry: Involvement of Immediate Early Genes

RAPHAEL PINAUD[1] AND LIISA A. TREMERE[2]

[1]Neurological Sciences Institute, OHSU, Portland, OR, USA, and
[2]CROET, OHSU, Portland, OR, USA

Introduction

As discussed in the preceding chapters, it is now very clear now that the adult vertebrate retina undergoes marked modifications in its synaptic and cellular architecture, primarily in response to pathological states. Remarkable gains have been obtained towards understanding the molecular, cellular and biochemical events underlying retinal rewiring in a number of experimental paradigms, as described in detail in Chapters 2, 3 and 4. A major question in the field of retinal plasticity, however, is the extent of circuit reorganization that occurs in response to normal visual experience in the retina. Pioneering experiments conducted between the late 1950's and early 1970's addressed this particular issue by investigating the effects of light deprivation and stimulation on the organization of the mammalian retina. One of the main advantages in using the paradigms of sensory deprivation and stimulation is that it is rarely associated with retinal cell damage. In 1958, Weiskrantz reported that cats that had been deprived of normal visual information (form-deprived) displayed a marked reduction in thickness of the inner plexiform layer (IPL) (Weiskrantz, 1958). A few years later, in 1961, Rasch and colleagues provided evidence for considerable modifications in retinal circuitry as a function of light deprivation: it was demonstrated that cats that had been dark-reared for an extended period of time also underwent a significant reduction in the thickness of the IPL. In addition, chimpanzees that had also been maintained in darkness exhibited marked retinal ganglion cell degeneration (Rasch et al., 1961).

Subsequent electron microscopic studies provided the first detailed account for the modifications in synaptic architecture observed in animals that had been light-deprived. This series of experiments provided evidence for significant increases in the number of amacrine cell synapses in the rat retina, particularly those between amacrine and ganglion cells, these possibly

R. Pinaud, L. A. Tremere, P. De Weerd (Eds.), Plasticity in the Visual System:
From Genes to Circuits, 79-95, ©2006 Springer Science + Business Media, Inc.

accounting for the gross alterations observed in IPL thickness in response to light-deprivation (Sosula and Glow, 1971; Fifkova, 1972a, b; Chernenko and West, 1976). Together these early anatomical data provided the first set of evidence suggesting that the circuitry of the intact mammalian retina is plastic and its cytoarchitecture exhibits malleability and likely undergoes reorganization based on a previous history of activation. In addition, these findings suggest that visual experience might play an active role in triggering the refinement of retinal connectivity.

Although these early works provided evidence that visual experience actively influenced the organization of the normal, adult, retinal circuitry, it became a commonly accepted view to contemporary neuroscientists that the adult retina is capable of extremely limited, if any, plasticity (Chernenko and West, 1976). Central to this argument was the notion that high levels of plasticity, and consequently the high possibility of circuit rewiring, was inversely correlated to the degree of fidelity of information processing and transfer of any given population of neurons. Given that accurate representations of the visual world are required for appropriate computations by neural ensembles in higher cortical areas, this dogma gained strength despite a complete lack of experimental data that supported (or refuted) these claims.

The dogma that the normal adult retina displays extremely limited plasticity was the reason for our interest in this field. Our interest in retinal plasticity occurred indirectly, in unrelated sets of experiments studying the rat striate cortex. We were interested in investigating genes that might be involved in triggering the cascade of genomic events leading to circuit rewiring in the mammalian brain, in a paradigm known to induce widespread cortical plasticity: the enriched environment protocol (EE; details below). At that time, we had preliminary evidence indicating that two immediate early genes were likely involved in the induction of the early stages of plasticity in the brain. Based on the highly plastic nature of the mammalian brain, we reasoned that if these genes were truly involved in the induction of plasticity in the adult central nervous system (CNS), they would exhibit a differential expression pattern in highly plastic CNS regions (e.g., primary sensory areas), as compared to regions where plasticity is very limited (e.g., retina, according to the fidelity hypothesis discussed above). Thus, in order to support our claims regarding the involvement of these genes in the induction of plasticity, we decided that we would use the retina as an internal control for our data collected in the brain. Our working hypothesis was that while brain sites would exhibit high induction of these genes in response to exposure to the EE, the retina of the same animals analyzed would display low gene expression levels. To our surprise, the results observed did not support our working hypothesis, as it will be discussed below.

Environmental Enrichment

The experimental conditions termed "enriched environment" refers to the creation of a housing and/or play facility where animals are expected to undergo an increase in the complexity and, perhaps, relative rates of sensory processing. In addition to changes in brain mass, studies on the effects of EE in laboratory animals, such as rats, revealed marked phenotypical and neurochemical changes, which have been forwarded as the physical signs of CNS plasticity, associated with the chronic or acute exposure to a complex environment (for reviews see Rosenzweig et al., 1972; van Praag et al., 2000). For example, rearing animals in an EE increased the number of dendritic arborizations, soma volume, levels of neurotrophic factors and the neuron-to-glia ratio in all main sensory systems (Rosenzweig et al., 1972; Volkmar and Greenough, 1972; Globus et al., 1973; Falkenberg et al., 1992; van Praag et al., 2000). Furthermore, this procedure led to a marked enlargement of synaptic terminals, a phenomenon that has been associated with increased numbers of quanta and, therefore, increased neurotransmitter release per action potential (Rosenzweig et al., 1972; van Praag et al., 2000).

Our research group has strongly advocated the use of the EE paradigm to invoke behaviorally-associated CNS plasticity (Pinaud et al., 2001; Pinaud et al., 2002a; Pinaud, 2004). One of the main advantages of the EE setup, as compared to other paradigms used to investigate plasticity, such as sensory deafferentation, is the complete absence of injury. In addition, visual stimulation associated with EE exposure is not associated with dramatic changes in light intensity, but rather involve, theoretically, changes in the complexity of the visual environment.

Even though it is widely accepted that experiences in an EE induce vast morphological and neurochemical modifications in central neurons, only a few studies have specifically addressed what functional benefits may correlate with time spent in a complex visual environment. One of these studies, conducted by Prusky and colleagues, showed that visual acuity in mice was increased by 18% following rearing in an EE from birth (Prusky et al., 2000). This finding supports the idea that exposure of animals to a complex visual environment leads to modifications in CNS neurons aimed at improved sensory system performance.

Gene Expression and Rewiring: The IEGs NGFI-A and *arc*

Phenotypic changes are dependent on gene expression. Changes in cell structure and physical interactions between neurons have been postulated to depend on a series of biochemical signals that first lead to the expression of immediate early genes (IEGs) (Hughes and Dragunow, 1995; Herdegen and Leah, 1998; Pinaud, 2004; see also chapter 8). Often the expression of IEGs is induced quickly after the cell stimulation and, in all cases, induction of members of this class of genes does not depend on *de-novo* protein synthesis (Hughes and Dragunow, 1995; Kaczmarek and Chaudhuri, 1997; Herdegen and Leah, 1998). Therefore, it has been argued that inducible IEGs provide the most immediate genomic response in central neurons to sensory input (Pinaud et al., 2002a; Pinaud, 2004).

As stated above, we were interested in studying the expression of IEGs in response to EE exposure as members of this family of genes are likely involved in the early phases of the induction of plasticity in the CNS. Two IEGs became the focus of our research efforts: the nerve growth factor-induced gene A (NGFI-A; also known as *zif-268*; *egr-1*; *krox-24* and *zenk*) (Milbrandt, 1987; Christy et al., 1988; Lemaire et al., 1988; Sukhatme et al., 1988; Mello et al., 1992) and the activity-regulated cytoskeletal gene (*arc*; also known as *arg 3.1*) (Lyford et al., 1995). The protein products encoded by these two genes shared interesting, and presented complimentary, characteristics that placed them well-positioned to mediate key aspects of the induction of plasticity in response to a complex sensory environment. For example, both NGFI-A and *arc* are activity-dependent genes (Murphy et al., 1991; Worley et al., 1991; Lyford et al., 1995; Herdegen and Leah, 1998; Steward et al., 1998). Furthermore, their expression depends on the activation of the NMDA receptors, which have been involved in various aspects of synaptic reinforcement and facilitation in a number of experimental systems (Cole et al., 1989; Wisden et al., 1990; Kunizuka et al., 1999; Steward and Worley, 2001b). Despite these critical similarities between NGFI-A and *arc*, the protein products encoded by these genes plays dramatically different roles for cell physiology. NGFI-A is a zinc-finger transcription factor, whose expression is coupled to the activation of the ERK/mitogen-activated protein kinase (MAP) kinase pathway, and thus plays a role in the control of the expression of other genes that possess the NGFI-A binding domain contained within their promoters (Herdegen and Leah, 1998; Dziema et al., 2003) (see also Chapter 8). For instance, it has been previously demonstrated that the NGFI-A protein is involved in the transcriptional regulation of synapsin I and synapsin II genes, as well as the gene that encodes for synaptobrevin II (Thiel et al., 1994; Petersohn et al., 1995; Petersohn and Thiel, 1996) in *in-vitro* essays. These proteins have been demonstrated to play a direct role in neuro-transmitter release. Other candidate proteins putatively regulated by NGFI-A include neurotransmitter-gated ion channels, the monoamine oxidase B gene, the alpha-7 subunit of the nicotinic acetylcholine receptor, neurofilament and the adenosine 5′-triphosphate binding cassette, sub-family A, transporter 2 (ABCA2)(Pospelov et al., 1994; Carrasco-Serrano et al., 2000; Wong et al., 2002; Davis et al., 2003; Mello et al., 2004). Thus, NGFI-A is well positioned to affect the expression of genes that are involved not only in neurotransmitter release and membrane transport, but also other fundamental neuronal functions such as excitability. The activity-dependent IEG *arc*, unlike NGFI-A, does not encode a transcription factor, but rather a growth factor (Lyford et al., 1995). Interestingly, after transcribed, *arc* mRNA is readily transported to dendrites where local protein synthesis takes place (Lyford et al., 1995; Steward et al., 1998; Steward and Worley, 2001a, b). This interesting feature of *arc* expression has led to the suggestion that the protein encoded by this gene might be involved in activity-dependent dendritic recon-figuration, such as spine rotation, retraction or elongation, in addition to dendrite sprouting or retraction (Pinaud et al., 2001; Pinaud, 2004), which typically occurs under conditions of altered sensory input, such as in the case of lesions or enhanced activity (Grutzendler et al., 2002; Nimchinsky et al., 2002; Trachtenberg et al., 2002).

Based on these expression characteristics, these two IEGs became highly attractive to our research group as possible regulators of some aspects of plasticity in CNS neurons. As mentioned above, assuming that these genes play a role in triggering the early phases of plasticity in CNS networks, it would be expected that regions that exhibit low plasticity levels, such as the adult retina, would also display low expression levels of both NGFI-A and *arc*. We have thus looked at IEG in the adult rat retina in response to a paradigm that reliably induces plasticity in the mammalian brain.

Plasticity-Driven Gene Expression in the Adult Mammalian Retina

NGFI-A Expression in Response to an EE

We have recently demonstrated that exposure of animals for 1 hour/day to an EE setting for 3 weeks, a procedure that induces plasticity in the cerebral cortex, is capable of inducing the IEGs NGFI-A and *arc* in discrete regions of the retina, a region of the CNS previously thought to be incapable, or have limited capabilities, of undergoing reorganizational changes (Chernenko and West, 1976; Pinaud, 2004).

We showed that undisturbed (UD) rats, which remained housed in standard laboratory cages, and handled-only (HO) controls displayed only a few immunolabeled cells in the inner nuclear layer (INL) and ganglion cell layer (GCL), when compared to animals exposed to an EE (Pinaud et al., 2002b) (Fig. 5.1). Animals in the latter group had a marked increase of NGFI-A labeling in these retinal layers. In addition, Arc immunoreactivity was found in an upregulated fashion in the inner and outer plexiform layers of EE animals, but not in UD and HO controls (discussed below) (Fig. 5.1). These findings suggest that enhanced complexity of the visual environment triggers a cascade of gene expression that might be correlated with circuit rewiring. We postulate that these putative wiring modifications may be associated with the optimization of information capture and processing and thus, increased functional capacity. In fact, it has been demonstrated that exposure of mice to an enriched environment does enhance visual acuity by 18%, as compared to control animals (Prusky et al., 2000). It remains to be investigated whether the enhanced functional performance of visual circuitry is related to optimization of neural networks at the level of the retina or higher brain areas.

In our experiments, increased NGFI-A immunoreactivity was also found in the GCL, even though it is possible that some of the labeled neurons were amacrine cells. Since the different classes of amacrine cells were not explored in these experiments, it is difficult to speculate about the type of neuromodulation

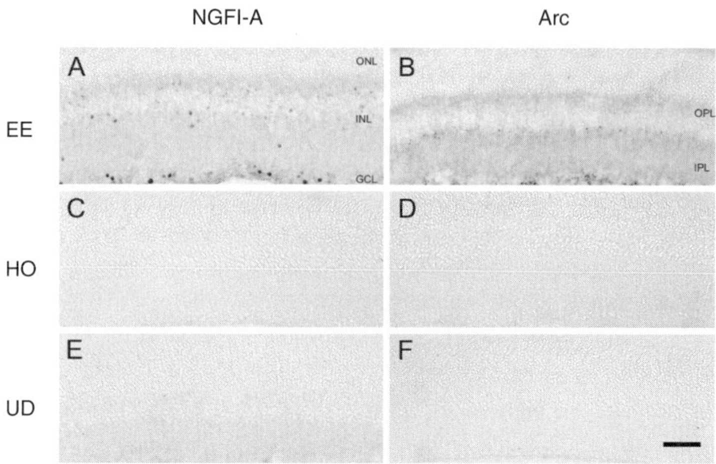

FIGURE 5.1. EE exposure drives the expression of NGFI-A and Arc. NGFI-A immunoreactivity was increased in the retinas of EE rats (A), but not in HO or UD controls (C and E, respectively). EE retinas also underwent increases in immunoreactivity for the IEG Arc in both IPL and OPL (B). In contrast, HO and UD animals exhibited significantly reduced Arc immunoreactivity in these layers (D and F, respectively). NGFI-A immunolabeled neurons were distributed evenly throughout the INL and GCL of EE animals. Increased Arc labeling was found in putative dendrites located within the IPL and OPL; those also appeared to be evenly distributed across these layers. GCL, ganglion cell layer; INL, inner nuclear layer; ONL, outer nuclear layer; IPL, inner plexiform layer; OPL, outer plexiform layer. Scale bar = 50µm.

that might have influenced visual responses and ultimately increased NGFI-A expression in the GCL of EE animals. Based on data collected from the cerebral cortex, noradrenergic influence is essential for NGFI-A expression (Cirelli et al., 1996; Pinaud et al., 2000). Should a similar situation apply in the retina, it is possible that adrenergic amacrine cells themselves might modulate NGFI-A expression as part of a putative neural plastic response.

Bipolar cells are believed to be the main class of NGFI-A expressing cells in response to the EE setting in the INL (Pinaud et al., 2002b). Should NGFI-A expression indicate sites of enhanced plasticity, it is possible that the complexity of information in the visual environment drives the expression of this activity-dependent gene that can ultimately alter cell structure and, subsequently, function. Although phenotypical changes of bipolar cells could be essential for the optimization of retinal function, a possibility also exits that horizontal and/or amacrine cells might also be undergoing plastic changes. Horizontal cells are believed to be involved in the regulation of information inflow through the OPL and into the second order retinal neurons, the bipolar cells (Dacey, 1999; Kamermans and Spekreijse, 1999; Lukasiewicz, 2005). More specifically, it has been proposed that horizontal cells play an important role in early spatial processing of visual input by developing the characteristic center-surround receptive fields of bipolar neurons. Given that NGFI-A immunoreactivity is localized to the nuclear compartment, it is not possible to differentiate the identity of these cells using strictly immunolabeling targeted at this protein. However, current efforts from our group are being directed at characterizing the neurochemical identity of NGFI-A-positive cells in the retinas of animals exposed to the EE setting. Detailing the cell types that exhibit NGFI-A induction will provide a better understanding of the key players involved in retinal rewiring that putatively occurs in response to exposure of animals to a complex visual environment.

Previous studies have indicated that the majority of amacrine cells use GABA as a neurotransmitter (Wu, 1992; Koulen et al., 1998). GABA's role in the OPL, at least where the terminals of horizontal cells oppose the bipolar cell process within the invaginated synapses of the cone system, may be to alter dynamically the center versus surround area as a means of preferentially funneling sensory information through the most biologically appropriate parts of retinal circuitry. Furthermore, it is possible that the molecular mechanisms that are sensitive to enhanced complexity of the environment were triggered in bipolar cells as a means of correcting for sub-optimal processing capabilities and, possibly, loss of environmental information that could potentially compromise the fidelity of visual information (Pinaud, 2004).

At present, there is currently only limited data concerning candidate plasticity gene expression in the retina of visual species. The great majority of data concerning retinal IEG expression is obtained in either rats or mice. We have, however, recently collected supporting evidence for a role of NGFI-A in the primate retina. More specifically, we have demonstrated that in a species of diurnal monkey, NGFI-A is expressed at high basal levels in undisturbed conditions (Pinaud et al., 2003). This was a surprising result given that UD rodents expressed NGFI-A at relatively low basal levels.

One possible explanation for this difference in basal expression between rodent and primate species is that primates rely more heavily on vision for normal behavior. Such a statement can be substantiated both by behavioral studies and by the proportion of cortex dedicated to visual processing. Thus, we have postulated that the expression of this gene, as well as other molecular machinery involved in retinal plasticity, might be primed in the adult monkey retina (Pinaud et al., 2003). Such mechanism could facilitate change in response to reorganizational pressures in these more visually-dependent animals (Pinaud et al., 2003; Pinaud, 2004).

Arc Expression in the Retina of EE animals

Arc immunoreactivity was enhanced strictly in EE animals for both the OPL and IPL (Pinaud et al., 2001) (Fig. 5.1). As stated above, the mRNA encoded by the IEG *arc* is rapidly delivered to dendrites upon cell stimulation and, therefore, it has been postulated that Arc protein is involved in activity-dependent dendritic reconfiguration in the cerebral cortex and hippocampus (Lyford et al., 1995; Steward et al., 1998; Pinaud et al., 2001). We observed a significantly higher Arc expression in the OPL and IPL of EE animals, but not HO and UD controls (Fig. 5.1). These findings are in agreement with the proposed role for the protein encoded by this IEG. In the experiments conducted by us (Pinaud et al., 2001; Pinaud et al., 2002b) the hypothesis is that this protein may be expressed in response to enhanced levels of visual complexity and translated locally in dendrites as part of a process underlying circuitry restructuring (Pinaud et al., 2001). The relay of visual information in the OPL depends primarily upon synapses between photoreceptors and bipolar cells. It is possible that the exposure of animals to an EE increased the absolute number of transduced visual events, or some other measure of visual activity such that altered cell structure was required. Furthermore, it is plausible that this structural reorganization could involve a rearrangement of the synaptic contacts between photoreceptors, bipolar cells, and horizontal cells.

Increased Arc immunoreactivity was also observed in the IPL, the retinal layer that primarily primarily contains connections between bipolar and ganglion cells (Fig. 5.1). Although the encoding of information across different retinal layers is dissimilar, it is possible that a similar need for reorganizational plasticity, possibly involving the same molecular players, would be found in both IPL and OPL. This possibility, however, remains to be experimentally tested.

GAP-43 and Synapsin I Upregulation in the Retina

GAP-43 and Synapsin I expression were used by us, in the same set of experiments, to confirm enhanced plasticity in the retina, in response to EE exposure (Pinaud et al., 2002b). The expression of both of these late response genes (LGs) has been repeatedly used as reliable markers for plasticity in the CNS. For example, optic nerve transection induces these LGs in the lateral geniculate nucleus (Baekelandt et al., 1994). In addition, several studies have demonstrated the upregulation of these LGs in the cortex after alterations in sensory drive and, more recently, after LTP induction in the hippocampus (Levin and Dunn-Meynell, 1993; Schauwecker et al., 1995; Bendotti et al., 1997; Sato et al., 2000; Suemaru et al., 2000). As previously described by other authors, basal levels of Synapsin I and GAP-43 were observed in the IPL of both HO and UD animals (Haas et al., 1990; Mandell et al., 1990; McIntosh and Blazynski, 1991; Mandell et al., 1992; Reh et al., 1993; Lopez-Costa et al., 2001). However, a modest but significant increase in Synapsin I and GAP-43 immunoreactivity was observed in the IPL of EE animals (Fig. 5.2). The most robust difference in the expression levels of these genes was, however, in the OPL, where HO and UD animals exhibited minimal GAP-43 and Synapsin I, while EE animals underwent a marked upregulation of the protein products of these LGs (Fig. 5.2).

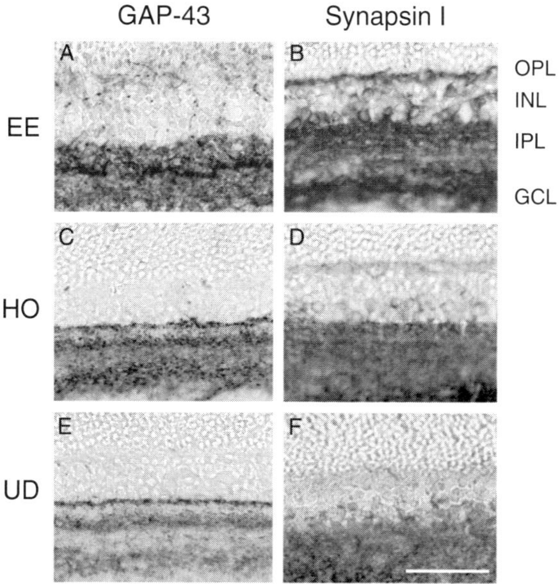

FIGURE 5.2. GAP-43 and Synapsin I are differentially expressed in EE retinas, as compared to HO and UD retinas. In EE retinas, GAP-43 labeling of beaded dendrites could be detected in the OPL (A), but was not detected in HO and UD retinas (C and E, respectively). A significant increase was also detected for Synapsin I immunolabeling in the OPL of EE animals (B). In HO and UD retinas, the OPL was unlabeled for GAP-43 (C and E, respectively) and Synapsin I (D and F, respectively). Strong bands of GAP-43 immunoreactive puncta were visible in the IPL of all experimental groups, with significantly higher intensity in EE animals (A, C and E). Synapsin I immunolabeling in the IPL was also prominent in all animal groups, with higher density in EE animals (B) when compared to HO and UD controls (D and F, respectively). OPL, outer plexiform layer; INL, inner nuclear layer; IPL, inner plexiform layer; GCL ganglion cell layer. Scale bar = 50μm.

The proteins encoded by LGs have been postulated to underlie more stable physical changes that ultimately alter function in CNS structures (Baekelandt et al., 1994; Han and Greengard, 1994; Melloni et al., 1994; Baekelandt et al., 1996; Yang et al., 1998; Aarts et al., 1999; Hilfiker et al., 1999). The upregulation of both classic plasticity markers discussed in this review lend further support to the hypothesis that retinal circuitry was altered in response to exposure to an EE setting.

Experience-Dependent Gene Expression Properties and Time Courses

The modifications in the expression levels for both IEGs and LGs detailed above were observed in the last experimental session (21st day), after the last exposure to the EE. As detailed above, we observed dramatic changes in the expression levels of NGFI-A, Arc, Synapsin I and GAP-43 in the retinas of animals exposed to this EE experience regimen for 1 hour per day. More recently, we have investigated in more detail the alterations in gene expression that result from EE exposure. In this new set of experiments, we investigated IEG and LG expression in animals that were exposed to the EE for 1 hour during 1 day, 2 days, 5 days, 1 week, 2 weeks, 3 weeks and animals that were exposed to the EE for 3 weeks and remained undisturbed for an additional month. All of these animals were sacrificed immediately after exposure to EE, however, in each timepoint, we also included animals that were sacrificed immediately prior to EE exposure. This experiment shed light into very important expression properties of these genes. We found, as expected, that both IEGs were rapidly and transiently induced in response to EE exposure. Both NGFI-A and Arc were induced by a single exposure to the EE. Interestingly, by the beginning of the subsequent session (day 2), the expression levels of both genes had returned to basal levels. These findings are not surprising taking into consideration the fact that the protein products of IEGs often exhibit a very short half-life (in a scale of minutes) (Hughes and Dragunow, 1995; Herdegen and Leah, 1998). However, exposure of animals to a second session (day 2) in the EE triggered a new wave of IEG expression in the retina. Irrespective of the number of exposures, expression of our genes of interest followed the same pattern as the one described previously in the original report (Pinaud et al., 2002b). Thus, daily 1 hour exposures of animals to the EE setting lead to daily upregulation of IEGs, followed by return to basal levels within a few hours after return of animals to lab home cages (Fig. 5.3).

Possibly the most interesting finding of this time-course study was revealed when investigating the expression profiles of the LGs GAP-43 and Synapsin I. We found that no alterations were found in the expression levels of these genes until the 5th session. After that point, a steady increase in the protein products for these LGs persisted until the 3rd week. Unlike IEG expression, LG expression was high prior to re-exposure of animals to the EE, suggesting that their protein expression is more long-lasting and stable than that of IEGs. Interestingly, animals that were studied a month after the last EE exposure still exhibited high LG expression (Fig. 5.3). Together, these findings suggest that although LG induction requires repeated exposures to the enriched visual environment, the modifications associated with these proteins are stable up until one month after the last session.

Immediate Early Genes as Detectors of Biological Repetition
and Relevance?

It is clear for decades now that enhanced complexity of the sensory environment leads to reorganization of CNS circuitry (Rosenzweig et al., 1972; Volkmar and Greenough, 1972; Globus et al., 1973; van Praag et al., 2000). Although the most compelling evidence that support this claim has been col-

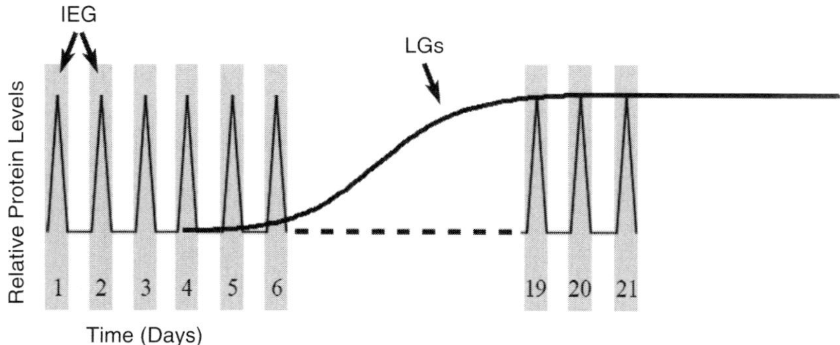

FIGURE 5.3. Schematic detailing the expression profiles of IEGs (NGFI-A and Arc) and LGs (Synapsin I and GAP-43) across time. IEG protein levels are found at basal levels prior to, but undergo a significant increase in response to, EE exposure. Due to the short half-life of these proteins, their levels return to baseline a few hours after EE experience is interrupted (return to home cages). IEG induction and subsequent return to basal levels appear to occur in response to each EE exposure episode, regardless of the number of occurrences. Conversely, LG induction exhibits a slow onset with a delay of approximately 4-5 days after the first EE exposure. LG protein levels are stable and appear to reach its peak after approximately 3 weeks of daily exposures to the EE. Interestingly, LG levels appear to be very stable: 1 month after the last EE experience, those levels appear to be unchanged from peak levels.

lected in studies conducted in the cerebral cortex, similar processes, possibly mediated through the same mechanisms, are also likely to occur in the adult mammalian retina.

We have proposed above that the putative reorganization of retinal circuits that occurs in response to complex environments has the objective of enhancing retinal information capture and processing capabilities. In addition, it is our position that in our paradigm (EE) any network reorganization is triggered by information (overload) contained in the visual environment. A paradoxical notion becomes obvious when evaluating the time course for these phenomena to occur. On the one hand, exposure of animals to the complex visual environment is an online, essentially instantaneous event. On the other hand, very little doubt also lies over the fact that circuitry rewiring involves gene expression and, consequently, protein synthesis, which is a very slow phenomenon. How can one conciliate these two notions? Assuming these concepts are truly accurate, the retina would always be "optimally" rewired for an event that has happened hours to days passed. In addition, subsequent visual events could also trigger further reorganization. If such situation was true, the retina would be under constant reorganization and virtually never "optimized". Further, reorganization is not only a time- but also a very energy-consuming process. To get around these problems, we propose that the retina contains a mechanism for detection of biological "repetition" and relevance.

In addition, we believe that IEG expression might shed light into the existence of this possibility. Figure 5.3 shows that one peak of IEG expression occurs following each exposure of animals to an EE. However, expression of LGs, which are putatively involved in circuit rewiring *per se*, require several exposures to the complex environment. In addition, LG accumulation occurs in an almost linear fashion, reaching its peak and plateau 2-3 weeks after the initial exposure to the EE. Thus, we propose that repeated IEG expression in response to an altered environment provides for a fast genetic mechanism involved in monitoring stimulus occurrences and relevance. This "accumulation" of IEG signal, at least in the case of transcription factors, crosses a threshold that then leads to the induction of LGs, which would begin the laborious task of circuit rewiring. This mechanism would prevent single and/or irrelevant occurrences from triggering widespread network modifications in the retina. Direct experimental evidence for such a hypothesis remains to be obtained. At present, research efforts are being directed at identifying genes that share the same regulatory domains, which could code for a coordinate gene expression wave potentially leading to retinal rewiring. In addition, it is currently important to establish with certainty the target genes regulated by each IEGs of interest likely involved in the circuit reorganization process (e.g., NGFI-A, c-fos, c-jun, etc). Once these questions are established, direct interventions targeted at IEG expression will be required, following examination of LG expression and morphological rewiring. Finally, physiological experiments will shed light onto the outcomes of this putative visually-driven retinal network reorganization.

Psychophysical Influences on Retinal IEG Expression

With the increased complexity of the sensory environment in the EE paradigm it becomes more difficult to understand exactly which psychophysical aspects, alone or in combination, may account for the increased expression of IEGs in EE subjects. Some of the more salient features of this condition are likely to be an increased frequency of borders, contrast and color changes. In addition, the greater range of depth over which animals can focus can explain upregulation of both *arc* and NGFI-A IEGs in the retina of EE animals. This hypothesis was supported previously in the chick retina using on-focus and de-focus by use of specific goggles for each condition (Fischer et al., 1999). The main finding of this work is that different classes of cells in the retina express NGFI-A in response to each kind of visual condition. In the experimental setup of the work by Pinaud and colleagues (Pinaud et al., 2002b), rats in the EE could detect visual objects over a much greater area, according to individual interests. Thus it is possible that NGFI-A and Arc are being expressed in different cell groups, in response to biologically relevant preselected characteristics of the environment. Sensory processing often appears to occur for both individual stimuli as well as in more integrative, and perhaps synergistic, forms. At present, it would be difficult to interpret what proportion or populations of cells belong to distinct classes that are individually stimulated by separate psychophysical aspects found in the EE, versus those cells that respond to more complex forms of sensory stimuli.

Finally, it appears that retinal cells carry molecular mechanisms that are sensitive to quantal values of information available in the environment. These mechanisms may be carefully regulated in the retina in order to optimize information collection and processing and these results implicate the IEGs NGFI-A and Arc in this process. In order to directly pinpoint the isolated contributions of these IEGs, studies using gene expression interference approaches should be directed at specific genes or mRNA species utilizing the same EE paradigm. Identifying specific cell types that are expressing NGFI-A and Arc is definitely an important next step in the current lines of investigations and, lastly, functional studies should address the physiological outcomes of such manipulations.

Acknowledgments: We thank Dr. William Currie and Harold Robertson for discussions and support. The work described here was supported in part by CIHR and H&SF of New Brunswick.

References

Aarts LH, Verkade P, van Dalen JJ, van Rozen AJ, Gispen WH, Schrama LH, Schotman P (1999) B-50/GAP-43 potentiates cytoskeletal reorganization in raft domains. Mol Cell Neurosci 14:85–97.

Baekelandt V, Eysel UT, Orban GA, Vandesande F (1996) Long-term effects of retinal lesions on growth-associated protein 43 (GAP-43) expression in the visual system of adult cats. Neurosci Lett 208:113–116.

Baekelandt V, Arckens L, Annaert W, Eysel UT, Orban GA, Vandesande F (1994) Alterations in GAP-43 and synapsin immunoreactivity provide evidence for synaptic reorganization in adult cat dorsal lateral geniculate nucleus following retinal lesions. Eur J Neurosci 6:754–765.

Bendotti C, Baldessari S, Pende M, Southgate T, Guglielmetti F, Samanin R (1997) Relationship between GAP-43 expression in the dentate gyrus and synaptic reorganization of hippocampal mossy fibres in rats treated with kainic acid. Eur J Neurosci 9:93–101.

Carrasco-Serrano C, Viniegra S, Ballesta JJ, Criado M (2000) Phorbol ester activation of the neuronal nicotinic acetylcholine receptor alpha7 subunit gene: involvement of transcription factor Egr-1. J Neurochem 74:932–939.

Chernenko GA, West RW (1976) A re-examination of anatomical plasticity in the rat retina. J Comp Neurol 167:49–62.

Christy BA, Lau LF, Nathans D (1988) A gene activated in mouse 3T3 cells by serum growth factors encodes a protein with "zinc finger" sequences. Proc Natl Acad Sci U S A 85:7857–7861.

Cirelli C, Pompeiano M, Tononi G (1996) Neuronal gene expression in the waking state: a role for the locus coeruleus. Science 274:1211–1215.

Cole AJ, Saffen DW, Baraban JM, Worley PF (1989) Rapid increase of an immediate early gene messenger RNA in hippocampal neurons by synaptic NMDA receptor activation. Nature 340:474–476.

Dacey DM (1999) Primate retina: cell types, circuits and color opponency. Prog Retin Eye Res 18:737–763.

Davis W, Jr., Chen ZJ, Ile KE, Tew KD (2003) Reciprocal regulation of expression of the human adenosine 5′-triphosphate binding cassette, sub-family A, transporter 2 (ABCA2) promoter by the early growth response-1 (EGR-1) and Sp-family transcription factors. Nucleic Acids Res 31:1097–1107.

Dziema H, Oatis B, Butcher GQ, Yates R, Hoyt KR, Obrietan K (2003) The ERK/MAP kinase pathway couples light to immediate-early gene expression in the suprachiasmatic nucleus. Eur J Neurosci 17:1617–1627.

Falkenberg T, Mohammed AK, Henriksson B, Persson H, Winblad B, Lindefors N (1992) Increased expression of brain-derived neurotrophic factor mRNA in rat hippocampus is associated with improved spatial memory and enriched environment. Neurosci Lett 138:153–156.

Fifkova E (1972a) Effect of light and visual deprivation on the retina. Exp Neurol 35:450–457.

Fifkova E (1972b) Effect of visual deprivation and light on synapses of the inner plexiform layer. Exp Neurol 35:458–469.

Fischer AJ, McGuire JJ, Schaeffel F, Stell WK (1999) Light- and focus-dependent expression of the transcription factor ZENK in the chick retina. Nat Neurosci 2:706–712.

Globus A, Rosenzweig MR, Bennett EL, Diamond MC (1973) Effects of differential experience on dendritic spine counts in rat cerebral cortex. J Comp Physiol Psychol 82:175–181.

Grutzendler J, Kasthuri N, Gan WB (2002) Long-term dendritic spine stability in the adult cortex. Nature 420:812–816.

Haas CA, DeGennaro LJ, Muller M, Hollander H (1990) Synapsin I expression in the rat retina during postnatal development. Exp Brain Res 82:25–32.

Han HQ, Greengard P (1994) Remodeling of cytoskeletal architecture of nonneuronal cells induced by synapsin. Proc Natl Acad Sci U S A 91:8557–8561.

Herdegen T, Leah JD (1998) Inducible and constitutive transcription factors in the mammalian nervous system: control of gene expression by Jun, Fos and Krox, and CREB/ATF proteins. Brain Res Brain Res Rev 28:370–490.

Hilfiker S, Pieribone VA, Czernik AJ, Kao HT, Augustine GJ, Greengard P (1999) Synapsins as regulators of neurotransmitter release. Philos Trans R Soc Lond B Biol Sci 354:269–279.

Hughes P, Dragunow M (1995) Induction of immediate-early genes and the control of neurotransmitter-regulated gene expression within the nervous system. Pharmacol Rev 47:133–178.

Kaczmarek L, Chaudhuri A (1997) Sensory regulation of immediate-early gene expression in mammalian visual cortex: implications for functional mapping and neural plasticity. Brain Res Brain Res Rev 23:237–256.

Kamermans M, Spekreijse H (1999) The feedback pathway from horizontal cells to cones. A mini review with a look ahead. Vision Res 39:2449–2468.

Koulen P, Malitschek B, Kuhn R, Bettler B, Wassle H, Brandstatter JH (1998) Presynaptic and postsynaptic localization of GABA(B) receptors in neurons of the rat retina. Eur J Neurosci 10:1446–1456.

Kunizuka H, Kinouchi H, Arai S, Izaki K, Mikawa S, Kamii H, Sugawara T, Suzuki A, Mizoi K, Yoshimoto T (1999) Activation of Arc gene, a dendritic immediate early gene, by middle cerebral artery occlusion in rat brain. Neuroreport 10:1717–1722.

Lemaire P, Revelant O, Bravo R, Charnay P (1988) Two mouse genes encoding potential transcription factors with identical DNA-binding domains are activated by growth factors in cultured cells. Proc Natl Acad Sci U S A 85:4691–4695.

Levin BE, Dunn-Meynell A (1993) Regulation of growth-associated protein 43 (GAP-43) messenger RNA associated with plastic change in the adult rat barrel receptor complex. Brain Res Mol Brain Res 18:59–70.

Lopez-Costa JJ, Goldstein J, Mangeaud M, Saavedra JP (2001) Expression of GAP-43 in the retina of rats following protracted illumination. Brain Res 900:332–336.

Lukasiewicz PD (2005) Synaptic mechanisms that shape visual signaling at the inner retina. Prog Brain Res 147:205–218.

Lyford GL, Yamagata K, Kaufmann WE, Barnes CA, Sanders LK, Copeland NG, Gilbert DJ, Jenkins NA, Lanahan AA, Worley PF (1995) Arc, a growth factor and activity-regulated gene, encodes a novel cytoskeleton-associated protein that is enriched in neuronal dendrites. Neuron 14:433–445.

Mandell JW, Czernik AJ, De Camilli P, Greengard P, Townes-Anderson E (1992) Differential expression of synapsins I and II among rat retinal synapses. J Neurosci 12:1736–1749.

Mandell JW, Townes-Anderson E, Czernik AJ, Cameron R, Greengard P, De Camilli P (1990) Synapsins in the vertebrate retina: absence from ribbon synapses and heterogeneous distribution among conventional synapses. Neuron 5:19–33.

McIntosh H, Blazynski C (1991) GAP-43-like immunoreactivity in the adult retina of several species. Brain Res 554:321–324.

Mello CV, Vicario DS, Clayton DF (1992) Song presentation induces gene expression in the songbird forebrain. Proc Natl Acad Sci U S A 89:6818–6822.

Mello CV, Velho TA, Pinaud R (2004) Song-induced gene expression: a window on song auditory processing and perception. Ann N Y Acad Sci 1016:263–281.

Melloni RH, Jr., Apostolides PJ, Hamos JE, DeGennaro LJ (1994) Dynamics of synapsin I gene expression during the establishment and restoration of functional synapses in the rat hippocampus. Neuroscience 58:683–703.

Milbrandt J (1987) A nerve growth factor-induced gene encodes a possible transcriptional regulatory factor. Science 238:797–799.

Murphy TH, Worley PF, Nakabeppu Y, Christy B, Gastel J, Baraban JM (1991) Synaptic regulation of immediate early gene expression in primary cultures of cortical neurons. J Neurochem 57:1862–1872.

Nimchinsky EA, Sabatini BL, Svoboda K (2002) Structure and function of dendritic spines. Annu Rev Physiol 64:313–353.

Petersohn D, Thiel G (1996) Role of zinc-finger proteins Sp1 and zif268/egr-1 in transcriptional regulation of the human synaptobrevin II gene. Eur J Biochem 239:827–834.

Petersohn D, Schoch S, Brinkmann DR, Thiel G (1995) The human synapsin II gene promoter. Possible role for the transcription factor zif268/egr-1, polyoma enhancer activator 3, and AP2. J Biol Chem 270:24361–24369.

Pinaud R (2004) Experience-dependent immediate early gene expression in the adult central nervous system: evidence from enriched-environment studies. Int J Neurosci 114:321–333.

Pinaud R, Tremere LA, Penner MR (2000) Light-induced zif268 expression is dependent on noradrenergic input in rat visual cortex. Brain Res 882:251–255.

Pinaud R, Penner MR, Robertson HA, Currie RW (2001) Upregulation of the immediate early gene arc in the brains of rats exposed to environmental enrichment: implications for molecular plasticity. Brain Res Mol Brain Res 91:50–56.

Pinaud R, Tremere LA, Penner MR, Hess FF, Robertson HA, Currie RW (2002a) Complexity of sensory environment drives the expression of candidate-plasticity gene, nerve growth factor induced-A. Neuroscience 112:573–582.

Pinaud R, De Weerd P, Currie RW, Fiorani Jr M, Hess FF, Tremere LA (2003) Ngfi-a immunoreactivity in the primate retina: implications for genetic regulation of plasticity. Int J Neurosci 113:1275–1285.

Pinaud R, Tremere LA, Penner MR, Hess FF, Barnes S, Robertson HA, Currie RW (2002b) Plasticity-driven gene expression in the rat retina. Brain Res Mol Brain Res 98:93–101.

Pospelov VA, Pospelova TV, Julien JP (1994) AP-1 and Krox-24 transcription factors activate the neurofilament light gene promoter in P19 embryonal carcinoma cells. Cell Growth Differ 5:187–196.

Prusky GT, Reidel C, Douglas RM (2000) Environmental enrichment from birth enhances visual acuity but not place learning in mice. Behav Brain Res 114:11–15.

Rasch E, Swift H, Riesen AH, Chow KL (1961) Altered structure and composition of retinal cells in darkreared mammals. Exp Cell Res 25:348–363.

Reh TA, Tetzlaff W, Ertlmaier A, Zwiers H (1993) Developmental study of the expression of B50/GAP-43 in rat retina. J Neurobiol 24:949–958.

Rosenzweig MR, Bennett EL, Diamond MC (1972) Brain changes in relation to experience. Sci Am 226:22–29.

Sato K, Morimoto K, Suemaru S, Sato T, Yamada N (2000) Increased synapsin I immunoreactivity during long-term potentiation in rat hippocampus. Brain Res 872:219–222.

Schauwecker PE, Cheng HW, Serquinia RM, Mori N, McNeill TH (1995) Lesion-induced sprouting of commissural/associational axons and induction of GAP-43 mRNA in hilar and CA3 pyramidal neurons in the hippocampus are diminished in aged rats. J Neurosci 15:2462–2470.

Sosula L, Glow PH (1971) Increase in number of synapses in the inner plexiform layer of light deprived rat retinae: quantitative electron microscopy. J Comp Neurol 141:427–451.

Steward O, Worley PF (2001a) A cellular mechanism for targeting newly synthesized mRNAs to synaptic sites on dendrites. Proc Natl Acad Sci U S A 98:7062–7068.

Steward O, Worley PF (2001b) Selective targeting of newly synthesized Arc mRNA to active synapses requires NMDA receptor activation. Neuron 30:227–240.

Steward O, Wallace CS, Lyford GL, Worley PF (1998) Synaptic activation causes the mRNA for the IEG Arc to localize selectively near activated postsynaptic sites on dendrites. Neuron 21:741–751.

Suemaru S, Sato K, Morimoto K, Yamada N, Sato T, Kuroda S (2000) Increment of synapsin I immunoreactivity in the hippocampus of the rat kindling model of epilepsy. Neuroreport 11:1319–1322.

Sukhatme VP, Cao XM, Chang LC, Tsai-Morris CH, Stamenkovich D, Ferreira PC, Cohen DR, Edwards SA, Shows TB, Curran T, et al. (1988) A zinc finger-encoding gene coregulated with c-fos during growth and differentiation, and after cellular depolarization. Cell 53:37–43.

Thiel G, Schoch S, Petersohn D (1994) Regulation of synapsin I gene expression by the zinc finger transcription factor zif268/egr-1. J Biol Chem 269:15294–15301.

Trachtenberg JT, Chen BE, Knott GW, Feng G, Sanes JR, Welker E, Svoboda K (2002) Long-term in vivo imaging of experience-dependent synaptic plasticity in adult cortex. Nature 420:788–794.

van Praag H, Kempermann G, Gage FH (2000) Neural consequences of environmental enrichment. Nat Rev Neurosci 1:191–198.

Volkmar FR, Greenough WT (1972) Rearing complexity affects branching of dendrites in the visual cortex of the rat. Science 176:1145–1147.

Weiskrantz L (1958) Sensory deprivation and the cat's optic nervous system. Nature 181:1047–1050.

Wisden W, Errington ML, Williams S, Dunnett SB, Waters C, Hitchcock D, Evan G, Bliss TV, Hunt SP (1990) Differential expression of immediate early genes in the hippocampus and spinal cord. Neuron 4:603–614.

Wong WK, Ou XM, Chen K, Shih JC (2002) Activation of human monoamine oxidase B gene expression by a protein kinase C MAPK signal transduction pathway involves c-Jun and Egr-1. J Biol Chem 277:22222–22230.

Worley PF, Christy BA, Nakabeppu Y, Bhat RV, Cole AJ, Baraban JM (1991) Constitutive expression of zif268 in neocortex is regulated by synaptic activity. Proc Natl Acad Sci U S A 88:5106–5110.

Wu SM (1992) Feedback connections and operation of the outer plexiform layer of the retina. Curr Opin Neurobiol 2:462–468.

Yang Y, Tandon P, Liu Z, Sarkisian MR, Stafstrom CE, Holmes GL (1998) Synaptic reorganization following kainic acid-induced seizures during development. Brain Res Dev Brain Res 107:169–177.

6

Attentional Activation of Cortico-Reticulo-Thalamic Pathways Revealed by Fos Imaging

VICENTE MONTERO

Department of Physiology and Waisman Center, University of Wisconsin,
Madison, WI, USA

Introduction

The brain of conscious animals and humans is constantly inundated by myriads of stimuli of different kinds (visual, acoustic, etc.); however only a fraction of them reaches awareness and control of behavior. Selective attention filters relevant from behaviorally irrelevant stimuli (Desimone and Duncan, 1995). There is considerable evidence in primates and humans that responses in several visual cortical areas, including the primary visual cortex, can be modulated by visual attention (Motter, 1993; Brefczynski and DeYoe, 1999; Ito and Gilbert, 1999). A role of the thalamic reticular nucleus in attentional processes was earlier hypothesized (Yingling and Skinner, 1977; Crick, 1984). However, attentional modulation of cortico-reticulo-thalamic pathways has only recently been addressed experimentally (Montero, 1999; 2000).

The GABAergic thalamic reticular nucleus (TRN) is strategically located to influence thalamocortical interactions by virtue of being in the pathway of glutamatergic thalamo-cortical and cortico-thalamic axons (Montero and Wenthold, 1989; Deschênes and Hu, 1990; Von Krosigk et al., 1999), from which it receives synapses (Ohara and Lieberman, 1981; Montero, 1989), and projecting back inhibitory synapses into specific thalamic nuclei (Montero and Scott, 1981; Montero, 1983; Peschanski et al., 1983). The TRN is subdivided into different sectors that are related to neocortical areas and their specific thalamic nuclei (Jones, 1975). In the rat, the visual, acoustic and somatic sectors are located in caudo-dorsal, caudo-ventral and central parts of the nucleus, respectively (Shosaku et al., 1989). The TRN receives topographically organized projections from primary cortical sensory areas (Montero et al., 1977) which, by forming the vast majority of synapses in this nucleus (Liu and Jones, 1999), suggest a topographically precise and predominant cortical control of its functions. The dependence of the visual sector of TRN

R. Pinaud, L. A. Tremere, P. De Weerd(Eds.), Plasticity in the Visual System:
From Genes to Circuits, 97-124, ©2006 Springer Science + Business Media, Inc.

on cortical inputs is supported by drastic diminution of its responses to visual stimuli, or to electrical stimulation of the optic nerve during inactivation of the visual cortex (Kayama et al., 1984). The TRN discharges tonically in wakefulness and in bursts in slow-wave sleep (Steriade et al., 1986), participating at this stage in the generation of spindles (Steriade and Llinas, 1988) by reciprocal interactions between TRN and their corresponding thalamo-cortical neurons (Bal et al., 1995).

The thalamic visual relay, the lateral geniculate nucleus (LGN), receives a massive glutamatergic projection from the primary visual cortex (V1) (Montero and Wenthold, 1989; Montero, 1991), acting on metabotropic glutamate receptors in distal dendrites of geniculate neurons (McCormick and von Krosigk, 1992). Despite multiple studies on the role of the cortico-geniculate pathway, determination of its function is still one of the main unresolved questions in neuroscience (Rivadulla et al., 2002).

To analyze the role of the visual TRN in attention mechanisms, the induction of the immediate early gene c-*fos* was used as marker of neuronal activity (Sheng and Greenberg, 1990; Kaczmarek and Chaudhuri, 1997; Clayton, 2000) in the visual pathways of the rat, including the visual (TRN), in different states of vigilance: a) In anesthetized rats that were stimulated with visual patterns in restricted regions of the visual field (Montero and Jian, 1995). b) in rats that freely explored a novel-complex (Montero, 1997), or a novel-simple environment. c) In control non-exploratory rats. d) In monocular amblyopic rats that explored the novel-complex environment (NCE) (Montero, 1999). e) In rats with LGN and visual TRN deprived of top-down inputs from V1 that explored the NCE (Montero, 2000). In addition, neurochemical analysis (high performance liquid chromatography, HPLC) of the excitatory neurotransmitter glutamate and inhibitory neurotransmitter GABA was determined in LGN and the medial geniculate nucleus (MGN) of exploratory rats, to test the hypothesis that attentional exploration of the NCE enhanced glutamatergic cortical inputs and GABAergic TRN inputs to LGN, but not to MGN.

Fos immunocytochemical mapping of neural activity is particularly suitable to detect activation of populations of neurons in visual centers of the rat for several reasons. The retino-geniculate, retino-collicular, geniculo-cortical, cortico-geniculate and cortico-reticular pathways are glutamatergic (Montero and Wenthold, 1989; Montero, 1990; Sakurai and Okada, 1992; Montero, 1994), whose actions are partially mediated by NMDA receptors (De Curtis et al., 1989; Funke et al., 1991; Sato et al., 1999), which by allowing Ca^{2+} entrance to the cell, trigger the intracellular cascade leading to rapid induction of the immediate-early gene c-*fos* (Ghosh et al., 1994; Chaudhuri, 1997). The time course of Fos induction, and its down-regulation, is well established for the rat visual cortex, with a maximum expression at 1.5-2 h post light stimuli (Chaudhuri et al., 2000). Fos is negligibly or weakly expressed in LGN, TRNv, Superior Colliculus (SC) and V1 of cage naive rats and in control non-exploring rats (Herdegen et al., 1993; Montero, 1997). Fos immunoreactive neurons can be double-labeled with antibodies to other neurotransmitters (Sakata et al., 2002). Finally, Fos mapping of neuronal activity is far superior to microelectrode mapping for the simple reason that Fos can detect activation of populations of neurons at single cell level in alert, freely-moving animals.

Fos induction by visual patterns stimulation in visual
centers of anesthetized rat

In these experiments (Montero and Jian, 1995) we analyzed the populations of neurons in central visual pathways of the rat that show Fos induction in restricted regions of visual field representation subsequent to localized visual patterned stimulation. In anesthetized rats, moving or stationary visual grids or dots patterns presented in a restricted region of the visual field, elicited Fos induction in corresponding retinotopic regions of visual centers which were previously mapped electrophysiologically or anatomically: in the ventral lateral geniculate nucleus (vLGN), LGN (Montero et al., 1968), the nucleus of optic tract (NOT) (Scalia and Arango, 1979), superficial gray of superior colliculus (Siminoff et al., 1966), primary visual cortex and extrastriate visual areas (Montero et al., 1973b; Montero, 1993). The advantage of using restricted visual stimulation versus total light stimulation of the eye, is that in these experiments it was possible to determine the correspondence of the populations of Fos-immunoreactive (Fos+) cells with the retinotopy of the stimulated region, in contrast to absence of Fos induction in non-stimulated regions.

Figure 6.1A shows the population of Fos+ cells in a coronal section of LGN of a rat that was stimulated with 20° moving patterns at the horizontal meridian and 60° elevation. The population of Fos+ cells are distributed in a retinotopically restricted field in LGN which translocates from central regions caudally in the nucleus, to ventral regions rostrally, following precisely the general orientation of retinotopic projection columns that were detected electrophysiologically in rat LGN (Montero et al., 1968). The population of Fos+ neurons in LGN was composed by geniculo-cortical relay cells and interneurons, as determined by analysis of somata size of Fos+ cells. Thus, the GABAergic interneurons in rat LGN (Montero and Singer, 1985; Gabbott et al., 1986) participate with inhibitory inputs in whatever activity was engendered by the visual stimuli in the receptive field of different types of relay cells (Montero et al., 1966; Montero and Brugge, 1969; Fukuda et al., 1979).

In the superior colliculus the population of Fos+ cells was equally retinotopically restricted to the visually stimulated region of the superficial gray layer, recipient of retinal inputs (Sefton and Dreher, 1985). However, there were no Fos+ cells in the deep gray layers (See Figs. 4 and 5 in Montero and Jian, 1995).

In contrast to LGN and SC cells, in anesthetized rats visual stimulation did not induce Fos in the visual TRN, despite activation of the glutamatergic geniculo-reticulo pathways by visual stimuli adequate to generate

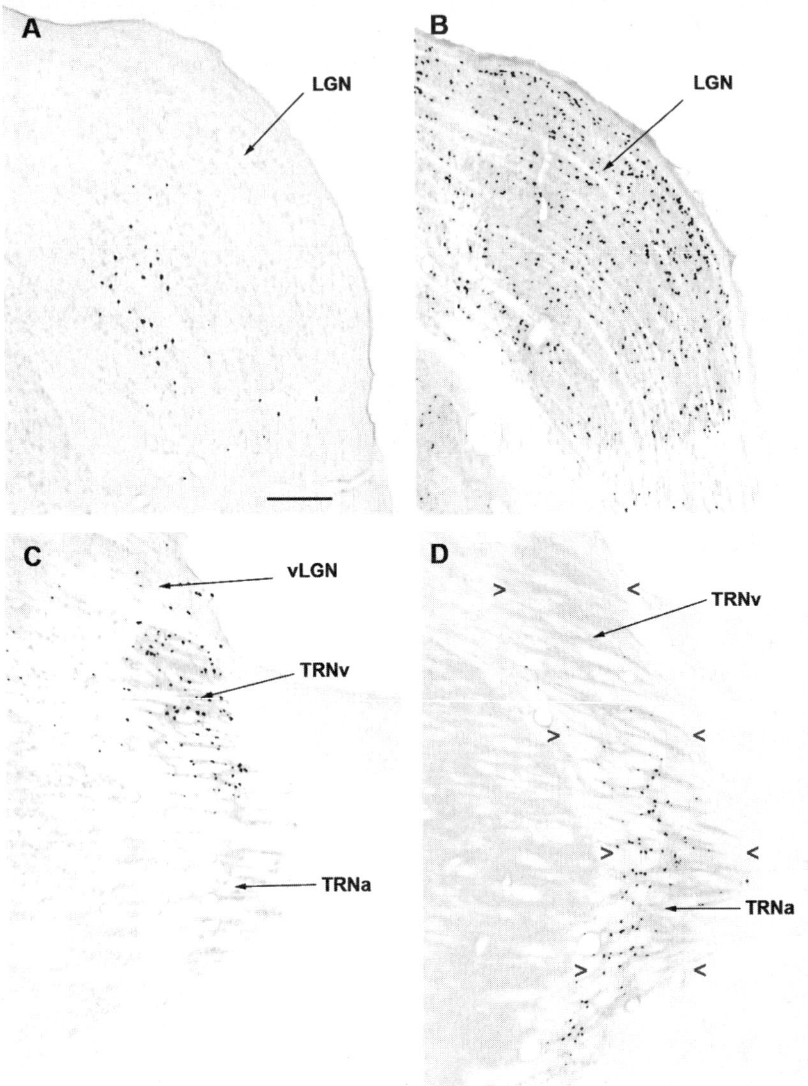

FIGURE 6.1. **A.** Localized population of Fos+ neurons in the lateral geniculate nucleus (LGN) of an anesthetized rat induced by visual patterns stimulation in a restricted region of the visual field. This coronal section is from the same animal, but rostral to those shown in figure 1 in (Montero and Jian, 1995), following the retinotopic lines of projection in the nucleus. **B.** Fos expression in whole extent of LGN of a rat that explored a novel-complex environment (NCE). **C.** Fos induction restricted to the caudo-dorsal visual sector the thalamic reticular nucleus (TRNv) of a rat that explored a NCE. The acoustic sector in the caudo-ventral thalamic reticular nucleus (TRNa) is devoid of Fos+ cells, despite intense noise of the rat's interaction with

trans-synaptic action potentials in visual TRN neurons (Hale et al., 1982). This is an essential difference of synaptic inputs requirements of TRN cells in comparison to LGN cells to express Fos, particularly in view of the fact that visual TRN cells are able to express Fos in alert, attentive rats, as discussed below.

In the primary visual cortex a restricted Fos+ field was induced in retinotopically corresponding region to the stimuli. Fos+ cells were present predominantly in layer 4, the main recipient of geniculo-cortical inputs, and in layer 6, the secondary recipient of these inputs (Peters and Feldman, 1976; Peters and Saldanha, 1976). Very few Fos+ cells were present in supragranular layers 2/3 and in layer 5 (Fig. 10A in Montero and Jian, 1995). This laminar distribution of visual stimuli induced Fos+ cells in V1 of anesthetized rats is drastically different, as shown below, to that in rats exploring a novel-complex environment, in which there is in addition Fos induction in layers 2/3 and 5.

It is important to emphasize that the same visual stimuli (stationary and moving visual patterns) that elicited trans-synaptic trains of action potentials in LGN of anesthetized rats (Montero et al., 1968), were able to induce Fos expression in the population of trans-synaptically stimulated neurons in LGN, and layers 4 and 6 of V1 in anesthetized rats. Altogether, these results are a direct demonstration that glutamatergic synaptic inputs that are known to induce action potentials in LGN neurons and in the granular and infragranular neurons of V1 in anesthetized animals are sufficient to induce Fos in these trans-synaptically activated neurons.

In the extrastriate visual areas (Montero, 1993), moving visual stimuli generated Fos induction more strongly in the anterolateral (AL) area than in the lateromedial (LM), posteromedial (PM) or anteromedial (AM) areas. The robust Fos induction in area AL by moving stimuli suggests that among extrastriate visual areas in the rat and other rodents (Coogan and Burkhalter, 1993; Schuett et al., 2002) this area might be specialized in motion detection, and consequently, be analogous to motion-sensitive areas PMLS in the cat (Spear and Baumann, 1975) and MT in the monkey (Zeki, 1974).

---

elements of the NCE that induced Fos label in brainstem acoustic centers. Coronal section at about bregma -3.6 mm in relation to a stereotaxic map (Paxinos and Watson, 1998). **D.** Selective Fos induction in the acoustic caudo-ventral TRNa sector a nursing rat in a dark NCE listening to pups sounds placed nearby. Notice absence of Fos label in the visual TRNv sector. Rats in the NCE listening to novel white noise or pure tones did not show Fos induction in the acoustic TRN. These results indicate that sensory inputs with behavioral significance to the animal, but not mere sensory inputs, even if novel, are necessary to induce Fos in sensory sectors of TRN. This is in contrast to Fos induction in LGN, where mere sensory inputs in an anesthetized animal are able to induce Fos, as shown in A. Coronal sections. Scales = 0.2 mm, applies to all panels.

The visual, auditory, and somatic sectors of TRN in the rat, as defined electrophysiologically (Shosaku and Simitomo, 1983), occupy the caudo-dorsal, caudo-ventral, and central regions of the nucleus, respectively, with almost no overlap. The physiologically defined visual sector extends from bregma -3.4 to -2.4 mm. This is in good agreement with our (unpublished) results in which injections of tracers in V1 labeled the caudo-dorsal sector of TRN from bregma -3.6 to 2.8 mm, with respect to a brain atlas (Paxinos and Watson, 1998). As originally described for the retinotopic projections from V1 to visual TRN in the rabbit (Montero et al., 1977), V1 cortical projections to the visual TRN in the rat (Coleman and Mitrofanis, 1996) are also distributed in medio-lateral flattened, and rostro-caudal elongated discoidal fields, that match the rostro-caudally extensive medio-lateral flattened discoidal shape of dendritic arborizations of TRN cells in the rat (Ohara and Havton, 1996).

Fos induction in rat visual pathways by exploration of a novel-complex environment

To evaluate the functional role of visual TRN in attention mechanisms we used a behavioral protocol to naturally elicit visual attention in alert rats. Since novelty attracts attention in humans (Anderson, 1994; Tiitinen et al., 1994; Daffner et al., 1998; Posner et al., 1999), as well as in rats (Oswald et al., 2001; Dias and Honey, 2002; Li et al., 2003), rats were exposed for the first time to a novel-complex or a novel-simple environment.

In rats that explored a novel-complex environment (Montero, 1997), the visual sector of TRN was consistently and strongly labeled with Fos+ cells (Figs. 6.1C, 6.3C, 6.3E); the somatic sector was restrictively labeled in central parts of the nucleus (Montero, 1999), probably in the region of vibrissae representation (Shosaku et al., 1984), while the auditory sector in the caudo-ventral TRN, was not labeled (Fig. 6.1C). Only the most rostral limbic sector of TRN (Cornwall et al., 1990; Gonzalo-Ruiz and Lieberman, 1995) was as intensely and consistently labeled as the TRNv. The strong Fos induction in TRNv of exploring rats was in striking contrast with absence of Fos+ neurons in the whole extent of TRN in control rats (cage naive, and awake control non-exploring rats), which is consistent with a previous report of absence of Fos+ neurons in the TRN of rats in undisturbed condition (Herdegen et al., 1995).

Although rats, being nocturnal rodents, depend mostly on their olfaction and vibrissae for recognition of the environment in darkness, under illumination rats orient themselves using distal visual cues (Lashley, 1938; Parron et al., 2004), and are capable of recognition of complex visual scenes (Simpson and Gaffan, 1999) despite their low visual acuity (Prusky et al., 2002) with respect to diurnal mammals. The multiple organization of extrastriate visual areas in the rat (Montero et al., 1973b; Coogan and Burkhalter, 1993) and the multiple variety of receptive field types in its primary visual cortex (Montero, 1981; Burne et al., 1984) testify to complex visual functions in these animals. In contrast to retinotopically restricted Fos activated region of LGN and absence of Fos induction in the intermediate layer of SC in visually stimulated anesthetized rats, in exploring rats there was Fos induction in the whole visual field representation in LGN (Fig. 6.1B, coronal section; Fig. 6.3C, sagittal section), and in the superficial and intermediate layers of the SC. The primary visual cortex of exploring rats showed strong Fos induction in all its extent and in all cortical layers (Fig. 6.3A). The fact that the supragranular layers 2/3 were strongly activated in exploring rats, in contrast to poor activation of these layers in visually stimulated anesthetized rats and in awake control non-exploring rats, suggests a specific engagement of the supragranular layers of V1 in mechanisms of attentive perception of the environment, as previously proposed (Montero, 1997).

Intense noises generated during exploration by rat's own interactions with objects of the NCE (as high as about 90 dB, Montero et al., 2001) induced several bands of Fos+ cells in brainstem acoustic centers (e.g. dorsal cochlear nucleus and inferior colliculus), but not in the acoustic sector of TRN. It should be noted that both, the dorsal cochlear nucleus and inferior colliculus in the rat and mouse express Fos tonotopically induced by acoustic stimuli in anesthetized and awake conditions (Ehret and Fischer, 1991; Friauf, 1992). To test for the possibility that self-generated noises could be actively inhibited in acoustic centers, rats were stimulated for the first time during exploration of the NCE (thus introducing the factor of "novelty" to the acoustic stimuli) with pure tones (16 KHz tone at 78 dB) or white noise (45 dB). In none of these rats there was Fos induction in the caudo-ventral acoustic sector of TRN, in contrast to activation of the visual TRN. To test for the possibility that the acoustic TRN could be activated by sounds that have behavioral significance, and that the acoustic TRN cells have the capability of expressing Fos, nursing rats in the NCE in darkness were exposed to noises and ultrasound vocalization of their pups (Farrell and Alberts, 2002) placed in an adjacent box. In these cases there was strong Fos induction in the caudo-ventral acoustic TRN in contrast to absence of Fos in the caudo-dorsal TRNv (Fig. 6.1D). Altogether, these observations indicate that mere sensory inputs to an awake, alert animal, which are not significant for the recognition of the environment or do not have behavioral significance, despite of being novel, do not induce Fos expression in the corresponding sector of TRN.

Attentional activation of visual TRN depends on top-down inputs from the primary visual cortex via glutamatergic cortico-reticulo-geniculate pathways

To analyze the neuronal circuits responsible for the Fos-detected activation of the visual TRN in rats that explored the NCE, we hypothesized that TRNv activation was primarily under the influence of the visual cortex. This hypothesis was tested in monocular amblyopic rats, and in rats with excito-toxic lesions in layer 6 of V1, that explored the NCE.

a) Asymmetry of visual TRN activation in monocular amblyopic rats that explored the NCE

The visual cortex is the main site of sensory loss after monocular deprivation (MD) during a critical postnatal period, on the basis that the physiology of the deprived geniculate laminae in MD cats is more or less normal (Wiesel and Hubel, 1963; Kiorpes and Movshon, 1996), even with respect to spatial resolution (Shapley and So, 1980), while the influence of the deprived eye on visual cortex responses is profoundly altered (Wiesel and Hubel, 1965; Kiorpes and Movshon, 1996). As in all mammals studied in this respect, there is a critical period in rats (from postnatal day 14 to 45) during which MD causes changes in responses of units in the visual cortex. There is a marked shift of ocular dominance distribution in favor of the open eye, and a drastic decrease in spatial resolution of responses to the deprived eye (Domenici et al., 1991; Fagiolini et al., 1994), amblyopia which can also be detected behaviorally (Rothblat et al., 1978). This, and the fact that retino-geniculate pathways in the rat are about 90–95% crossed (Hayhow et al., 1962), offered the opportunity to test the effects of monocular amblyopia on the exploration-dependent Fos-detected activation of the visual sector of TRN contralateral to the deprived eye. The main question was whether there is a reduction in the number of Fos+ cells in the TRNv contralateral to the deprived, amblyopic eye in comparison to the TRNv contralateral to the non-deprived normal eye.

In MD rats (n = 5) that explored the NCE there was diminution of the number of Fos+ cells in the visual sector of TRN in the side contralateral to the deprived eye (Fig. 6.2A) in comparison to the side contralateral to the normal eye (Fig. 6.2B), which was significant ($P < 0.01$) in four cases. In contrast, the number of Fos+ cells per area (Fos cells density, FCD) in the monocular lateral region of LGN (Hayhow et al., 1962) in the sides con-tralateral to the deprived and normal eyes was not significantly different in

FIGURE 6.2. **A** and **B:** Fos induction in the left and right visual TRN of a right eye monocular amblyopic rat that explored the NCE. (**A**) Shows diminution in number of Fos+ cells in the left TRNv contralateral to the right amblyopic eye, in compar-ison to (**B**) where numerous Fos+ cells in the right TRNv contralateral to the left normal eye are present. **C** and **D:** Fos induction in the left and right visual TRN of a rat without the cortico-reticulo-geniculate pathway from V1 in the right side, that explored the NCE. Note drastic diminution in the number of Fos+ cell in the TRNv of the right, lesioned side (**D**) with respect to the TRN v in the left, normal side (**C**). Coronal sections at about bregma -3.6 mm. Scale = 0.2 mm, applies to all panels.

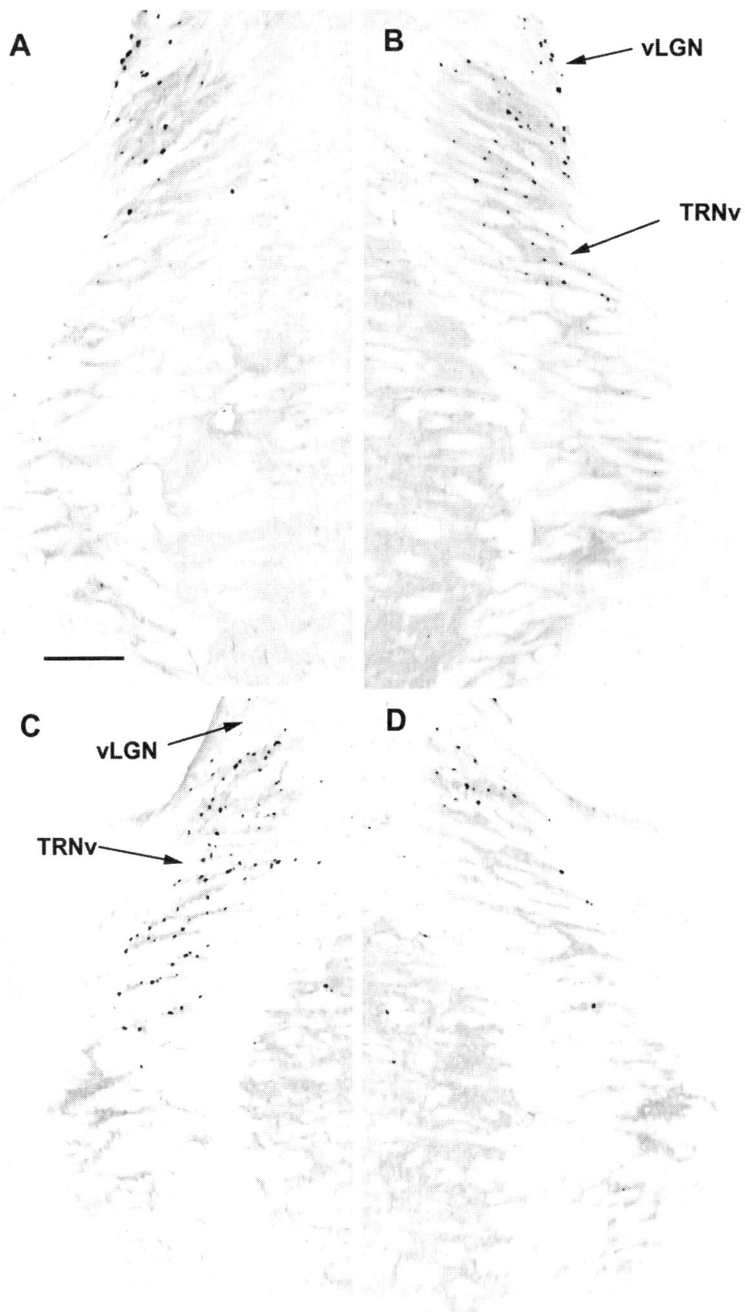
A
B
vLGN
TRNv
C
vLGN
TRNv
D

these animals ($P > 0.05$). In the monocular medial segment of layer 4 of V1 (Montero, 1973), recipient of geniculate axons conveying visual inputs exclusively from the contralateral eye, there was a significant diminution of FCD in the side contralateral to the deprived eye in two of the animals, while in layer 6 of the monocular V1, giving origin to cortico-reticulo-geniculate axons, there was significant ($P < 0.05$) diminution of FCD in the side contralateral to the deprived eye in all animals.

An index of activation asymmetry in the structures analyzed showed that activation in the right and left LGN was virtually identical. In the other structures there was a gradient of activation asymmetry which was less pronounced in layer 4 of V1, stronger in layer 6 of V1, and strongest in the visual TRN. A scatter plot of exploration time *versus* the index of activation asymmetry in TRNv for individual rats (Fig. 4 in Montero, 1999) showed a significant positive correlation between these two variables (Spearmans non-parametric correlation $r = 0.9$, $P = 0.05$). Thus, the longer the rats explored the stronger was the activation asymmetry in TRNv. By contrast, non-significant correlations were found between exploration time *versus* right/left FCD ratios in layers 4 and 6 of V1 ($P > 0.05$).

In summary, these results showed that in monocular amblyopic rats that explored a NCE there was a gradient of diminished Fos-detected activation in the analyzed cortical and subcortical visual structures contralateral to the deprived eye in comparison to those contralateral to the normal eye. The activation asymmetry was strongest in the TRNv, less strong in layer 6 of the monocular segment of the visual cortex, and weakest in layer 4 of the monocular visual cortex, while the activation of right and left LGN was virtually similar. In addition, there was a positive correlation between the time spent in exploration of the NCE and the degree of activation asymmetry in the TRNv. These results suggest that the asymmetric TRNv activation in monocular amblyopic exploring rats is a consequence of a similar activation asymmetry in layer 6 of the visual cortex, on the rationale that this layer provides the cortical input to TRNv (Bourassa and Deschênes, 1995; Murphy and Sillito, 1996), and since activation of the other visual input to TRNv, the LGN (Ahlsén et al., 1978; Montero, 1989) was virtually equal in both sides. Thus, the results imply that in the awake, attentive animal, TRNv activation is predominantly under visual cortex control. Furthermore, the correlation between the intensity of TRNv activation and exploration time found in this study, and the fact that TRNv activation in rats is triggered by exploration of a novel-complex environment, in contrast to absence of TRNv activation in control, awake non-exploring rats (Montero, 1997), suggest a link between visual attention to the environment and TRNv activation.

The symmetry in Fos-detected activation in LGN contralateral to the normal and deprived eyes is consistent with normal physiological parameters found in LGN of amblyopic animals (Wiesel and Hubel, 1963; Kiorpes and Movshon, 1996) and suggests that retinal inputs from the normal and deprived eye in the alert, attentive animal, are sufficient to elicit Fos induction in LGN. In contrast, TRNv activation requires additional inputs other than the LGN, which as the results suggest, are top-down inputs from the visual cortex.

b) Effects of excitotoxic lesions of layer 6 of V1 on activation
of visual TRN in rats that explored the novel-complex environment

To obtain direct evidence that Fos-detected activation of the visual TRN in rats that explored the NCE is under control of the primary visual cortex, unilateral excitotoxic lesions (ibotenic acid) were placed with multiple microinjections restricted to layer 6 of most of V1, to determine whether the absence of the cortico-reticulo-geniculate pathway originating in this layer would affect activation of TRNv (Montero, 2000). Since these pathways are only ipsilateral, it was possible to compare qualitatively and quantitatively in the same animal, and in the same sections, the number of Fos+ neurons induced by exploration of the NCE in the TRNv of the lesioned and normal sides. The advantage of this internal control is that it abolished the effects of behavioral variability among animals, as well as variability in the immunocytochemical stain.

In three animals with lesions in layer 6 of V1 (N=3) there was diminished Fos-detected activation in the right lesioned-side TRNv, after the exploration test, in comparison to the left, normal-side TRNv. This asymmetric activation is demonstrated qualitatively in Fig. 6.2, where coronal sections show clear diminution of Fos+ neurons in the TRNv of the right, lesioned side (Fig. 6.2D), in comparison with the number of Fos+ neurons in the left, normal side (Fig. 6.2C). Quantitative analysis of the number of Fos+ neurons in TRNv per section in the three lesioned rats, as counted in 12 sections of the right and left TRNv in each animal, showed a highly significant (P **0.001, t-test) reduction of the number of Fos+ neurons in the right TRNv of the lesioned side in comparison with that in the left TRNv of the nonlesioned side. In contrast, in TRNv of both sides of a normal control animal that explored the NCE, the number of Fos+ neurons was not significantly different ($P = 0.92$).

The main objective of this study was to determine the effects of lack of visual cortex inputs to TRNv in its activation induced by attentive exploration of a novel environment. Therefore, a concern was that the excitotoxic lesion of layer 6 in V1 could also have provoked retrograde degeneration in LGN, which would implicate lack of geniculate inputs as an additional factor in diminished TRNv activation in these animals. Thus, it was necessary to determine whether the excitotoxic lesions in layer 6 of V1 provoked retrograde degeneration in LGN, either as a consequence of unintended compromise of layer 4, which is the main target of geniculo-cortical axons (Peters and Feldman, 1976), or as a consequence of the lesion in layer 6, which is a minor target of these axons (Peters and Feldman, 1976). This evaluation was done by measuring the cross-sectional area of 200 neuronal cell bodies in Nissl-stained sections of the right and left LGN in each of the lesioned animals. The results of this analysis showed no significant differences in the cross-sectional areas of LGN neuronal perikaria in the right and left sides ($P > 0.05$), indicating that the lesions in layer 6 of V1 in the experimental animals did not induce retrograde degeneration in LGN. If the ibotenate lesions had provoked retrograde degeneration in LGN, at the survival time used (11–15 days) the cross-sectional area of LGN cells would have been reduced to about 60% of their original size (Agarwala and Kalil, 1998). Conversely, the lack of retrograde degeneration in LGN after excitotoxic lesions in layer 6 of V1, traversed by geniculo-cortical axons, provides direct evidence that ibotenate does not destroy fibers of passage.

To assess the effects of the ibotenate lesions in layer 6 of V1 on Fos labeling in LGN, which necessarily implied lack of excitatory glutamatergic cortico-geniculate inputs, and diminution of GABAergic inhibitory inputs from the less active TRNv into LGN, the number of Fos+ neurons in LGN of both sides was assessed qualitatively and quantitatively. This quantitative analysis showed no statistical difference (t-test) in the mean number of Fos+ neurons per mm^2 in LGN of the right and left sides.

These results provided direct evidence that Fos-detected activation of the TRNv in rats that explored a NCE is dependent primarily on top-down cortico-reticular inputs from V1 via axon collaterals of cortico-geniculate axons originating in layer 6 (Bourassa and Deschênes, 1995). The normal morphology and Fos label of LGN cells in the lesioned side indicates that geniculate inputs to TRNv were viable, and consequently not responsible for the diminished activation in TRNv. The similarity of Fos label in LGN of the lesioned side with that in the normal side might be explained by mutual cancellation of lack of excitatory cortico-geniculate inputs and diminution of inhibitory inputs from the TRNv into LGN cells.

Altogether, these results suggest that the observed diminution of Fos label in TRNv induced by lesions in layer 6 of V1, or by amblyopia, is not caused by lack of geniculate inputs to TRNv, but is provoked primarily by diminution of V1 inputs into TRNv. The observed dependence of attentional activation of TRNv on cortical inputs from V1, necessarily reflects attentional activation of the cortico-geniculate pathway originating in layer 6 and innervating both the TRNv and LGN (Bourassa and Deschênes, 1995; Murphy and Sillito, 1996) forming the vast majority of synapses in both nuclei (Montero, 1991; Liu and Jones, 1999). The retinotopic organization of V1 innervation of TRNv and LGN (Montero et al., 1977), and of TRNv projections to LGN (Pinault et al., 1995), is ideally suitable to influence geniculo-cortical transmission in a spatially discrete manner. The results are thus consistent with the hypothesis (Montero, 1999) that a focus of attention in V1 generates a core of enhanced geniculo-cortical transmission via direct cortico-geniculate inputs, and surround regions of depressed geniculo-cortical transmission via cortico-reticulo-geniculate pathways (Fig. 6.4).

Visual attention facilitates detection and discrimination of features in the visual scene, independent of foveal vision or eye movements (Posner, 1980). There is substantial evidence that activity in the primary visual cortex of monkeys and humans can be modulated by attention (Motter, 1993; Roelfsema et al., 1998; Vidyasagar, 1998; Watanabe et al., 1998; Brefczynski and DeYoe, 1999 ; McAdams and Maunsell, 1999). The results suggest that top-down inputs from V1 to TRNv and LGN via the cortico-reticulo-geniculate pathway are a consequence of attentional activation in V1, and provide supportive evidence to the proposed notion (Montero, 1999) that attentional modulation of thalamo-cortical transmission is a main function of cortico-thalamic pathways to sensory relay nuclei.

Neurochemical analysis of attentional activation of cortical pathways to LGN but not to the MGN in rats that explored the novel-complex environment

The previous results showed that in rats that explored a NCE there is Fos-detected activation of the visual TRN, but not of the acoustic TRN, which is dependent on top-down cortical glutamatergic inputs to LGN and to TRNv. Thus, with respect to awake non-exploring controls, the LGN of exploring rats should receive increased glutamatergic inputs from the direct cortico-geniculate pathway, and increased GABAergic inputs from the activated visual TRN. By contrast, since the acoustic TRN is not activated in exploring rats, despite intense noises generated during exploration, suggestive of absence of glutamatergic top-down from the primary auditory cortex (A1) to TRN and to the medial geniculate nucleus (MGN), the levels of glutamate (Schwarz et al., 2000) and GABA in the MGN (Montero, 1983) of exploring rats should be less than in LGN of the same rats, and equal to the levels of these neurotransmitters in MGN of non-exploring controls.

These predictions were tested in a neurochemical analysis (HPLC) of levels of glutamate and GABA in the LGN and MGN of rats that explored the NCE for 30 min and in non-exploring awake control rats (Montero et al., 2001). The results showed that levels of glutamate and GABA in LGN of exploring rats were significantly higher than in LGN of control animals. In contrast, the levels of these neurotransmitters in MGN of the exploring and control rats were not significantly different. In addition, the levels of glutamate and GABA in LGN of control rats were not significantly different to the levels of these neurotransmitters in MGN of exploring and control rats.

The results confirmed the predictions that in exploring rats there is increased activity of glutamatergic and GABAergic pathways to LGN but not to MGN, and are consistent with the Fos experiments indicating that in exploring rats there is increased glutamatergic and GABAergic inputs from V1 and visual TRN, respectively, to LGN, but that there is no increased glutamatergic inputs from A1 to MGN, and of GABAergic acoustic TRN to MGN.

Fos induction in rats visual pathways by exploration of a novel-simple environment

Rats placed in a novel-simple environment (NSE), an empty 50 cm-sided translucent plexiglass box with vertical and horizontal black lines in the walls, perform significantly more rearings during the first 30 min of the test than rats in the NCE (117 ± 12 vs 21 ± 10; $P < 0.001$; N = 8). Rearings are elevation of head and trunk with release of forelegs from the floor which are considered unconditioned orienting and scanning phases of attention. In contrast to rats in the NCE, in these animals there is retinotopically restricted Fos induction confined to the upper visual field regions in dorsal parts of LGN (Montero et al., 1968) (Figs. 6.3C and 6.3D), in caudal parts of V1 (Montero et al., 1973b) (Fig. 6.3B), and in medial parts of the SC (Siminoff et al., 1966). Of particular interest is the fact that Fos label in TRNv is also restricted, but to the reciprocal, lower visual field representation in its caudo-ventral region (Figs. 6.3D and 6.3F). The visual field representation in the rat TRNv is similar to that found in the rabbit TRNv (Montero et al., 1977), as confirmed in tracer studies (Lozsadi et al., 1996), and is also similar to the retinotopy of rat LGN (Montero et al., 1968), where upper to lower visual field is represented dorso-ventrally, and nasal to temporal visual field is represented medio-laterally in LGN. Fos induction restricted to the upper visual field regions in the intermediate layers of the SC layer is indicative of predominant eye movements towards the upper visual field. Neurons in the intermediate layer represent a retinotopically organized motor map which respond to eye movements towards their receptive fields (Walker et al., 1995; Ozen et al., 2000), and are activated by attentional shifts associated with eye movements (Kustov and Robinson, 1996; Ignashchenkova et al., 2004; Krauzlis et al., 2004). Contrary to common belief, rats are capable of ample eye movements including saccades (fast eye movements Montero and Robles, 1971; Strata et al., 1990). Thus, Fos induction in the whole visual field representation in the SC indicates that in rats that explore the NCE eye movements are directed to all regions of the visual field, while in rats in the NSE, retinotopically restricted Fos label in the SC indicates that there are predominant eye movements towards the upper visual field.

As in LGN and SC, Fos induction in V1 of rats that explored the NSE was retinotopically restricted to the upper visual field representation in the caudal region with negligible label in the rest of V1 (Fig. 6.3B). In the retinotopically restricted Fos field in caudal V1, all layers were labeled, but the label was particularly strong and wider in layer 2/3 than in layer 6. In this respect, to a certain extent, the retinotopically restricted Fos label in V1 was similar to that found in this cortex in anesthetized rats stimulated with localized visual stimuli (Montero and Jian, 1995), with the significant difference that the Fos label in rats that explored the NSE was more intense in all cortical layers, and particularly in layer 2/3, which was not labeled in anesthetized rats.

The retinotopically asymmetrical Fos induction found in LGN, TRNv, SC and V1 of rats that explored the NSE opened the question whether this asymmetry was already present in the retina, and therefore was simply the result of stronger activation of upper visual representation in the retina. To check for this possibility, quantification of the number of Fos+ cells per area in the retinal ganglion cell layer was done in rats that explored the NSE or the NCE. It has been previously shown that spontaneous activity that occurs in the retina in darkness does not induce Fos expression in the retina of dark-adapted rabbits, in contrast to strong Fos induction in the retina of light stimulated animals (Sagar and Sharp, 1990). The number of Fos+ cells per area in the ganglion cell layer in the upper and lower hemiretinas of rats that explored the NSE or the NCE was not significantly different ($P > 0.05$). These results suggest that the retinotopically asymmetrical activation found in the analyzed visual centers in rats that explored the NSE is a consequence of neural interactions beyond the retina.

A significant result of these experiments is the demonstration of activation of reciprocal regions of the visual field in LGN versus TRNv in rats that explored the NSE. A plausible explanation of this retinotopic complementarity is that activation of the lower visual field in TRNv induces inhibition in the lower visual field in LGN, such that retinal inputs from the lower visual field are blocked in geniculo-cortical transmission.

Fos mapping in these experiments represents the temporal integration of neural activity engendered in visual centers of rats at different times during the 2-2.5 h exploration tests. Expression of the Fos protein in the rat visual cortex peaks at 1.5-2 h after light stimuli onset (Chaudhuri et al., 2000). Thus, Fos induction retinotopically restricted to the upper visual in the SC indicates predominant eye movements towards the upper visual field in rats that explored the NSE, while Fos induction in the whole visual field representation in the SC of rats that explored the NCE indicates temporal integration of neural activity engendered by eye movements to all regions of the visual field. Fos label is also preferentially restricted to the upper visual field in LGN and V1, and -significantly- to the lower visual field in TRNv, in rats that explored the NSE. Thus, conceivably, Fos induction in the whole visual field representation in LGN, TRNV, and V1 in rats that explored the NCE corresponds to the temporal integration of neuronal activation in these centers representing different, segmental regions of the visual field, particularly during the first 30 min of the behavioral test, when rats perform more active exploration.

Discussion

In these studies Fos imaging was used to detect neural activity at single cell level in populations of neurons in the visual pathways of the rat that was elicited: a) by restricted visual stimulation in anesthetized animals; b) in

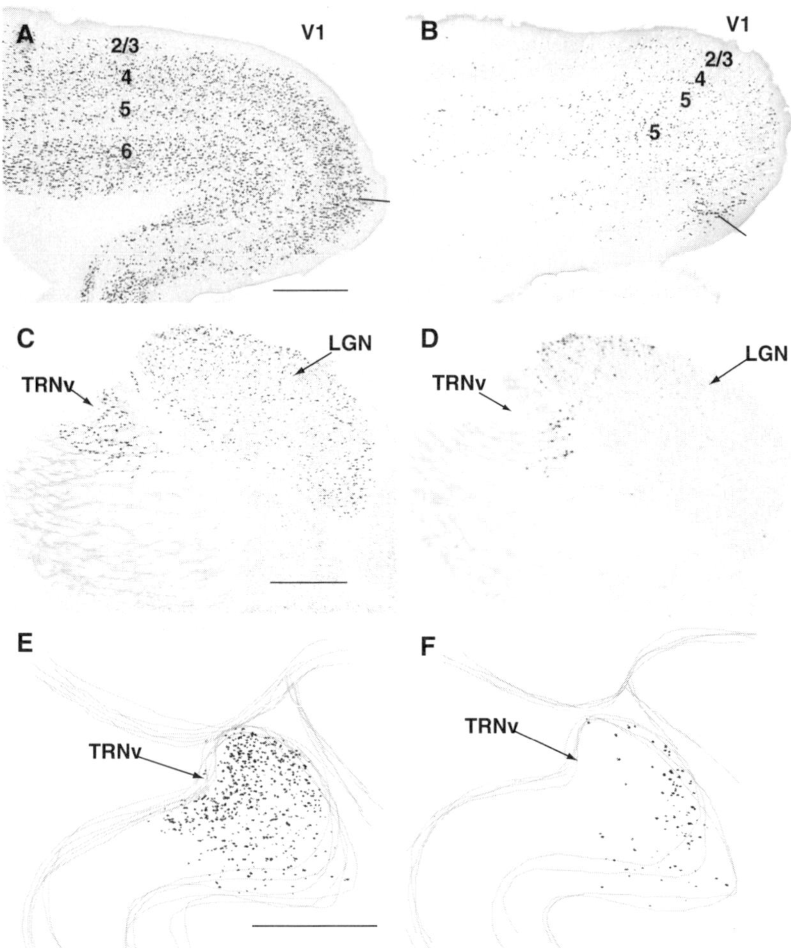

FIGURE 6.3. Comparison of Fos label in V1, LGN and TRNv of a rat (RC) that explored the NCE (**A, C, E**) and of a rat (RS) that explored the NSE (**B, D, F**). **A:** Sagittal section showing intense Fos label in all the rostro-caudal extent of V1 and in all cortical layers (indicated by small numbers) except layer 1. Caudal limit of V1 is indicated by a black line at the occipital pole. This section is representative of Fos induction obtained in the whole visual field representation in V1 of rats that explored the NCE. In contrast in V1 of rats that explored the NSE, as shown in the sagittal section of rat RC in **B**, Fos label was restricted to the upper visual field representation in the most caudal part of V1, while the rest of V1 showed much less Fos label, similar to that seen in V1 of controls non-exploring rats. In the restricted Fos field at the occipital pole, layers 2/3 were most extensively labeled, while Fos label in layer 6 was more narrow. **C:** Sagittal section of LGN and TRNv of rat RC that explored the NCE, showing intense Fos label in the whole extent of LGN, and consequently, in the whole visual field representation in this nucleus, similar to that seen in coronal sections of LGN of rats that explored the NCE, as shown in Fig. 1B. Immediately rostral

behavioral paradigms in which rats are naturally paying attention to visual features of two different types of novel environments; c) in control awake non-exploring rats; and d) in exploring rats after monocular amblyopia or after excitotoxic lesion of layer 6 of the primary visual cortex

Fos induction of TRNv in attentive rats is dependent on top-down cortical inputs from V1

The dependence of Fos induction in visual TRN in attentive exploring rats on top-down inputs from layer 6 of the primary visual cortex is a significant finding of these experiments since it reveals command by the cortex of TRN inhibitory filtering of sensory inputs in thalamo-cortical pathways. The purely excitatory nature of glutamatergic cortical inputs to TRN has been demonstrated *in vivo* (Ahlsén and Lindström, 1983); and in an *in vitro* study the excitatory cortical inputs to TRN were demonstrated to be mediated by NMDA and non-NMDA glutamate receptors (De Curtis et al., 1989). GluR4 receptors occur at cortical synapses in the TRN, and are vastly more frequent than in cortical synapses on thalamic relay cells (Golshani et al., 2001). Iontophoresis of glutamate in TRN of anesthetized animals shifts oscillatory firing of its neurons to high frequency tonic discharges (Marks and Roffwarg, 1991; Warren and Jones, 1994); for review, see (Steriade et al., 1993; McCormick and Bal, 1997). The TRN in the cat discharges in bursts during sleep in contrast to tonic discharges during wakefulness (Steriade et al., 1986). More pertinent to the present results, similar discharge patterns in the sleep-wake cycle were corroborated in the TRN of freely moving rats by Marks and

---

to LGN, the visual sector of TRN in the most caudal region of this nucleus at this sagittal plane (about lateral 3.9 mm (Paxinos and Watson, 1998)) is fully labeled with Fos+ cells for about 0.5 mm. **D:** Sagittal section of LGN and TRNv of rat RS that explored the NSE. Most of LGN is devoid of Fos label, except the most rostral and dorsal part representing the upper visual field (Montero et al., 1968). In contrast to Fos label in the whole extent of TRNv in rat RC, Fos label in TRNv of rat RS is confined to the most caudo-ventral lower visual field representation in the visual TRN. Significantly, Fos label in the lower visual field of TRNv reveals neural activation that is reciprocal to the upper visual field activation in LGN. **E:** Diagram of thresholded Fos+ cells in TRNv of rat RC compiled from several sections. The whole extent of TRNv is labeled, with stronger Fos label in the dorsal, upper visual field region of TRNv. **F:** Diagram of thresholded Fos+ cells in TRNv of rat RS compiled from several sections. Fos+ cells are restricted to the caudo-ventral lower visual field in TRNv, being almost absent in the caudo-dorsal upper visual field representation in TRNv. Scales = 0.5 mm, in left panels apply to right panels.

Roffwarg, 1993 (Marks and Roffwarg, 1993) who demonstrated that TRN discharges in bursts during slow wave and rapid eye movement sleep, while it discharges tonically in waking state. Since rat TRN does not express Fos during sleep (unpublished observation), bursting discharges *per se* do not induce Fos in TRN. And vice-versa, since rat TRN discharges tonically in wakefulness, tonic discharges contribute to Fos induction in this nucleus.

The excitatory center of receptive field size in thalamo-cortical relay cells looses discrimination of sensory inputs after neurotoxic lesion of the respective sector of TRN, as demonstrated by drastic increase in the number of whiskers able to activate a neuron in the somatic relay nucleus after somatic TRN lesions (Lee et al., 1994), indicating an important role of TRN in the discriminatory ability of thalamic relay cells, and consequently, of cortical cells. Neuronal circuits that may be involved in these TRN effects on receptive field discrimination of thalamic relay cells are open loop inhibitory connections between adjacent thalamic relay cells via TRN cells (Pinault and Deschênes, 1998), and extensive collaterals of geniculo-cortical and cortico-geniculate axons in the visual TRN (Ahlsén et al., 1978; Murphy and Sillito, 1996). These circuits may be involved in results (Tsumoto et al., 1978) demonstrating that excitation of a restricted locus in the cat visual cortex by glutamate elicited a central core of excitation in LGN of about 3.4° in diameter, within which the geniculate cells had over-lapping receptive fields with the stimulated cortical cells, and surround inhibition that extended for several degrees. More recently, it has been shown that top-down cortical inputs from V1 to LGN not only increase the inhibitory surround of receptive fields of relay cells in LGN (Cudeiro and Sillito, 1996), but also the strength of the excitatory center of these cells (Rivadulla et al., 2002). Altogether, these results suggest that the excitability of the receptive field center of thalamic relay cells is under the direct control of glutamatergic cortical inputs from primary sensory cortices, while the inhibitory receptive field surround is also under cortical control, via disynaptic inhibition mediated by the TRN. Thus, as previously proposed (Montero, 1999), attention-dependent modulation of thalamo-cortical transmission is one of the main functions of cortico-thalamic pathways to thalamic relay nuclei.

Recent studies have confirmed attentional modulation of neural activity in the rat TRN (McAlonan et al., 2000) and in LGN of the monkey (Vanduffel et al., 2000) and humans (O'Connor et al., 2002). However, in none of these reports the influence of top-down inputs from the visual cortex was considered a causal mechanism for TRN or LGN attentional activation. In the present results, the combination of Fos mapping of neural activity in rats exploring different novel environments and restricted excitotoxic lesion of layer 6 of the visual cortex, elucidated an essential role of top-down inputs from the primary visual cortex on attentional activation of the visual TRN, and consequently, on LGN transfer to the cortex.

Different neural mechanisms required for Fos induction in LGN and TRN

In these experiments, neurons in LGN and in TRNv showed an essential difference with respect to synaptic inputs required for Fos induction. In anesthetized rats, the same type of visual stimulation that generated neuronal discharges recorded with microelectrodes in LGN (Montero et al., 1968) and TRNv (Hale et al., 1982), was sufficient to induce Fos expression in the populations of stimulated neurons in LGN (Montero and Jian, 1995), but not in the TRNv. Thus, Fos induction in TRNv represents a clear example of Clayton's 'genomic action potential' (Clayton, 2000) requiring integration of representational inputs from the retina, and additional modulatory inputs, in this case from the cortex and brainstem which are triggered in the awake animal by behaviorally significant stimuli. In control awake non-exploring rats (Montero, 1997), as well as in exploring rats, there is strong Fos induction in the brainstem neuromodulators centers: noradrenergic locus coeruleus (LC), serotonergic dorsal raphe nucleus (DR) and cholinergic laterodorsal tegmental nucleus (LDTg) (McCormick, 1989; McCormick and Bal, 1994). The rich noradrenergic input from LC to TRN (Asanuma, 1992) is further activated during exploration of a novel environment (Vankov et al., 1995). However, since there is significantly less Fos induction in TRNv of exploring rats as a consequence of lack of top-down inputs from V1 versus the non-lesioned, normal side (Montero, 2000), these results indicate that increased noradrenergic modulatory inputs by exploration *per se* are insufficient to induce Fos in TRN. In fact, our results suggest that Fos induction in the visual TRN requires the combination of topographically organized glutamatergic inputs from LGN, diffuse modulatory noradrenergic inputs from LC, and topographically organized, attentionally triggered, glutamatergic inputs from the visual cortex. Thus, Fos induction in sensory sectors of TRN in alert, freely moving animals, reflects attentional activation of cortical inputs to TRN and to sensory relay nuclei, and could be used as a gauge to reveal attention mechanisms in cortico-thalamic circuits

Fos induction in supragranular layers of the visual cortex V1 in attentive exploring rats

Strong Fos induction in the supragranular layers 2/3 in V1 of rats that explored the NCE or NSE, which is not present in V1 of visually stimulated anesthetized rats or in control rats, suggests as previously proposed (Montero, 1997), that supragranular layers in V1, or in other primary cortical areas, are involved in cortical attentional mechanisms. Similarly, Sakata et al., (2000) reported attention-dependent Fos induction in all layers of V1 or A1 of rats attending to visual or acoustic stimuli in an audio-visual discrimination task. Fos induction was strongest in layers 2/3, and was observed preferentially in excitatory neurons but not in inhibitory neurons. It should be mentioned that in rats that explored the novel environments there is Fos induction not only in the primary visual cortex, but also in the several extrastriate visual areas (Montero et al., 1973b; Montero, 1993) and in the frontal eye field (Guandalini, 1998). The primary visual cortex in the rat is directly interconnected with all extrastriate visual areas (Montero et al., 1973a) and with the frontal eye field (Sanderson et al., 1991; Guandalini, 1998), receiving from extrastriate areas excitatory feedback connections that are distributed mainly outside granular layer 4 (Coogan and Burkhalter, 1993; Johnson, 1994), and from the frontal eye fields connections that are distributed in all laminae (Guandalini, 1998). In addition, supragranular layers in rat and cat V1 send intra-areal excitatory projections to layers 5 and 6 (Johnson and Burkhalter, 1992; Gilbert, 1993). Thus, the strong activation of supragranular layers of V1 in the alert, attentive rats demonstrated in these experiments, reflects integration in the primary visual cortex of inputs from the different types of cortical neural functions performed in extrastriate visual areas and frontal eye fields.

A 'focal attention' hypothesis

Altogether, the results of these experiments are consistent with the previously postulated 'focal attention' hypothesis (Montero, 1999), expressed schematically in Fig. 6.4. In this view, a focus of attention in V1 (Motter, 1993) transmits top-down excitatory influences to LGN via retinotopically organized glutamatergic corticogeniculate pathway (Montero and Wenthold, 1989). Concomitantly, collaterals of the activated cortico-geniculate axons innervate more diffusely the visual TRN (Bourassa and Deschênes, 1995; Murphy and Sillito, 1996), which in turn sends axons making inhibitory synapses on LGN relay cells (Montero and Scott, 1981) that represent surrounding regions of the focus of attention. The axons of LGN relay cells innervate TRNv cells which project to neighboring regions of the nucleus in open loop circuits (Ahlsén et al., 1978; Pinault and Deschenes, 1998). The net result of this cortical influence on LGN and visual TRN is increased geniculo-cortical transmission of visual inputs from a focus of attention (Fig. 6.4, large arrow), and decreased geniculo-cortical transmission of visual inputs from surrounding regions of the focus of attention (Fig. 6.4, small arrows). Consistent with this hypothesis, attentional activation of the visual TRN in rats that explored the NCE is dependent on top-down inputs from V1 (Montero, 2000). Rats that explored the NSE did significantly more rearings than rats exploring the NCE, and showed restricted activation of upper visual field regions in superficial and deep layers of the SC, both phenomena indicative of shifts of visual attention to the upper visual field. The reciprocal retinotopic activation in LGN and TRNv detected by Fos in rats that explored the NSE, i.e., the lower visual field is active in TRNv but not in LGN, and the upper visual field is active in LGN but not in TRNv (Figs. 6.3D, 6.3F), indicates inhibitory effects of TRNv on regions of the lower visual field in LGN, outside of the focus of attention in the upper visual field

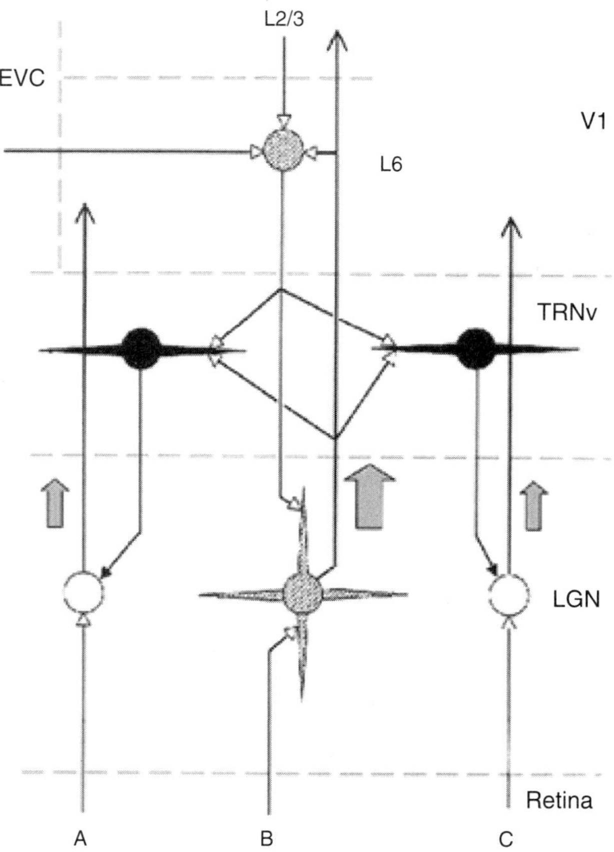

FIGURE 6.4. Diagram of the 'focal attention' hypothesis. A focus of attention in V1, generated by inputs to layers 2/3 and 6 of V1 from extrastriate visual cortical areas (EVC) and frontal eye fields, transmits top-down excitatory influences to LGN via retinotopically organized glutamatergic corticogeniculate pathway. Concomitantly, collaterals of the activated cortico-geniculate axons innervate more diffusely the visual TRN, which in turn sends axons establishing inhibitory synapses in LGN in regions surrounding the focus of attention. In turn, axons of LGN relay cells innervate TRNv cells which project to neighboring regions of the nucleus in open loop circuits. The net result of this cortical influence on LGN and visual TRN is increased geniculo-cortical transmission of visual inputs from a focus of attention (large arrow), in contrast to decreased geniculo-cortical transmission of inputs from surrounding regions of the focus of attention (small arrows).

of V1. Altogether, these results indicate that attentional modulation of thalamo-cortical transmission is a main function of cortico-thalamic pathways to sensory relay nuclei and to TRN.

Acknowledgements: Supported by NIH grants EY02877, MH57558 HD35885 and from the Whitehall Foundation.

References

Agarwala S, Kalil RE (1998) Axotomy-induced neuronal death and reactive astrogliosis in the lateral geniculate nucleus following a lesion of the visual cortex in the rat. J Comp Neurol 392:252–263.

Ahlsén G, Lindström S (1983) Corticofugal projection to perigeniculate neurons in the cat. Acta Physiol Scand 118:181–184.

Ahlsén G, Lindström S, Sybirska E (1978) Subcortical axon collaterals of principal cells in the lateral geniculate body of the cat. Brain Res 156:106–109.

Anderson B (1994) the volume of the cerebellar molecular layer predicts attention to novelty in rats. Brain Research 641:160–162.

Asanuma C (1992) Noradrenergic innervation of the thalamic reticular nucleus: A light and electron miscroscopic immunohistochemical stydy in rats. JCompNeurol 319:299–311.

Bal T, M. vK, McCormick D (1995) Role or the ferret perigeniculate nucleus in the generation of synchronized oscillations *in vitro.* J Physiol 483:665–685.

Bourassa J, Deschênes M (1995) Corticothalamic projections from the primary visual cortex in rats: A single fiber study using biocytin as an anterograde tracer. Neuroscience 253:253–263.

Brefczynski JA, DeYoe EA (1999) A physiological correlate of the 'spotlight' of visual attention. Nature Neuroscience 2:370–374.

Burne RA, Parnavelas JG, Lin C-S (1984) Response properties of neurons in the visual cortex of the rat. ExpBrain Res 53:374–383.

Chaudhuri A (1997) Neural activity mapping with inducible transcription factors. Neuroreport 8:R3–R7.

Chaudhuri A, Zangenehpour S, Rahbar-Dehgan F, Ye F (2000) Molecular maps of neural activity and quiescence. Acta Neurobiol Exp (Wars) 60:403–410.

Clayton DF (2000) The Genomic Action Potential. Neurobiology of Learning and Memory 74:185–216.

Coleman KA, Mitrofanis J (1996) Organization of the visual reticular thalamic nucleus of the rat. EurJNeurosci 8:388–404.

Coogan TA, Burkhalter A (1993) Hierarchical organization of areas in rat visual cortex. JNeurosci 13:3749–3772.

Cornwall J, Cooper JD, Phillipson OT (1990) Projections to the rostral reticular thalamic nucleus in the rat. ExpBrain Res 80:157–171.

Crick F (1984) Function of the thalamic reticular complex: The searchlight hypothesis. ProcNatlAcadSciUSA 81:4586–4590.

Cudeiro J, Sillito AM (1996) Spatial frequency tuning of orientation-discontinuity-sensitive corticofugal feedback to the cat lateral geniculate nucleus. J Physiol 490 (Pt 2):481–492.

Daffner KR, Mesulam MM, Scinto LFM, Cohen LG, Kennedy BP, West WC, Holcomb PJ (1998) Regulation of attention to novel stimuli by frontal lobes: an event-related potential study. Neuroreport 9:787–791.

De Curtis M, Spreafico R, Avanzini G (1989) Excitatory amino acids mediate responses elicited *in vitro* by stimulation of cortical afferents to reticularis thalamic neurons in the rat. Neuroscience 33:275–284.

Deschênes M, Hu B (1990) Electrophysiology and pharmacology of the corticothalamic input to lateral thalamic nuclei: an intracellular study in the cat. EuropJNeurosci 2:140–152.

Desimone R, Duncan J (1995) Neural mechanisms of selective visual attention. AnnuRevNeurosci 18:193–222.

Dias R, Honey RC (2002) Involvement of the rat medial prefrontal cortex in novelty detection. Behav Neurosci 116:498–503.

Domenici L, Berardi N, Carmignoto G, Vantini G, Maffei L (1991) Nerve growth factor prevents the amblyopic effects of monocular deprivation. Proc Natl Acad Sci USA 88:8811–8815.

Ehret G, Fischer R (1991) Neuronal activity and tonotopy in the auditory system visualized by c-fos gene expression. Brain Res 567:350–354.

Fagiolini M, Pizzorusso T, Berardi N, Domenici L, Maffei L (1994) Functional postnatal development of the rat primary visual cortex and the role of visual experience: Dark rearing and monocular deprivation. Vision Res 34:709–720.

Farrell WJ, Alberts JR (2002) Maternal responsiveness to infant Norway rat (Rattus norvegicus) ultrasonic vocalizations during the maternal behavior cycle and after steroid and experiential induction regimens. J Comp Psychol 116:286–296.

Friauf E (1992) Tonotopic order in the adult and developing auditory system of the rat as shown by c-fos immunocytochemistry. EuropJNeurosci 4:798–812.

Fukuda Y, Sumitomo I, Sugitani M, Iwama K (1979) Receptive-field properties of cells in the dorsal part of the albino rat's lateral geniculate nucleus. JpnJPhysiol 29: 283–307.

Funke K, Eysel UT, FitzGibbon T (1991) Retinogeniculate transmission by NMDA and non-NMDA receptors in the cat. Brain Res 547:229–238.

Gabbott PLA, Somogyi J, Stewart MG, Hamori J (1986) Neurons in the dorsal lateral geniculate nucleus of the rat: characterisation using a combination of Golgi-impregnation and GABA-immunocytochemistry. ExplBrain Res 61:311–322.

Ghosh A, Ginty DD, Bading H, Greenberg ME (1994) Calcium regulation of gene expression in neuronal cells. J Neurobiol 25:294–303.

Gilbert CD (1993) Circuitry, architecture, and functional dynamics of visual cortex. Cereb Cortex 3:373–386.

Golshani P, Liu X-B, Jones EG (2001) Differences in quantal amplitude reflect GluR4-subunit number at corticothalamic synapses on two populations of thalamic neurons. Proc Natl Acad Sci USA 98:4172–4177.

Gonzalo-Ruiz A, Lieberman AR (1995) Topographic organization of projections from the thalamic reticular nucleus to the anterior thalamic nuclei in the rat. Brain Res Bull 37:17–35.

Guandalini P (1998) The corticocortical projections of the physiologically defined eye field in the rat medial frontal cortex. Brain Res Bull 47:377–385.

Hale PT, Sefton AJ, Baur LA, Cottee LJ (1982) Interrelations of the rat's thalamic reticular and dorsal lateral geniculate nuclei. ExpBrain Res 45:217–229.

Hayhow WR, Sefton A, Webb C (1962) Primary optic centers in the rat in relation to the terminal distribution of the crossed and uncrossed optic nerve fibers. J Comp Neurol 118:295–322.

Herdegen T, Sandkuhler J, Gass P, Kiessling M, Bravo R, Zimmermann M (1993) Jun, fos, krox, and creb transcription factor proteins in the rat cortex: basal expression and induction by spreading depression and epileptic seizures. JCompNeurol 333:271–288.

Herdegen T, Kovary K, Buhl A, Bravo R, Zimmermann M, Gass P (1995) Basal expression of the inducible transcription factors c-jun, junB, junD, c-fos, fosB and krox-24 in the adult rat brain. JCompNeurol 354:39–56.

Ignashchenkova A, Dicke PW, Haarmeier T, Thier P (2004) Neuron-specific contribution of the superior colliculus to overt and covert shifts of attention. Nat Neurosci 7:56–64.

Ito M, Gilbert CD (1999) Attention modulates contextual influences in the primary visual cortex of alert monkeys. Neuron 22:593–604.

Johnson RR, Burkhalter A (1992) Evidence for excitatory amino acid neurotransmitters in the geniculo-cortical pathway and local projections within rat primary visual cortex. ExpBrain Res 89:20–30.

Jones EG (1975) Some aspects of the organization of the thalamic reticular complex. JCompNeurol 162:285–308.

Kaczmarek L, Chaudhuri A (1997) Sensory regulation of immediate-early gene expression in mammalian visual cortex: implications for functional mapping and neural plasticity. Brain Res Brain Res Rev 23:237–256.

Kayama Y, Shosaku A, Doty R, W. (1984) Cryogenic blockade of the visual corticothalamic projection in the rat. ExpBrain Res 54:157–165.

Kiorpes L, Movshon JA (1996) Amblyopia: A developmental disorder of the central visual pathways. Cold Spring Harbor Symposia on Quantitative Biology 41:39–48.

Krauzlis RJ, Liston D, Carello CD (2004) Target selection and the superior colliculus: goals, choices and hypotheses. Vision Res 44:1445–1451.

Kustov AA, Robinson DL (1996) Shared neural control of attentional shifts and eye movements. Nature 384:74–77.

Lashley KS (1938) The mechanism of vision. XV. Preliminary studies of the rat's capacity for detailed vision. J gen Psychol 18:123–193.

Lee SM, Friedberg MH, Ebner FF (1994) The role of GABA-mediated inhibition in the rat ventral posterior medial thalamus. JNeurophysiol 71:1702–1715.

Li S, Cullen WK, Anwyl R, Rowan MJ (2003) Dopamine-dependent facilitation of LTP induction in hippocampal CA1 by exposure to spatial novelty. Nat Neurosci 6:526–531.

Liu XB, Jones EG (1999) Predominance of corticothalamic synaptic inputs to thalamic reticular nucleus neurons in the rat. J Comp Neurol 414:67–79.

Lozsadi DA, Gonzalez-Soriano J, Guillery RW (1996) The course and termination of corticothalamic fibres arising in the visual cortex of the rat. Eur J Neurosci 8:2416–2427.

Marks GA, Roffwarg HP (1991) Cholinergic modulation of responses to glutamate in the thalamic reticular nucleus of the anesthetized rat. Brain Res 557:48–56.

Marks GA, Roffwarg HP (1993) Spontaneous activity in the thalamic reticular nucleus during the sleep/wake cycle of the freely-moving rat. Brain Research 623:241–248.

McAdams CJ, Maunsell JHR (1999) Effects of attention on orientation-tuning functions of single neurons in macaque cortical area V4. J Neurosci 19:431–441.

McAlonan K, Brown VJ, Bowman EM (2000) Thalamic reticular nucleus activation reflects attentional gating during classical conditioning. J Neurosci 20:8897–8901.

McCormick DA (1989) Cholinergic and noradrenergic modulation of thalamocortical processing. Trends in Neurosci 12:215–221.

McCormick DA, von Krosigk M (1992) Corticothalamic activation modulates thalamic firing through glutamate "metabotropic" receptors. ProcNatlAcadSciUSA 89:2774–2778.

McCormick DA, Bal T (1994) Sensory gating mechanisms of the thalamus. Current Opinion Neurobiol 4:550–556.

McCormick DA, Bal T (1997) Sleep and arousal: Thalamocortical mechanisms. Annu Rev Neurosci 20:185.

Montero VM (1973) Evoked responses in the rat's visual cortex to contralateral, ipsilateral, and restricted photic stimulation. Brain Res 53:192–196.

Montero VM (1981) Comparative studies on the visual cortex. In: Cortical Sensory Organization. Vol. 2 Multiple Visual Areas (Woolsey CN, ed), pp 33–81. Clifton, N.J.: The Humana Press.

Montero VM (1983) Ultrastructural identification of axon terminals from the thalamic reticular nucleus in the medial geniculate body in the rat: An EM autoradiographic study. ExpBrain Res 51:338–342.

Montero VM (1989) Ultrastructural identification of synaptic terminals from cortical axons and from collateral axons of geniculo-cortical relay cells in the perigeniculate nucleus of the cat. ExpBrain Res 75:65–72.

Montero VM (1990) Quantitative immunogold analysis reveals high glutamate levels in synaptic terminals of retino-geniculate, cortico-geniculate, and geniculo-cortical axons in the cat. Visual Neurosci 4:437–443.

Montero VM (1991) A quantitative study of synaptic contacts on interneurons and relay cells of the cat lateral geniculate nucleus. ExpBrain Res 86:257–270.

Montero VM (1993) Retinotopy of cortical connections between the striate cortex and extrastriate visual areas in the brain. ExpBrain Res 94:1–15.

Montero VM (1994) Quantitative immunogold evidence for enrichment of glutamate but not aspartate in synaptic terminals of retino-geniculate, geniculo-cortical, and cortico-geniculate axons in the cat. VisNeurosc 11:675–681.

Montero VM (1997) C-*fos* induction in sensory pathways of rats exploring a novel environment: Shifts of active thalamic reticular sectors by predominant sensory cues. Neuroscience 76:1069–1081.

Montero VM (1999) Amblyopia decreases activation of the corticogeniculate pathway and visual reticularis in attentive rats: A 'focal attention hypothesis'. Neuroscience 91:805–817.

Montero VM (2000) Attentional activation of the visual thalamic reticular nucleus depends on 'top-down' inputs from the primary visual cortex via corticogeniculate pathways. Brain Res 864:95–104.

Montero VM, Brugge JF (1969) Direction of movement as the significant stimulus parameter for some lateral geniculate cells in the rat. Vision Res 9:71–88.

Montero, VM, Robles, L (1971) Saccadic modulation of cell discharges in the lateral geniculate nucleus. Vision Res.11(Suppl.3): 253–268.

Montero VM, Scott GL (1981) Synaptic terminals in the dorsal lateral geniculate nucleus from neurons of the thalamic reticular nucleus: A light and electron microscope autoradiographic study. Neuroscience 6:2561–2577.

Montero VM, Singer W (1985) Ultrastructural identification of somata and neural processes immunoreactive to antibodies against glutamic acid decarboxylase (GAD) in the dorsal lateral geniculate nucleus of the cat. ExpBrain Res 59: 151–165.

Montero VM, Wenthold RJ (1989) Quantitative immunogold analysis reveals high glutamate levels in retinal and cortical synaptic terminals in the lateral geniculate nucleus of the macaque. Neuroscience 31:639–647.

Montero VM, Jian S (1995) Induction of c-*fos* protein by patterned visual stimulation in central visual pathways of the rat. Brain Res 690:189–199.

Montero VM, Brugge JF, Beitel RE (1966) Visual field representation and receptive field characteristics of neurons in the dorsal lateral geniculate nucleus of the albino rat. Anat Rec 154:388.

Montero VM, Brugge JF, Beitel RE (1968) Relation of the visual field to the lateral geniculate body of the albino rat. JNeurophysiol 31:221–236.

Montero VM, Bravo H, Fernandez V (1973a) Striate-peristriate cortico-cortical connections in the albino and gray rat. Brain Res 53:202–207.

Montero VM, Rojas A, Torrealba F (1973b) Retinopic organization of the striate and peristriate visual cortex in the albino rat. Brain Res 53:197–201.

Montero VM, Guillery RW, Woolsey CN (1977) Retinotopic organization within the thalamic reticular nucleus demonstrated by a double label autoradiographic technique. Brain Res 138:407–421.

Montero VM, Wright LS, Siegel F (2001) Increased glutamate, GABA and glutamine in lateral geniculate nucleus but not in medial geniculate nucleus caused by visual attention to novelty. Brain Res 916:152–158.

Motter BC (1993) Focal attention produces spatially selective processing in visual cortical areas V1, V2, and V4 in the presence of competing stimuli. JNeurophysiol 70:909–919.

Murphy PC, Sillito AM (1996) Functional morphology of the feedback pathway from area 17 of the cat visual cortex to the lateral geniculate nucleus. J Neurosci 16:1180–1192.

O'Connor DH, Fukui MM, Pinsk MA, Kastner S (2002) Attention modulates responses in the human lateral geniculate nucleus. Nature Neurosci online 15 October 2002:1–7.

Ohara PT, Lieberman AR (1981) Thalamic reticular nucleus: anatomical evidence that cortico-reticular axons establish monosynaptic contact with reticulo-geniculate projection cells. Brain Res 207:153–156.

Ohara PT, Havton LA (1996) Dendritic arbors of neurons from different regions of the rat thalamic reticular nucleus share a similar orientation. Brain Res 731:236–240.

Oswald CJ, Yee BK, Rawlins JN, Bannerman DB, Good M, Honey RC (2001) Involvement of the entorhinal cortex in a process of attentional modulation: evidence from a novel variant of an IDS/EDS procedure. Behav Neurosci 115:841–849.

Ozen G, Augustine GJ, Hall WC (2000) Contribution of superficial layer neurons to premotor bursts in the superior colliculus. J Neurophysiol 84:460–471.

Parron C, Poucet B, Save E (2004) Entorhinal cortex lesions impair the use of distal but not proximal landmarks during place navigation in the rat. Behav Brain Res 154:345–352.

Paxinos G, Watson C (1998) The rat brain in stereotaxic coordinates, Fourth Edition Edition. New York, NY: Academic Press.

Peschanski M, Ralston HJ, Roudier F (1983) Reticularis thalami afferents to the ventrobasal complex of the rat thalamus: an electron microscope study. Brain Res 270:325–329.

Peters A, Saldanha J (1976) The projection of the lateral geniculate nucleus to area 17 of the rat cerebral cortex. III. Layer VI. Brain Res 105:533–537.

Peters A, Feldman ML (1976) The projection of the lateral geniculate nucleus to area 17 of the rat cerebral cortex. I. General description. J Neurocytol 5:63–84.

Pinault D, Deschênes M (1998) Anatomical evidence for a mechanism of lateral inhibition in the rat talamus. Eur J Neurosci 10:3462–3469.

Pinault D, Bourassa J, Deschênes M (1995) Thalamic reticular input to the rat visual thalamus: a single fiber study using biocytin as an anterograde tracer. Brain Res 670:147–152.

Posner MI (1980) Orienting of attention. Q J Exp Psychol 32:3–25.

Posner MI, Pothbart MK, Digirolamo GJ (1999) Develpment of brain networks for orienting to novelty. Zh Vyssh Nerv Deyat Imen Pavlov 49:715–722.

Prusky GT, Harker KT, Douglas RM, Whishaw IQ (2002) Variation in visual acuity within pigmented, and between pigmented and albino rat strains. Behav Brain Res 136:339–348.

Rivadulla C, Martinez LM, Varela C, Cudeiro J (2002) Completing the corticofugal loop: A visual role for the corticogeniculate type 1 metabotropic glutamate receptor. J Neurosci 22:2956–2962.

Roelfsema PR, Lamme VAF, Spekreijse H (1998) Object-based attention in the primary visual cortex of the macaque monkey. Nature 395:376–381.

Rothblat LA, Schwartz ML, Kasdan PM (1978) Monocular deprivation in the rat: evidence for an age-related defect in visual behavior. Brain Research 158:456–460.

Sagar SM, Sharp FR (1990) Light induces a Fos-like nuclear antigen in retinal neurons. Brain Res Mol Brain Res 7:17–21.

Sakata S, Kitsukawa T, Kaneko T, Yamamori T, Sakurai Y (2002) Task-dependent and cell-type-specific Fos enhancement in rat sensory cortices during audio-visual discrimination. Eur J Neurosi 15:735–743.

Sakurai T, Okada Y (1992) Selective reduction of glutamate in the rat superior colliculus and dorsal lateral geniculate nucleus after contralateral enucleation. Brain Res 573:197–203.

Sanderson KJ, Dreher B, Gayer N (1991) Prosencephalic connections of striate and extrastriate areas of rat visual cortex. ExpBrain Res 85:324–334.

Sato H, Hata Y, Tsumoto T (1999) Effects of blocking non-N-methyl-D-aspartate receptors on visual responses of neurons in the cat visual cortex. Neuroscience 94: 697–703.

Scalia F, Arango V (1979) Topographic organization of the projections of the retina to the pretectal region in the rat. JCompNeurol 186:271–292.

Schuett S, Bonhoeffer T, Hubener M (2002) Mapping retinotopic structure in mouse visual cortex with optical imaging. J Neurosci 22:6549–6559.

Schwarz DWF, Tennigkeit F, Puil E (2000) Metabotropic transmitter actions in auditory thalamus. Acta Otolaryngol 120:251–254.

Sefton AJ, Dreher B (1985) Visual system. Sydney, Australia: Academic Press.

Shapley R, So YT (1980) Is there an effect of monocular deprivation on the proportions of X and Y cells in the cat lateral geniculate nucleus? ExpBrain Res 39:41–48.

Sheng M, Greenberg ME (1990) The regulation and function of c-fos and other immediate early genes in the nervous system. Neuron 4:477–485.

Shosaku A, Simitomo I (1983) Auditory neurons in the rat thalamic reticular nucleus. ExpBrain Res 49:432–442.

Shosaku A, Kayama Y, Sumitomo I (1984) Somatotopic organization in the rat thalamic reticular nucleus. Brain Res 311:57–63.

Shosaku A, Kayama Y, Sumitomo I, Sugitani M, Iwama K (1989) Analysis of recurrent inhibitory circuit in rat thalamus: Neurophysiology of the thalamic reticular nucleus. Prog Neurobiol 32:77–102.

Siminoff R, Schwassmann HO, Kruger L (1966) An electrophysiological study of the visual projection to the superior colliculus of the rat. JCompNeurol 127:435–444.

Simpson EL, Gaffan EA (1999) Scene and object vision in rats. Q J Exp Psychol B 52:1–29.

Spear PD, Baumann TP (1975) Receptive field characteristics of single neurons in lateral suprasylvian visual area of the cat. JNeurophysiol 38:1403–1420.

Steriade M, Llinas RR (1988) The functional states of the thalamus and the associated neuronal interplay. PhysiolRev 68:649–742.

Steriade M, Domich L, Oakson G (1986) Reticularis thalami neurons revisited: activity changes during shifts in states of vigilance. JNeurosci 6:68–81.

Steriade M, McCormick DA, Sejnowski TJ (1993) Thalamocortical oscillations in the sleeping and aroused brain. Science 262:679–688.

Tiitinen H, May P, Reinikainen K, Naatanen R (1994) Attentive novelty detection in humans is governed by pre-attentive sensory memory. Nature 372:90–92.

Tsumoto T, Creutzfelt OD, Legendy CR (1978) Functional organization of the corticofugal system from the visual cortex to the lateral geniculate nucleus in the cat (with an appendix on geniculo-cortical mono-synaptic connections). ExplBrain Res 32:345–364.

Vanduffel W, Tootell RBH, Orban GA (2000) Attention-dependent suppression of metabolic activity in the early stages of the macaque visual system. Cereb Cortex 10:109–126.

Vankov A, Herve-Minvielle A, Sara.S.J. (1995) Response to novelty and its rapid habituation in locus coeruleus neurons of the freely exploring rat. EurJNeurosci 7:1180–1187.

Vidyasagar TR (1998) Gating of neuronal responses in macaque primary visual cortex by attentional spotlight. Neuroreport 9:1947–1952.

Von Krosigk M, Monckton JE, Reiner PB, McCormick DA (1999) Dynamic properties of corticothalamic excitatory postsynaptic potentials and thalamic reticular inhibitory postsynaptic potentials in thalamocortical neurons of the guinea-pig dorsal lateral geniculate nucleus. Neuroscience 91:7–20.

Walker MF, Fitzgibbon EJ, Goldberg ME (1995) Neurons in the monkey superior colliculus predict the visual result of impending saccadic eye movements. J Neurophysiol 73:1988–2003.

Warren RA, Jones EG (1994) Glutamate activation of cat thalamic reticular nucleus: effects on response properties of ventroposterior neurons. ExpBrain Res 100:215–226.

Watanabe T, Sasaki Y, Miyauchi S, Putz B, Norio F, Nielsen M, Takino R, Miyakawa S (1998) Attention-regulated activity in human primary visual cortex. J Neurophysiol 79:2218–2221.

Wiesel TN, Hubel DH (1963) Effects of visual deprivation on morphology and physiology of cells in the cat's lateral geniculate body. JNeurophysiol 26:978–993.

Wiesel TN, Hubel DH (1965) Comparison of the effects of unilateral and bilateral eye closure on cortical unit responses in kittens. J Neurophysiol 28:1029–1040.

Yingling CD, Skinner JE (1977) Gating of thalamic input to cerebral cortex by nucleus reticularis thalami. In: Attention, Voluntary Contraction and Event-Related Cerebral Potentials. Prog. clin. Neurophysiol. Vol 1 (Desmedt JE, ed), pp 70–96. Basel: Karger.

Zeki SM (1974) Functional organization of a visual area in the posterior bank of the superior temporal sulcus of the rhesus monkey. JPhysiolLondon 236:549–573.

Part II

Cortical Plasticity

7

Neuromodulatory Transmitters in Sensory Processing and Plasticity in the Primary Visual Cortex

RAPHAEL PINAUD[1], THOMAS A. TERLEPH[2]
AND LIISA A. TREMERE[3]

[1]Neurological Sciences Institute, OHSU, Portland, OR, USA, [2]Department of Psychology, Rutgers University, Piscataway, NJ, USA and [3]CROET, OHSU, Portland, OR, USA

Introduction

In this chapter we discuss the roles of four neurotransmitter systems, noradrenaline (NA), acetylcholine (ACh), serotonin (5-HT) and dopamine (DA) in processing at the level of the primary visual cortex (V1), and review their participation in experience-dependent plasticity paradigms. For each of these neuromodulators, two parts are presented: First we discuss the anatomical distribution of neuromodulatory fibers and terminals, and post-synaptic receptors in V1. Second, we review the effects of altering neuromodulatory transmission to receptive field (RF) properties, sensory processing and experience-dependent plasticity, with an emphasis on plasticity of ocular dominance column paradigms.

Postsynaptic receptor subtypes determine which of several modulatory signal transduction systems will be activated, thus influencing how a given transmitter will act: with excitatory effects, inhibitory effects, or both. As discussed below, neuromodulatory transmitters may contribute to plasticity by increasing the signal-to-noise ratio of cell populations, changing the threshold for activity-dependent synaptic modifications, and gating behavioral states. Receptor classes and even receptor subtypes generally show distinct laminar distributions within the visual cortex that vary across species. The distribution patterns and densities of afferent inputs also differ across cortical layers, and in a species-specific manner. Whenever possible, and wherever relevant to the issues discussed, we outline inter-species differences in both fiber and receptor distributions. Knowledge of the anatomical-functional organization of these neurotransmitter systems allows for a more complete understanding of visual

R. Pinaud, L. A. Tremere, P. De Weerd (Eds.), Plasticity in the Visual System:
From Genes to Circuits, 127-151, ©2006 Springer Science + Business Media, Inc.

cortical processing dynamics and facilitates a more thorough understanding of the principles that underlie plastic changes in the mammalian V1.

Noradrenaline

Noradrenergic projections and receptors in V1

Noradrenaline (NA) cortical afferents are derived primarily from the locus coeruleus (LC) (Freedman et al., 1975; Foote et al., 1983). Tract-tracing studies have demonstrated that the NA projections from LC to the visual cortex are robust and bilateral across a number of species (Gatter and Powell, 1977; Dursteler et al., 1979; Tigges et al., 1982; Sato et al., 1989).

NA synthesis is catalyzed by the enzyme dopamine-beta-hydroxylase (DBH), using dopamine as a precursor; this enzyme therefore can serve as a cytochemical marker for the presence of NA in axons and cell bodies. Axons containing this enzyme are found in all cortical layers. Unlike other types of afferents that enter specific cortical regions radially, NA-containing axons innervate all six cortical layers in an unusual way. Fibers that have reached the rostral cortical poles extend caudally, innervating longitudinal slabs of cortex from the frontal to occipital pole (Morrison et al., 1981).

The density of DBH-immunoreactive fibers differs by cortical area and layer, depending on the species studied (Kosofsky et al., 1984; Gu, 2002). There are highly differentiated laminar orderings of NA throughout the cortex, including V1 (Gatter and Powell, 1977; Dursteler et al., 1979; Tigges et al., 1982; Kosofsky et al., 1984). Descriptions of intracortical NA fiber trajectories and distributions obtained from the rat only weakly predict what is found in the primate cortex (Foote et al., 1983) and NA innervations of the primate cortex have a far greater regional variation in density and pattern relative to that of the rat.

In the rat, DBH-immunoreactive fibers are dense in the superficial layers (Morrison et al., 1978). To the contrary, in the cat V1, fibers are abundant in both supragranular and infragranular layers, with a low density of DBH-positive inputs observed in the thalamorecipient layer IV (Liu and Cynader, 1994). NA fiber innervation of V1 in New World Guyanan squirrel monkeys (*Saimiri sciureus*) is most dense in infragranular layers, moderate in supragranular layers, and virtually absent in layer IV (Morrison et al., 1982b; Morrison et al., 1982a). Squirrel monkeys have a laminar distribution of NA innervation in V1 that is dissimilar from that found in the Old World cynomolgus monkey (*Macaca fascicularis*) (Foote and Morrison, 1984). In the V1 of adult cynomolgus monkeys, noradrenergic fibers are sparse relative to those of other neurotransmitters systems such as the serotonergic system (details below), and laminar differences in NA axon density are not detectable (Kosofsky et al., 1984). NA axons in the cynomolgus monkey form two broad bands of moderate density throughout deep and superficial layers, and intermediate layer IVc shows very few NA fibers (Kosofsky et al., 1984).

NA innervation has also been mapped via DBH-immunoreactivity in the human visual cortex (Gaspar et al., 1989). In humans, NA axons are present in all cortical layers. DBH-immunoreactive fibers are most dense in layer V, followed by VI, and NA innervations are sparse in all other cortical layers.

Adrenoreceptors are coupled to G proteins; activation of these receptors therefore influences a number of second messengers systems. Adrenoreceptors are pharmacologically differentiated into the alpha and beta receptor families, each of these families being further subdivided into a number of subtypes, based upon the results of pharmacological and cloning studies (for reviews see Hoffman and Lefkowitz, 1980; Jonowsky and Sulser, 1987; Civantos Calzada and Aleixandre de Artinano, 2001). The beta receptors are differentiated from alpha receptors by their sensitivity to isoproterenol. All three of the beta receptor subtypes (beta$_1$, beta$_2$, and beta$_3$) facilitate synthesis of cAMP, but only beta$_1$ and beta$_2$ are found in the cortex (for review see Stiles et al., 1984).

The beta$_1$ and beta$_2$ subtypes are linked to a stimulatory G protein that leads to enhanced catalytic activity of adenylyl cyclase, an enzyme that catalyzes the conversion of adenosine triphosphate (ATP) to cyclic adenosine monophosphate (cAMP). The alpha$_1$ and alpha$_2$ subtypes are also found in the cerebral cortex. Activation of alpha$_1$ receptors stimulates the phospholipase C pathway, leading to mobilization of intracellular Ca^{2+} via increased levels of the second messengers IP$_3$ and DAG; activation of alpha$_2$ receptors inhibits adenylyl cyclase through an inhibitory G protein, reducing intracellular cAMP levels (Berridge, 1998).

Primary visual cortex has a similar laminar distribution of alpha$_1$, beta$_1$, and beta$_2$ subunits with high densities of receptors in superficial layers, low densities in the middle layers and intermediate densities in the infragranular layers. This overall distribution pattern is seen in the cat (Aoki et al., 1986; Parkinson et al., 1988a; Jia et al., 1994), the New World common marmoset (*Callithrix jacchus*) (Gebhard et al., 1993), and the Old World rhesus monkey (*Macaca mulatta*) (Bigham and Lidow, 1995). The alpha$_2$ receptor shows a different distribution: in cats alpha$_2$ receptor density is lowest in layer VI (Jia et al., 1994), and in one study this receptor was only observed in a single band occupying layers II and III (Parkinson et al., 1988a). The laminar distribution pattern of NA fibers in the cat visual cortex resembles that of beta-adrenergic receptors, and differs from the distribution pattern of alpha-adrenergic receptors (Liu and Cynader, 1994), suggesting that beta-receptors are anatomically well positioned to participate in key functional roles in V1 (discussed below). In rhesus monkeys, alpha$_2$ receptor densities are highest in layers III, IVa, IVc, and VI (Bigham and Lidow, 1995).

Noradrenergic effects on physiological properties of V1 neurons
and plasticity

Earlier reports that characterized the contributions of NA to the modulation of response properties within sensory cortices used a number of strategies.

They have included the local delivery of NA into sites of interest via iontophoresis, electric stimulation of the noradrenergic locus coeruleus (LC) and pharmacological intervention directed at specific noradrenergic receptors.

Noradrenaline has been shown to markedly affect neuronal properties within V1 of both adult and juvenile animals; age differences do not appear to influence noradrenergic modulation on visual cortical neurons (Kasamatsu and Heggelund, 1982; Videen et al., 1984; Ego-Stengel et al., 2002).

For example, pharmacological antagonism of both alpha and beta receptors are known to alter spontaneous response properties of cat V1 neurons, suggesting that endogenous NA contributes to regulating normal cell activity (Sato et al., 1989). NA effects on cortical neurons, however, appear to be expressed in a bi-directional fashion, promoting both excitation and inhibition (Videen et al., 1984; Sato et al., 1989). For example, Sato and colleagues (1989) have shown that electrical stimulation of the LC suppressed V1 neuronal activity in about half of the cells recorded in their experiments. Other cells exhibited a facilitation of activity under conditions of enhanced noradrenergic tone. Interestingly, the nature of this differential effect of NA on V1 activity appears to be related to cortical layer, where facilitation was observed in neurons within the granular and supragranular layers, while depressive effects appeared to prevail in the infragranular layers (Sato et al., 1989). It has been proposed that this disparity in the neuronal response under enhanced NA reflects different patterns of LC projections across layers of V1 (Sato et al., 1989). Differential distribution of NA receptors across cortical layers could also account for these seemingly opposite effects of noradrenergic input. This hypothesis is supported by findings where antagonism of beta receptors affected response suppression, while alpha-specific antagonism interfered with the facilitatory effects triggered by enhanced levels of NA (Sato et al., 1989). Recent experiments conducted in a slice preparation have provided evidence for a role for NA in suppressing horizontal propagation of activity in rat V1 by decreasing excitatory and increasing inhibitory transmission directed at supragranular pyramidal neurons (Kobayashi et al., 2000). Studies conducted in the rat (Kolta et al., 1987; Waterhouse et al., 1990) and cat (Kasamatsu and Heggelund, 1982) visual cortex also suggested that NA acts to either modulate depression of baseline activity or to enhance visually-driven responses. For example, it has been reported that iontophoretic application of NA within rat V1 unmasked visually-driven responses originally suppressed prior to NA application. In these experiments, the authors also observed that local NA application enhanced RF border contrast (Waterhouse et al., 1990).

It also appears that more complex visual cortex network properties are modulated by noradrenergic input, although the reported findings have been contradictory. For example, it has been reported that local iontophoretic application of NA in V1 increases the selectivity of neurons for direction and speed of moving visual stimuli (McLean and Waterhouse, 1994), while other research groups have reported effects for NA on both orientation and directional selectivity (Madar et al., 1983). Experiments performed in adult rats, where pharmacological depletion of LC was conducted, also provided evidence for a role of NA in regulating V1 neuronal properties. These results show that orientation selectivity is severely impaired by decreased noradrenergic input. In addition, it was reported that receptive-field areas were expanded under these conditions, while the signal-to-noise ratio of neuronal responses appear to decrease (Siciliano et al., 1999). To the best of our knowledge there is no report to date that reconciles such discrepant results, possibly indicating that arousal and/or attentional levels might directly impact sensory processing at the level of V1. In fact, it has been shown that enhancement of noradrenergic LC activity contributes to attentional shifts in the rat during visual discrimination tasks (Devauges and Sara, 1990) and other learning paradigms (Carli et al., 1983; Everitt et al., 1983; Pisa et al., 1988; Selden et al., 1990b; Selden et al., 1990a). Currently we are not certain whether such effects of altered LC neuronal firing associated with behavioral states translates into differential NA modulation of visual cortical cells, a hypothesis that remains to be investigated.

Roles for noradrenergic inputs have also been demonstrated during ocular dominance (OD) plasticity. Pioneering experiments primarily conducted by Kasamatsu and colleagues have investigated the effects of eyelid suture coupled with intraventricular 6-OHDA treatment, a common chemical methodology used to deplete forebrain catecholaminergic transmission (Kasamatsu and Pettigrew, 1979). In these experiments, it was observed that animals treated with 6-OHDA exhibited an elevated percentage of cells with binocular responses, whereas control kittens, which underwent eyelid suture in the absence of 6-OHDA intervention, exhibited marked proportions of monocularly-driven neurons (Kasamatsu and Pettigrew, 1979). These experiments have suggested that catecholamines are major participants in the induction and/or maintenance of plastic states within the kitten V1. Direct evidence for the participation of NA in this process was obtained by the same group, where the same experiments were performed as outlined above, but in conjunction with local V1 NA infusion (Kasamatsu et al., 1979). As observed in their previous experiment (Kasamatsu and Pettigrew, 1979), 6-OHDA prevented OD shifts triggered by monocular deprivation. However, when NA was infused locally in combination with 6-OHDA treatment, plasticity was restored within V1, providing direct evidence for the participation of NA in maintaining V1 in a plastic state. Subsequent experiments have revealed that at least part of this effect was mediated through beta receptors in a concentration-dependent fashion (Kasamatsu and Shirokawa, 1985; Shirokawa and Kasamatsu, 1986). These data are in accordance with recent experiments that showed that monocular deprivation leads to an increase in $beta_1$-receptor, but not $alpha_1$-receptor, immunoreactivity in the visual cortex ipsilateral to the deprived eye in kittens (Nakadate et al., 2001). Interestingly, experiments conducted in the adult cat, where OD shifts do not occur (Hubel and Wiesel, 1970), using NA infusion or LC stimulation coupled with monocular deprivation, showed a significantly decreased number of binocularly-driven V1 neurons (Kasamatsu et al., 1979; Kasamatsu et al., 1985).

Noradrenaline and the control of plasticity-regulated gene expression

Beta receptor activation via NA binding leads to stimulatory G-protein (Gs) activation which, in turn, enhances catalytic activity of adenylyl cyclase (AC). AC catalyzes the conversion of ATP into cAMP, and increased cytoplasmic cAMP levels activate cAMP-dependent protein kinase A (PKA). The PKA pathway has been repeatedly implicated in a number of cellular plasticity mechanisms, including OD plasticity, and has been shown to involve the regulation of ERK/MAP Kinase regulation and CREB phosphorylation pathways (see Chapter 8). For example, it has been demonstrated that local pharmacological intervention that mimics enhanced strength of the PKA contributes to the restoration of OD plasticity in adult cats (Imamura et al., 1999). These results suggest that cAMP-sensitive pathways are involved in the maintenance of plasticity in V1. A critical question that arises from these experiments is whether these biochemical cascades involved in V1 plasticity trigger a gene expression program that is involved in regulating activity-dependent changes in the visual cortex. If so, it is important to define what genes participate in the neural plastic response. We have explored this possibility by focusing our own investigations on an activity-dependent, candidate-plasticity gene, the nerve-growth factor induced gene-A (NGFI-A). NGFI-A has been the focus of extensive study for the past few years as it is well positioned to orchestrate waves of gene expression involved ultimately in structural and functional rewiring of central nervous system (CNS) networks (reviewed in Knapska and Kaczmarek, 2004; Pinaud, 2004) (see also Chapter 8). NGFI-A encodes a transcriptional regulator that is sensitive to neuronal depolarization and has high affinity for a specific DNA motif found in the promoter region of a large number of genes expressed in the CNS (Milbrandt, 1987; Christy and Nathans, 1989). It has been shown that NGFI-A plays a role in the regulation of a number of genes that are putatively involved in CNS plasticity. For example, NGFI-A regulates the expression of the synapsin I and synapsin II genes, in addition to the gene that encodes synaptobrevin II (Thiel et al., 1994; Petersohn et al., 1995; Petersohn and Thiel, 1996). These genes have been shown to regulate the size of the readily releasable pool of neurotransmitter vesicles, and transmitter release (Hilfiker et al., 1999). NGFI-A has also been involved in the putative regulation of the monoamine oxidase B gene, neurofilament, the adenosine 5′-triphosphate binding cassette, sub-family A, transporter 2 (ABCA2) and the alpha-7 subunit of the nicotinic acetylcholine receptor (Pospelov et al., 1994; Carrasco-Serrano et al., 2000; Wong et al., 2002; Davis et al., 2003; Mello et al., 2004), thus likely to directly impact neuronal cell physiology, from neurotransmitter release and membrane transport to the regulation of cell excitability and neurotransmitter degradation rates. NGFI-A expression has been repeatedly correlated with classic plasticity inducing paradigms such as exposure to enriched visual environments (Wallace et al., 1995; Pinaud et al., 2002; Pinaud, 2004), which have been demonstrated to trigger widespread structural modifications in V1 connectivity (Volkmar and Greenough, 1972). In addition, NGFI-A expression is dependent on the activation of the NMDA-type glutamatergic receptors, which have been repeatedly implicated in various forms of enhanced synaptic efficacy, such as long-term potentiation (Cole et al., 1989; Wisden et al., 1990). Finally, NGFI-A expression is dramatically down-regulated in response to dark-adaptation and induced strongly by subsequent light-stimulation, another paradigm associated with visual system plasticity (reviewed in Kaczmarek and Chaudhuri, 1997; Herdegen and Leah, 1998; and Pinaud et al., 2000; Pinaud et al., 2003). We used the latter paradigm to investigate the contributions of noradrenergic input to the regulation of NGFI-A in the adult rat visual cortex (Pinaud et al., 2000). In our control experiments we found that dark adaptation significantly decreased the number of NGFI-A positive cells in the rodent V1, while a brief subsequent visual experience (light exposure) triggered a marked increase in NGFI-A expression across all cortical layers, with the exception of layer I, a finding that is in accordance with previously published results (Kaczmarek and Chaudhuri, 1997; Herdegen and Leah, 1998; Pinaud et al., 2000). We next ablated noradrenergic projections to V1 by locally infusing 6-OHDA within the LC, conducted the dark adaptation/light stimulation protocol and evaluated the patterns of NGFI-A expression in V1. Interestingly, 6-OHDA treatment prevented light-induced NGFI-A expression, providing direct evidence that noradrenergic input is required for NGFI-A induction (Pinaud et al., 2000). These experiments show, therefore, that NA directly modulates the activation of genetic mechanisms that are sensitive to sensory stimulation and possibly involved in the machinery required for CNS plasticity. Given that NGFI-A expression depends on increased cAMP levels and PKA activation, and NA potentiates this pathway through beta-adrenoreceptors, future work with selective antagonists targeted at each step of this biochemical cascade will allow a more detailed characterization of the second messenger systems involved in the induction of candidate-plasticity genes associated with experience-dependent cortical plasticity. An alternative role for NA in NGFI-A induction could be to regulate appropriate intracellular calcium levels required to trigger gene expression, along with other sources of calcium. NMDA receptor activation leads to calcium influx required for NGFI-A induction, however, a concomitant activation of beta-adrenoceptors could also potentially mobilize calcium from intracellular stores through a rise in IP_3 levels. Should this hypothesis prove to be correct, ablation of noradrenergic projections could significantly reduce intracellular calcium levels to prevent NGFI-A expression from occurring. This hypothesis, however, remains to be tested experimentally.

Acetylcholine

Cholinergic projections and receptors in V1

Depending on the types of target neurons, variations in postsynaptic receptor subtypes, and the different subcellular sites targeted, ACh have been shown to have diverse effects upon visual cortical neurons.

Early studies documented the laminar distribution pattern of cholinergic axons throughout the cortex based on electron microscopic visualization of choline acetyl transferase (ChAT) immunoreactive synapses (Mesulam et al., 1983; Houser et al., 1985). ChAT catalyzes ACh synthesis using acetyl CoA and choline as substrates. Given that ChAT is the acetylcholine-synthesizing enzyme, it serves as a marker for cholinergic neurons. The cerebral cortex has been found to have a diffuse cholinergic innervation, mostly from neurons in the nucleus Basilis of Meynert of the basal forebrain (Mesulam et al., 1983; Rye et al., 1984). In V1, ChAT immunoreactive fiber density varies across species. Humans show the largest number of immunoreactive fibers in layers I-III$_A$ (Mesulam et al., 1992), while the highest levels found in the adult macaque monkey are in layers I and IV (Mrzljak and Goldman-Rakic, 1993). ChAT immunoreactive fiber density also shows unique species-specific patterns in V1 for the rat (Parnavelas et al., 1986), mouse (Kitt et al., 1994), cat (Stichel and Singer, 1987a, b), and ferret (Henderson, 1987). For example, in rats and mice, although ChAT-positive fibers were distributed throughout all layers, highest densities were detected in cortical layers IV and V. Conversely, in the cat V1, highest fiber density was detected for layer I, while low numbers of ChAT immunolabeled fibers were found in the infragranular layers (Parnavelas et al., 1986; Stichel and Singer, 1987a; Kitt et al., 1994). It is therefore clear that there is considerable variability in the distribution patterns of cholinergic fibers, however, it is not currently known if and how these disparate patterns might contribute to species-specific visual processing.

Both muscarinic and nicotinic ACh receptor classes are found in the visual cortex (Zilles et al., 1989). Some muscarinic subtypes show a similar laminar distribution across V1 in several species, while the distribution pattern of other subtypes differs in a species-specific manner (Gu, 2003). Of the muscarinic receptors, anatomical localization in the cortex is best described for M$_1$ and M$_2$ receptor subtypes, largely because there are no specific labels for other subtypes (Gu, 2002). The M$_1$ muscarinic receptor, for example, is most dense in supragranular layers and least in middle cortical layers of the rat (Schliebs et al., 1989; Zilles et al., 1989), cat (Prusky and Cynader, 1990), and primates (Lidow et al., 1989a; Tigges et al., 1997). The M$_2$ subtype distribution, on the other hand, differs widely across these species. Even across individuals of the same species, M$_2$ receptor distribution can vary greatly. Peak density occurs in superficial layers and in lower layers IVc and IVa of the rhesus monkey (Lidow et al., 1989a; Tigges et al., 1997). In the cat V1, M$_2$ receptors are highly concentrated in layers I and V, although significant levels of this receptor are found in both supra- and infragranular layers (Prusky and Cynader, 1990). In the rat, although M_2 expression was found throughout all cortical layers, highest levels were found in the thalamorecipient layer IV, while lowest expression was detected in supragranular layers (Rossner et al., 1994).

Unlike muscarinic receptors, which are coupled to G proteins and show a long latency of action, all nicotinic ACh receptors (nAChRs) belong to the superfamily of ligand-gated ion channels. Most cortical nAChRs are thought to exist as heteromers of $alpha_4$ and $beta_2$ subunits or as $alpha_7$ homomers, and are found throughout the cortex (Metherate, 2004). The nAChRs can also be classified according to their affinity for alpha-bungarotoxin: low affinity nAChRs are alpha-bungarotoxin sensitive and high affinity nAChRs are not (reviewed in Gotti and Clementi, 2004).

These different nAChR subtypes not only serve distinct functions, but they show a different cortical distribution. Low-affinity nAChRs occur mainly within supragranular and infragranular layers, while high-affinity nAChRs are densest in the middle cortical layers (Clarke et al., 1985). Some nAChRs are located on thalamocortical afferent terminals primarily within cortical layer IV, but most receptors appear to be postsynaptic (Prusky et al., 1987; Parkinson et al., 1988b; Lavine et al., 1997). Most nAChRs appear on pyramidal cells and spiny stellate cells and, at lower frequencies, on inhibitory interneurons. Subcellular localization on pyramidal cells emphasized receptor placement on cell bodies and apical dendrites (Zilles et al., 1989).

Cholinergic effects on physiological properties of V1 neurons and plasticity

Like studies on the noradrenergic system, the effects of acetylcholine (ACh) on the properties of V1 neurons have primarily been investigated using chemical lesioning of cholinergic neurons, iontophoretic application of ACh and delivery of selective antagonists for particular types of cholinergic receptors. As detailed above, cholinergic basal forebrain neurons innervate the mammalian visual cortex profusely, suggesting a putative role for ACh in modulating visually-driven response properties in V1 (Saper, 1984; Mesulam et al., 1992). Studies conducted in cat primary visual cortex with electrical (via lateral geniculate nucleus stimulation) and visual stimuli have directly investigated the role of ACh in V1 physiology.

Local ACh application triggers both facilitation as well as depression of neuronal responses, results that are comparable at some levels with those obtained with noradrenaline (Sillito and Kemp, 1983; Sato et al., 1987a). Facilitatory effects however are more prevalent when compared with ACh-mediated suppression (Sillito and Kemp, 1983; Sato et al., 1987a).

Local ACh application has been reported to markedly influence V1 response properties, including enhanced directional and orientation selectivity (Sillito and Kemp, 1983). Unlike GABAergic (discussed in Chapter 11), and to a lesser degree, noradrenergic influences, cholinergic influences do not appear to regulate receptive field (RF) area, suggesting that this neurotransmitter does not, to any degree, determine the functional tuning of RFs. Furthermore, no differential actions of ACh on simple versus complex cells have been observed (Sillito and Kemp, 1983).

Another approach that has proven useful in investigating the role of ACh in visual processing has been the use of cytotoxic lesions directed at sites involved in ACh synthesis. For example, Sato and colleagues (1987b) conducted kainic-acid-induced unilateral lesions in the cat nucleus basalis magnocellularis (NBM), the major source of ACh projections to the visual cortex. Neurons located ipsilaterally to the lesion appeared to have undergone a substantial decrease in visually-evoked responses, while contralateral neurons were unaffected. Interestingly, direction and orientation selectivity were unaffected by depletion of cholinergic neurons in the NBM. However, rescue of normal cholinergic transmission for the ipsilateral side by local ACh application triggered both the response of originally silent neurons, and the facilitation of neuronal responses in the large majority of cells recorded (Sato et al., 1987b).

It has been reported that the effects of ACh on V1 neurons are blocked by infusion of pharmacological agents directed at the muscarinic (but not nicotinic) cholinergic receptors (Sato et al., 1987a), suggesting that ACh-mediated facilitation might be correlated with disinhibition. This effect was likely mediated through a reduction of the potassium current. In experiments conducted by another group (Sillito and Kemp, 1983), where the facilitatory effects of ACh were compared to disinhibition triggered by bicuculline application, it was concluded that the effects of both neurotransmitter systems are dissimilar.

As reported for the noradrenergic system, the effects of ACh-mediated facilitation or suppression appear to be layer-specific, with neuronal response suppression predominantly occurring for cells located in the granular cell layer and layer III, while facilitation was detected across all cortical layers (Sillito and Kemp, 1983; Sato et al., 1987a). This differential impact of ACh on V1 neurons might be correlated with the asymmetric expression of different receptor subtypes in sub-populations of visual cortical neurons. In fact, it has been shown that M_1 receptors are more prevalent in the granular and supragranular layers of V1 across a number of experimental models (Lidow et al., 1989a; Schliebs et al., 1989; Zilles et al., 1989; Prusky and Cynader, 1990). The highest concentration of M_2 receptors, however, appears to be regulated in a species-specific manner (Lidow et al., 1989a; Zilles et al., 1989; Prusky and Cynader, 1990). To our knowledge, no systematic analysis of the contributions of each receptor subtype to either facilitation or suppression behavior has been conducted to date.

A classic paradigm for postnatal visual system plasticity involves shifts in ocular dominance induced by monocular deprivation (reviewed in Miller, 1994; Bear and Rittenhouse, 1999; Katz and Crowley, 2002). The contributions of cholinergic transmission to this form of plasticity were originally investigated by Bear and Singer (1986), who used excitotoxic lesions directed at the cholinergic basal forebrain of kittens. These authors observed that by selectively targeting either the cholinergic or the noradrenergic system (by infusion of 6-hydroxydopamine; 6-OHDA), shifts of ocular dominance were preserved. Conversely, dual ablation of ACh and NA transmission severely delayed this form of plasticity, suggesting a synergistic effect between acetylcholine and noradrenaline enabling this form of visual cortical plasticity (Bear and Singer, 1986). Subsequent experiments using this same plasticity paradigm coupled with local infusion of specific cholinergic receptor antagonists elegantly showed that ACh plays a direct role in ocular dominance plasticity, mediated through muscarinic M_1 receptors (Gu and Singer, 1993). Antagonism of nicotinic ACh receptors does not seem to play a role in ocular dominance shifts but rather has been shown to enhance visually-driven responses in V1 (Parkinson et al., 1988b; Gu and Singer, 1993). As discussed above, M_1 receptor activation has been proposed to participate in visual cortical plasticity by enhancing activity via a decrease in potassium conductance. Another alternative for M_1 participation in the plasticity response is in the process that leads to the cleavage of phosphatidylinositol biphosphate (PIP_2), a membrane phospholipid that is an important component of several signal transduction pathways. PIP_2 cleavage leads to the generation of two products, diacylglycerol (DAG; the polar domain of the original PIP_2 structure) and inositol triphosphate (IP_3; the apolar domain) (Berridge, 1998). Given that IP_3 is apolar, it diffuses into the cytoplasm where it can then bind to specific IP_3 receptors in the endoplasmic reticulum, a major source of intracellular calcium. Mobilized calcium from intracellular stores appears to directly increase NMDA-receptor conductance in hippocampal neurons (Markram and Segal, 1992). NMDA receptors are repeatedly implicated in a number of experience-dependent synaptic modifications (for a review see Collingridge and Singer, 1990). Finally, ACh is also suggested to increase the signal-to-noise ratio for processing within V1, given that its iontophoretic application does not lead to enhancement of spontaneous neuronal activity (Sato et al., 1987a).

A recent study has demonstrated that pharmacological intervention directed at the cholinergic system within cat V1 affects stimulus-driven gamma oscillations and the associated neuronal synchronization via muscarinic receptors (Rodriguez et al., 2004). In addition, ACh agonist application along with the presentation of visual stimuli that are known to trigger synchronization lead to long-lasting enhancement of gamma oscillations and synchrony in visual cortical neurons (Rodriguez et al., 2004). Oscillatory behavior in the visual cortex may correlate strongly with higher order functions such as attentional processing (Muller et al., 2000; Fries et al., 2001), learning (Miltner et al., 1999; Gruber et al., 2001; Gruber et al., 2002) and memory (Tallon-Baudry et al., 1998), therefore directly implicating ACh in the regulation of these processes.

Serotonin

Serotonergic projections and receptors in V1

Serotonin synthesis occurs through two enzymatic steps: First the enzyme tryptophan hydroxylase catalyzes the conversion of the amino acid tryptophan into

L-5-hydroxytryptophan (5-HTP), using tetrahydrobiopterin as a co-factor. Second, 5-HTP is decarboxylated by the enzyme aromatic L-amino acid decarboxylase to form 5-hydroxytryptamine (5-HT; serotonin), using pyridoxal phosphate as a co-factor.

Serotonin has a wide cortical distribution, with high levels in the visual cortex (Azmitia and Segal, 1978; Qu et al., 2000). In fact, the extent of cortical arborization and innervation is such that 5-HT neurons are in contact with most rat cortical cells (Lidov et al., 1980). Early studies employing immunohistochemical methods to target 5-HT fibers in the rat show that labeled cell groups projecting to the cortex are found primarily in the dorsal and median raphe nuclei (Lidov et al., 1980). Anterograde labeling of axons from the dorsal raphe nucleus shows that these fibers are very fine and have granular or fusiform varicosities, while thicker fibers derived from the median raphe nucleus have large spherical varicosities. Axons from the dorsal raphe nucleus are much more common in the cortex (Kosofsky and Molliver, 1987).

The distribution density and laminar arrangement of 5-HT fibers throughout the cortex varies amongst species. These fibers show a relatively homogeneous distribution in the rat cortex (Lidov et al., 1980), although the orientation of 5-HT axons does vary according to laminar position (Papadopoulos et al., 1987). Throughout the ferret cortex, 5-HT fibers are distributed in a laminar fashion with density decreasing with increased depth below the pial surface: fibers are very dense in the supragranular layers and the granular and infragranular layers only exhibit sparse innervation. In addition, V1 of the ferret cortex shows a marked reduction of innervation in the transition from layer II to III (Voigt and de Lima, 1991a, b). Within the ferret cortex most innervations are made by fine fusiform axons containing small ovoid varicosities. Fewer innervations come from thick fibers (Voigt and de Lima, 1991a, b). In the cat visual cortex 5-HT fibers also exhibit a differential laminar distribution (Gu et al., 1990). These fibers are dense in the supragranular layers, less dense in layer V, and sparse in layers IV and VI. As with other species, the vast majority of 5-HT fibers in the cat cortex consist of fine axons with small varicosities, and a much smaller number of thick axons (Gu et al., 1990). Serotonergic innervation of the marmoset cortex similarly consists of two fiber types: one with small ovoid varicosities and with a laminar distribution throughout the cortex, and the other type sparsely distributed and containing large, spheroidal varicosities (Hornung et al., 1990). In the new world squirrel monkey, serotonergic fibers generally occur in the upper laminae and are especially dense in layer IV (Morrison et al., 1982b; Morrison et al., 1982a). In the V1 of cynomolgus monkeys (*Macaca fascicularis*) serotonergic innervation is similar to that of the squirrel monkey, yet shows a more robust pattern of lamination (Foote and Morrison, 1987).

The laminar distribution of cortical 5-HT fibers in these new and old world primates is complementary to the innervation of noradrenergic fibers, and serotonergic innervation is generally denser (Morrison et al., 1982b). For example, in the squirrel monkey, layer IV receives a very dense serotonergic projection (primarily innervating the spiny stellate cells of IVa and IVc) and a sparse noradrenergic projection, while in layers V and VI the opposite pattern of serotonergic and noradrenergic density is observed (Morrison et al., 1982b).

Serotonergic synapses are both symmetric and asymmetric, and found primarily on spines and dendritic shafts (Papadopoulos et al., 1987). In the primary visual cortex of the Japanese macaque (*Macaca fuscata*), synaptic contacts are made onto both stellate and pyramidal cells (Takeuchi and Sano, 1984).

There are currently seven recognized classes of 5-HT receptors, each with a distinct pharmacology and most exhibiting a number of subtypes (for reviews see Hoyer et al., 1994; Hoyer and Martin, 1997). Serotonin neuromodulatory action through 5-HT receptors, like ACh and NA, can potentially influence second messenger systems and cellular excitability throughout the cortex (Foote and Morrison, 1987).

Separate receptor subtypes show different distribution patterns within the cortex, both within and across species. For example, autoradiographic mapping of the human brain has revealed a high density of 5-HT-1$_A$ receptors in cortical layers I and II, a low density of 1$_C$ receptors throughout the cortex, and no 1$_B$ binding sites at all (Pazos et al., 1987). In the primary visual cortex of the rhesus monkey, both 5-HT1 and 5-HT2 receptors are found in high concentrations in sublayer IVc beta, and 5-HT1 receptor density is also high in layers III, V, VI and subdivisions of layer IV (Lidow et al., 1989b).

The general distribution of 5-HT1$_C$ sites in human and rat brain is similar (Hoyer et al., 1986a, b). 5-HT-2 receptor densities also have been localized in high concentrations throughout the human cortex, largely in layers III and V (Hoyer et al., 1986a). The anatomical distribution of this receptor subtype is also similar in human and rat. Even though the receptor's pharmacological profile does differ between rodents and humans, a comparatively larger difference may exist between humans and pigs (Hoyer et al., 1986a). Taken together, these findings indicate that there are not only many different 5-HT receptor subtypes but also a multitude of different patterns of cortical distribution, both within and across species, contributing to multiple putative functional roles in the mammalian visual cortex.

Serotonergic effects on physiological properties of V1 neurons
and plasticity

Different serotonin receptor subtypes are associated with a number of intracellular responses. In vivo experiments conducted in rat V1 have demonstrated that iontophoretic application of serotonin primarily triggers a suppression of neuronal activity (possibly by enhancing GABAergic transmission; see below), but can also increase stimulus evoked-responses (Waterhouse et al., 1990; Gu, 2002). Local serotonin treatment in the visual cortex has been reported to affect spike threshold and contrast of RF boundaries.

For example, serotonin reduces amplitude differences between spontaneous activity and stimulus-driven responses at RF borders, suggesting a role for this neurotransmitter in decreasing contrast. Interestingly, this effect appears to be the opposite of NA treatment, where RF contrast is enhanced (see above), leading authors to suggest that NA and serotonin act in complimentary ways to regulate visually-driven responses and possibly play a role in gain control (Waterhouse et al., 1990).

Experiments conducted in a slice preparation of the ferret V1 have revealed that agonist treatment leading to activation of the 5-HT3 receptor, (a ligand-gated ion channel) led to a restriction in the lateral extent of excitation, as well as a significant decrease in response amplitude. Electrically-driven responses have been shown to be significantly decreased by 5-HT3 agonist treatment, with a concomitant increase in spontaneous GABAergic events, suggesting that 5-HT3 activation leads to network inhibition (Roerig and Katz, 1997). Furthermore, the strength of spontaneous inhibition mediated by the activation of 5-HT3 receptors increases as a function of eye-opening, suggesting that activity-dependent modulation of serotonergic tone via 5-HT3 receptors might contribute to the development of visual cortical connections (Roerig et al., 1997). Serotonin application to rat pyramidal V1 neurons has also been shown to result in asymmetric responses, depending on the type of receptors activated. For example, activation of 5-HT3 receptors with the specific agonist 1-m-chlorophenyl biguanide has been shown to enhance the frequency of spontaneous GABAergic post-synaptic currents, while 5-HT1 activation with a specific agonist triggered a decrease of these sIPCs (Xiang and Prince, 2003). These results led the authors to propose that complex circuitry interactions affected by serotonin might occur within V1, allowing for intricate synaptic computations in this cortical region.

One of the classical forms of synaptic plasticity is long-term potentiation (LTP). LTP has been shown to occur in V1 synapses and a great deal of effort has been targeted at identifying the mechanisms that contribute to these experience-dependent synaptic strength alterations. In a recent and very elegant report, Edagawa and colleagues have demonstrated a critical role for serotonin in the induction of LTP in the rat visual cortex (Edagawa et al., 2001). These authors show that tetanic stimulation (a typical stimulus used to trigger LTP), when applied to the white matter, triggered LTP in V1 supragranular layers of young (2-3 weeks old), but not older (5 weeks old) rats. These findings suggest that plasticity mechanisms are fully active by 2-3 weeks, but are greatly reduced, if not ended, by 5 weeks in this preparation.

During normal development it has also been shown that levels of serotonin increase, suggesting that this neurotransmitter plays a role in suppressing visual cortex plasticity. Interestingly, treatment of 5-week old V1 slices with a serotonin receptor antagonist allowed for LTP induction in the supragranular layers. Cytotoxic lesions directed at serotonergic projections also allowed slices from 5-week old rats to undergo LTP induction (Edagawa et al., 2001).

These results strongly suggest that enhanced serotonergic transmission, which appears to progressively occur throughout the critical period, might be a major factor in the restriction of plasticity mechanisms within the rat V1.

The role of serotonin in OD plasticity has also been investigated in a number of studies. Monocular deprivation associated with depletion of serotonergic fibers has been shown to prevent OD shifts. In this paradigm, a high proportion of cells with binocular responses were detected after eyelid suture in experimental animals, while in the control condition, a shift to monocular responsiveness was detected for a large population of cells (Gu and Singer, 1995). Subsequent experiments using specific pharmacological agents directed at particular serotonin receptors revealed that OD shifts were significantly decreased by blockade of 5-HT2c receptors (Wang et al., 1997). It has been proposed that the role of serotonin receptor activation in cortical plasticity might be a contribution to the gating of NMDA receptors (Gu, 2002). Alternatively, activation of 5-HT2c receptors could mediate activation of the phospholipase C pathway, which is involved in the hydrolysis of PIP_2. This mechanism, via increases in IP_3 concentrations, could potentially lead to enhanced levels of intracellular calcium that would summate with calcium influx via NMDA receptor activation. This calcium signal could then directly mediate and/or trigger mechanisms of plasticity within V1. To our knowledge, however, the contribution of these biochemical pathways to visual cortical plasticity has not yet been satisfactorily addressed.

Dopamine

Dopaminergic projections and receptors in V1

Dopamine is synthesized by the action of the enzyme aromatic L-amino acid decarboxylase on precursor L-3,4-dihydroxyphenylalanine (L-DOPA). Dopamine (DA) axons innervating the cortex arise primarily from cells in the ventral tegmental area of the midbrain (Phillipson et al., 1987). A number of techniques have been used to identify and describe DA innervations of primary visual cortex in rats, including immunocytochemistry (Phillipson et al., 1987; Papadopoulos et al., 1989), retrograde axonal transport and HPLC (Phillipson et al., 1987). DA fibers are generally sparse in rat visual cortex. The principal innervations are in infragranular layers, with a minor innervation occurring in layer I (Phillipson et al., 1987). At the ultrastructural level, DA axon terminals and axonal varicosities form primarily asymmetrical synaptic contacts (Papadopoulos et al., 1989). DA innervation in the primate (human and non-human) cortex is much greater than that of rodents (Goldman-Rakic et al., 1992).

DA axons have been visualized in the neocortex of cynomolgus and squirrel monkeys, and show a different pattern of innervation from that of rats (Lewis et al., 1987). These axons are distributed throughout the entire neocortex and are only sparsely distributed in V1, with labeled fibers primarily located in layer I.

There are five dopaminergic receptors that are classified in two families, the D1-like family (D1- and D5-receptor subtypes) and the D2-like family (D2-, D3- and D4-receptor subtypes) (for reviews see Creese et al., 1983a, b; Gingrich and Caron, 1993; Bordet, 2004). All DA receptor subtypes belong to the G protein-coupled receptor family. The D1 and D5 receptor activation stimulates adenylyl cyclase (resulting in increased intracellular cAMP levels), while D2, 3, and 4 activation inhibits adenylyl cyclase (for review see Sokoloff and Schwartz, 1995).

The D1 receptor is present in all regions and laminae of the rat, cat, and monkey cortices, with higher regional densities in the cat and monkey (Richfield et al., 1989). D1 antibody labeling shows a trilaminar distribution in V1 of the rhesus monkey and human (Smiley et al., 1994). This trilaminar pattern is associated with bands in layers Ib-II and V-VI in both species, an additional band in layer IVb of the macaque and IVa in the human (Smiley et al., 1994). D1 receptor immunoreactivity has been documented primarily in dendritic spines and shafts (Smiley et al., 1994).

The D2 receptor is also distributed throughout the cerebral cortex of rats, cats, and monkeys, and is more homogeneous in regional distribution and laminar pattern than the D1 receptor (Richfield et al., 1989). D2 receptors are found in all cortical layers of the rhesus monkey, with the highest density of binding found in layer V (Lidow et al., 1991). However, D2 receptor concentrations decrease in the cortex along a rostro-caudal gradient, so that lowest levels are found in the occipital cortex of monkeys and, to a lesser extent, rats (Lidow et al., 1989c; Lidow et al., 1990; Lidow et al., 1991; Goldman-Rakic et al., 1992). In addition, D2 receptors are less common than D1 receptors in primate cortex (Goldman-Rakic et al., 1992).

Dopaminergic effects on physiological properties of V1 neurons

Experiments conducted in rats have revealed that microiontophoretic application of dopamine was similar to other catecholamines, such as NA. DA both enhanced and depressed the activity of spontaneously active cortical neurons (Bevan et al., 1978). The large majority (~75%) of these recorded neurons, however, exhibited increased activity, while the rest underwent a significant depression of neuronal firing (Bevan et al., 1978). The results from these early experiments could reflect a differential distribution of D1- versus D2-like receptors in the cortical regions evaluated.

The role of dopaminergic transmission in the regulation of visual cortical neuronal physiology and plasticity has not been extensively investigated to date. Work conducted in the cat visual cortex has revealed that iontophoretic application of DA leads to a significant and long-lasting depression of light-evoked activity (Reader, 1978). Contributions of DA to plasticity paradigms within V1 have not been investigated. However, a potential role for DA in modulating some forms of plasticity, such as potentiation and long-term depression in the pre-frontal cortex has been discussed in a recent review by

Gu (2002). Future work will be directed at investigating the direct contributions of DA to visual cortical plasticity (including OD shifts), and should dissect out the contributions of different families of DA receptors (D1-like and D2-like) to these processes.

Final Comments

We are now in the second generation of research aimed at characterizing how the classical neuromodulators NA, ACh, 5-HT and DA act as mediators of neural plasticity. Some of the earliest work in this field revealed synaptic changes such as facilitation (or depression) of neuronal activity in the presence of each of these neuromodulators. Various research groups have also shown that there are dramatic consequences of neuromodulator depletion on CNS response towards plasticity/reorganization following sensory deafferentation (e.g., monocular deprivation). More specifically, the absence of normal neuromodulatory action alone or in combination was shown to prevent the expression of plasticity and countered cortical reorganization.

In most cases the complete pathways and molecular consequences of neuromodulatory action at the cell membrane are not known. It is therefore difficult at this point to assemble a complete picture of how these neurochemical signals might invoke or enable plasticity. One biochemical component that appears fairly consistently in discussions of neuromodulator-mediated plasticity is calcium; intracellular calcium levels are often heavily regulated by the action of neuromodulator transmitter action at the cell membrane. This divalent cation has powerful and complex effects both on neuronal membranes and, in the cytoplasm, as a second messenger (see Chapter 8). Future research in this area will focus heavily on the precise intracellular effects of neuromodulatory action during normal sensory processing and during plasticity-inducing conditions. Furthermore, an emphasis on calcium handling, not only in regulating neuronal activity, but in determining plasticity-states and/or as remote transcriptional regulators appears to be warranted based on the evidence provided in this chapter.

References

Aoki C, Kaufman D, Rainbow TC (1986) The ontogeny of the laminar distribution of beta-adrenergic receptors in the visual cortex of cats, normally reared and dark-reared. Brain Res 392:109–116.

Azmitia EC, Segal M (1978) An autoradiographic analysis of the differential ascending projections of the dorsal and median raphe nuclei in the rat. J Comp Neurol 179:641–667.

Bear MF, Singer W (1986) Modulation of visual cortical plasticity by acetylcholine and noradrenaline. Nature 320:172–176.

Bear MF, Rittenhouse CD (1999) Molecular basis for induction of ocular dominance plasticity. J Neurobiol 41:83–91.

Berridge MJ (1998) Neuronal calcium signaling. Neuron 21:13–26.

Bevan P, Bradshaw CM, Pun RY, Slater NT, Szabadi E (1978) Responses of single cortical neurones to noradrenaline and dopamine. Neuropharmacology 17:611–617.

Bigham MH, Lidow MS (1995) Adrenergic and serotonergic receptors in aged monkey neocortex. Neurobiol Aging 16:91–104.

Bordet R (2004) [Central dopamine receptors: general considerations (Part 1)]. Rev Neurol (Paris) 160:862–870.

Carli M, Robbins TW, Evenden JL, Everitt BJ (1983) Effects of lesions to ascending noradrenergic neurones on performance of a 5-choice serial reaction task in rats; implications for theories of dorsal noradrenergic bundle function based on selective attention and arousal. Behav Brain Res 9:361–380.

Carrasco-Serrano C, Viniegra S, Ballesta JJ, Criado M (2000) Phorbol ester activation of the neuronal nicotinic acetylcholine receptor alpha7 subunit gene: involvement of transcription factor Egr-1. J Neurochem 74:932–939.

Christy B, Nathans D (1989) DNA binding site of the growth factor-inducible protein Zif268. Proc Natl Acad Sci U S A 86:8737–8741.

Civantos Calzada B, Aleixandre de Artinano A (2001) Alpha-adrenoceptor subtypes. Pharmacol Res 44:195–208.

Clarke PB, Schwartz RD, Paul SM, Pert CB, Pert A (1985) Nicotinic binding in rat brain: autoradiographic comparison of [3H]acetylcholine, [3H]nicotine, and [125I]-alpha-bungarotoxin. J Neurosci 5:1307–1315.

Cole AJ, Saffen DW, Baraban JM, Worley PF (1989) Rapid increase of an immediate early gene messenger RNA in hippocampal neurons by synaptic NMDA receptor activation. Nature 340:474–476.

Collingridge GL, Singer W (1990) Excitatory amino acid receptors and synaptic plasticity. Trends Pharmacol Sci 11:290–296.

Creese I, Sibley DR, Hamblin MW, Leff SE (1983a) The classification of dopamine receptors: relationship to radioligand binding. Annu Rev Neurosci 6:43–71.

Creese I, Sibley DR, Hamblin MW, Leff SE (1983b) Dopamine receptors in the central nervous system. Adv Biochem Psychopharmacol 36:125–134.

Davis W, Jr., Chen ZJ, Ile KE, Tew KD (2003) Reciprocal regulation of expression of the human adenosine 5′-triphosphate binding cassette, sub-family A, transporter 2 (ABCA2) promoter by the early growth response-1 (EGR-1) and Sp-family transcription factors. Nucleic Acids Res 31:1097–1107.

Devauges V, Sara SJ (1990) Activation of the noradrenergic system facilitates an attentional shift in the rat. Behav Brain Res 39:19–28.

Dursteler MR, Blakemore C, Garey LJ (1979) Projections to the visual cortex in the golden hamster. J Comp Neurol 183:185–204.

Edagawa Y, Saito H, Abe K (2001) Endogenous serotonin contributes to a developmental decrease in long-term potentiation in the rat visual cortex. J Neurosci 21:1532–1537.

Ego-Stengel V, Bringuier V, Shulz DE (2002) Noradrenergic modulation of functional selectivity in the cat visual cortex: an in vivo extracellular and intracellular study. Neuroscience 111:275–289.

Erisir A, Levey AI, Aoki C (2001) Muscarinic receptor M(2) in cat visual cortex: laminar distribution, relationship to gamma-aminobutyric acidergic neurons, and effect of cingulate lesions. J Comp Neurol 441:168–185.

Everitt BJ, Robbins TW, Gaskin M, Fray PJ (1983) The effects of lesions to ascending noradrenergic neurons on discrimination learning and performance in the rat. Neuroscience 10:397–410.

Foote SL, Morrison JH (1984) Postnatal development of laminar innervation patterns by monoaminergic fibers in monkey (Macaca fascicularis) primary visual cortex. J Neurosci 4:2667–2680.

Foote SL, Morrison JH (1987) Development of the noradrenergic, serotonergic, and dopaminergic innervation of neocortex. Curr Top Dev Biol 21:391–423.

Foote SL, Bloom FE, Aston-Jones G (1983) Nucleus locus ceruleus: new evidence of anatomical and physiological specificity. Physiol Rev 63:844–914.

Freedman R, Foote SL, Bloom FE (1975) Histochemical characterization of a neocortical projection of the nucleus locus coeruleus in the squirrel monkey. J Comp Neurol 164:209–231.

Fries P, Reynolds JH, Rorie AE, Desimone R (2001) Modulation of oscillatory neuronal synchronization by selective visual attention. Science 291:1560–1563.

Gaspar P, Berger B, Febvret A, Vigny A, Henry JP (1989) Catecholamine innervation of the human cerebral cortex as revealed by comparative immunohistochemistry of tyrosine hydroxylase and dopamine-beta-hydroxylase. J Comp Neurol 279:249–271.

Gatter KC, Powell TP (1977) The projection of the locus coeruleus upon the neocortex in the macaque monkey. Neuroscience 2:441–445.

Gebhard R, Zilles K, Schleicher A, Everitt BJ, Robbins TW, Divac I (1993) Distribution of seven major neurotransmitter receptors in the striate cortex of the New World monkey Callithrix jacchus. Neuroscience 56:877–885.

Gingrich JA, Caron MG (1993) Recent advances in the molecular biology of dopamine receptors. Annu Rev Neurosci 16:299–321.

Goldman-Rakic PS, Lidow MS, Smiley JF, Williams MS (1992) The anatomy of dopamine in monkey and human prefrontal cortex. J Neural Transm Suppl 36:163–177.

Gotti C, Clementi F (2004) Neuronal nicotinic receptors: from structure to pathology. Prog Neurobiol 74:363–396.

Greuel JM, Luhmann HJ, Singer W (1988) Pharmacological induction of use-dependent receptive field modifications in the visual cortex. Science 242:74–77.

Gruber T, Keil A, Muller MM (2001) Modulation of induced gamma band responses and phase synchrony in a paired associate learning task in the human EEG. Neurosci Lett 316:29–32.

Gruber T, Muller MM, Keil A (2002) Modulation of induced gamma band responses in a perceptual learning task in the human EEG. J Cogn Neurosci 14:732–744.

Gu Q (2002) Neuromodulatory transmitter systems in the cortex and their role in cortical plasticity. Neuroscience 111:815–835.

Gu Q (2003) Contribution of acetylcholine to visual cortex plasticity. Neurobiol Learn Mem 80:291–301.

Gu Q, Singer W (1993) Effects of intracortical infusion of anticholinergic drugs on neuronal plasticity in kitten striate cortex. Eur J Neurosci 5:475–485.

Gu Q, Singer W (1995) Involvement of serotonin in developmental plasticity of kitten visual cortex. Eur J Neurosci 7:1146–1153.

Gu Q, Patel B, Singer W (1990) The laminar distribution and postnatal development of serotonin-immunoreactive axons in the cat primary visual cortex. Exp Brain Res 81:257–266.

Henderson Z (1987) Cholinergic innervation of ferret visual system. Neuroscience 20:503–518.

Herdegen T, Leah JD (1998) Inducible and constitutive transcription factors in the mammalian nervous system: control of gene expression by Jun, Fos and Krox, and CREB/ATF proteins. Brain Res Brain Res Rev 28:370–490.

Hilfiker S, Pieribone VA, Czernik AJ, Kao HT, Augustine GJ, Greengard P (1999) Synapsins as regulators of neurotransmitter release. Philos Trans R Soc Lond B Biol Sci 354:269–279.

Hoffman BB, Lefkowitz RJ (1980) Alpha-adrenergic receptor subtypes. N Engl J Med 302:1390–1396.

Hornung JP, Fritschy JM, Tork I (1990) Distribution of two morphologically distinct subsets of serotonergic axons in the cerebral cortex of the marmoset. J Comp Neurol 297:165–181.

Houser CR, Crawford GD, Salvaterra PM, Vaughn JE (1985) Immunocytochemical localization of choline acetyltransferase in rat cerebral cortex: a study of cholinergic neurons and synapses. J Comp Neurol 234:17–34.

Hoyer D, Martin G (1997) 5-HT receptor classification and nomenclature: towards a harmonization with the human genome. Neuropharmacology 36:419–428.

Hoyer D, Pazos A, Probst A, Palacios JM (1986a) Serotonin receptors in the human brain. II. Characterization and autoradiographic localization of 5-HT1C and 5-HT2 recognition sites. Brain Res 376:97–107.

Hoyer D, Pazos A, Probst A, Palacios JM (1986b) Serotonin receptors in the human brain. I. Characterization and autoradiographic localization of 5-HT1A recognition sites. Apparent absence of 5-HT1B recognition sites. Brain Res 376:85–96.

Hoyer D, Clarke DE, Fozard JR, Hartig PR, Martin GR, Mylecharane EJ, Saxena PR, Humphrey PP (1994) International Union of Pharmacology classification of receptors for 5-hydroxytryptamine (Serotonin). Pharmacol Rev 46:157–203.

Hubel DH, Wiesel TN (1970) The period of susceptibility to the physiological effects of unilateral eye closure in kittens. J Physiol 206:419–436.

Imamura K, Kasamatsu T, Shirokawa T, Ohashi T (1999) Restoration of ocular dominance plasticity mediated by adenosine 3′,5′-monophosphate in adult visual cortex. Proc Biol Sci 266:1507–1516.

Jia WW, Liu Y, Lepore F, Ptito M, Cynader M (1994) Development and regulation of alpha adrenoceptors in kitten visual cortex. Neuroscience 63:179–190.

Jonowsky A, Sulser A (1987) Alpha and Beta adrenergic receptors in the brain. In: Psychopharmacology, The Third Generation of Progress (Meltzer HT, ed), pp 249–256. New York: Raven Press.

Kaczmarek L, Chaudhuri A (1997) Sensory regulation of immediate-early gene expression in mammalian visual cortex: implications for functional mapping and neural plasticity. Brain Res Brain Res Rev 23:237–256.

Kasamatsu T, Pettigrew JD (1979) Preservation of binocularity after monocular deprivation in the striate cortex of kittens treated with 6-hydroxydopamine. J Comp Neurol 185:139–161.

Kasamatsu T, Heggelund P (1982) Single cell responses in cat visual cortex to visual stimulation during iontophoresis of noradrenaline. Exp Brain Res 45:317–327.

Kasamatsu T, Shirokawa T (1985) Involvement of beta-adrenoreceptors in the shift of ocular dominance after monocular deprivation. Exp Brain Res 59:507–514.

Kasamatsu T, Pettigrew JD, Ary M (1979) Restoration of visual cortical plasticity by local microperfusion of norepinephrine. J Comp Neurol 185:163–181.

Kasamatsu T, Watabe K, Heggelund P, Scholler E (1985) Plasticity in cat visual cortex restored by electrical stimulation of the locus coeruleus. Neurosci Res 2:365–386.

Katz LC, Crowley JC (2002) Development of cortical circuits: lessons from ocular dominance columns. Nat Rev Neurosci 3:34–42.

Kitt CA, Hohmann C, Coyle JT, Price DL (1994) Cholinergic innervation of mouse forebrain structures. J Comp Neurol 341:117–129.

Knapska E, Kaczmarek L (2004) A gene for neuronal plasticity in the mammalian brain: Zif268/Egr-1/NGFI-A/Krox-24/TIS8/ZENK? Prog Neurobiol 74:183–211.

Kobayashi M, Imamura K, Sugai T, Onoda N, Yamamoto M, Komai S, Watanabe Y (2000) Selective suppression of horizontal propagation in rat visual cortex by norepinephrine. Eur J Neurosci 12:264–272.

Kolta A, Diop L, Reader TA (1987) Noradrenergic effects on rat visual cortex: single-cell microiontophoretic studies of alpha-2 adrenergic receptors. Life Sci 41:281–289.

Kosofsky BE, Molliver ME (1987) The serotonergic innervation of cerebral cortex: different classes of axon terminals arise from dorsal and median raphe nuclei. Synapse 1:153–168.

Kosofsky BE, Molliver ME, Morrison JH, Foote SL (1984) The serotonin and norepinephrine innervation of primary visual cortex in the cynomolgus monkey (Macaca fascicularis). J Comp Neurol 230:168–178.

Lavine N, Reuben M, Clarke PB (1997) A population of nicotinic receptors is associated with thalamocortical afferents in the adult rat: laminal and areal analysis. J Comp Neurol 380:175–190.

Lewis DA, Campbell MJ, Foote SL, Goldstein M, Morrison JH (1987) The distribution of tyrosine hydroxylase-immunoreactive fibers in primate neocortex is widespread but regionally specific. J Neurosci 7:279–290.

Lidov HG, Grzanna R, Molliver ME (1980) The serotonin innervation of the cerebral cortex in the rat—an immunohistochemical analysis. Neuroscience 5:207–227.

Lidow MS, Gallager DW, Rakic P, Goldman-Rakic PS (1989a) Regional differences in the distribution of muscarinic cholinergic receptors in the macaque cerebral cortex. J Comp Neurol 289:247–259.

Lidow MS, Goldman-Rakic PS, Gallager DW, Rakic P (1989b) Quantitative autoradiographic mapping of serotonin 5-HT1 and 5-HT2 receptors and uptake sites in the neocortex of the rhesus monkey. J Comp Neurol 280:27–42.

Lidow MS, Goldman-Rakic PS, Rakic P, Innis RB (1989c) Dopamine D2 receptors in the cerebral cortex: distribution and pharmacological characterization with [3H]raclopride. Proc Natl Acad Sci U S A 86:6412–6416.

Lidow MS, Goldman-Rakic PS, Rakic P, Gallager DW (1990) Autoradiographic comparison of D1-specific binding of [3H]SCH39166 and [3H]SCH23390 in the primate cerebral cortex. Brain Res 537:349–354.

Lidow MS, Goldman-Rakic PS, Gallager DW, Rakic P (1991) Distribution of dopaminergic receptors in the primate cerebral cortex: quantitative autoradiographic analysis using [3H]raclopride, [3H]spiperone and [3H]SCH23390. Neuroscience 40:657–671.

Liu Y, Cynader M (1994) Postnatal development and laminar distribution of noradrenergic fibers in cat visual cortex. Brain Res Dev Brain Res 82:90–94.

Markram H, Segal M (1992) Activation of protein kinase C suppresses responses to NMDA in rat CA1 hippocampal neurones. J Physiol 457:491–501.

McLean J, Waterhouse BD (1994) Noradrenergic modulation of cat area 17 neuronal responses to moving visual stimuli. Brain Res 667:83–97.

Mello CV, Velho TA, Pinaud R (2004) Song-induced gene expression: a window on song auditory processing and perception. Ann N Y Acad Sci 1016:263–281.

Mesulam MM, Mufson EJ, Levey AI, Wainer BH (1983) Cholinergic innervation of cortex by the basal forebrain: cytochemistry and cortical connections of the septal area, diagonal band nuclei, nucleus basalis (substantia innominata), and hypothalamus in the rhesus monkey. J Comp Neurol 214:170–197.

Mesulam MM, Hersh LB, Mash DC, Geula C (1992) Differential cholinergic innervation within functional subdivisions of the human cerebral cortex: a choline acetyltransferase study. J Comp Neurol 318:316–328.

Metherate R (2004) Nicotinic acetylcholine receptors in sensory cortex. Learn Mem 11:50–59.

Milbrandt J (1987) A nerve growth factor-induced gene encodes a possible transcriptional regulatory factor. Science 238:797–799.

Miller KD (1994) Models of activity-dependent neural development. Prog Brain Res 102:303–318.

Miltner WH, Braun C, Arnold M, Witte H, Taub E (1999) Coherence of gamma-band EEG activity as a basis for associative learning. Nature 397:434–436.

Morrison JH, Grzanna R, Molliver ME, Coyle JT (1978) The distribution and orientation of noradrenergic fibers in neocortex of the rat: an immunofluorescence study. J Comp Neurol 181:17–39.

Morrison JH, Molliver ME, Grzanna R, Coyle JT (1981) The intra-cortical trajectory of the coeruleo-cortical projection in the rat: a tangentially organized cortical afferent. Neuroscience 6:139–158.

Morrison JH, Foote SL, O'Connor D, Bloom FE (1982a) Laminar, tangential and regional organization of the noradrenergic innervation of monkey cortex: dopamine-beta-hydroxylase immunohistochemistry. Brain Res Bull 9:309–319.

Morrison JH, Foote SL, Molliver ME, Bloom FE, Lidov HG (1982b) Noradrenergic and serotonergic fibers innervate complementary layers in monkey primary visual cortex: an immunohistochemical study. Proc Natl Acad Sci U S A 79:2401–2405.

Mrzljak L, Goldman-Rakic PS (1993) Low-affinity nerve growth factor receptor (p75NGFR)- and choline acetyltransferase (ChAT)-immunoreactive axons in the cerebral cortex and hippocampus of adult macaque monkeys and humans. Cereb Cortex 3:133–147.

Mrzljak L, Levey AI, Goldman-Rakic PS (1993) Association of m1 and m2 muscarinic receptor proteins with asymmetric synapses in the primate cerebral cortex: morphological evidence for cholinergic modulation of excitatory neurotransmission. Proc Natl Acad Sci U S A 90:5194–5198.

Muller MM, Gruber T, Keil A (2000) Modulation of induced gamma band activity in the human EEG by attention and visual information processing. Int J Psychophysiol 38:283–299.

Nakadate K, Imamura K, Watanabe Y (2001) Effects of monocular deprivation on the expression pattern of alpha-1 and beta-1 adrenergic receptors in the kitten visual cortex. Neurosci Res 40:155–162.

Papadopoulos GC, Parnavelas JG, Buijs RM (1987) Light and electron microscopic immunocytochemical analysis of the serotonin innervation of the rat visual cortex. J Neurocytol 16:883–892.

Papadopoulos GC, Parnavelas JG, Buijs RM (1989) Light and electron microscopic immunocytochemical analysis of the dopamine innervation of the rat visual cortex. J Neurocytol 18:303–310.

Parkinson D, Coscia E, Daw NW (1988a) Identification and localization of adrenergic receptors in cat visual cortex. Brain Res 457:70–78.

Parkinson D, Kratz KE, Daw NW (1988b) Evidence for a nicotinic component to the actions of acetylcholine in cat visual cortex. Exp Brain Res 73:553–568.

Parnavelas JG, Kelly W, Franke E, Eckenstein F (1986) Cholinergic neurons and fibres in the rat visual cortex. J Neurocytol 15:329–336.

Pazos A, Probst A, Palacios JM (1987) Serotonin receptors in the human brain—III. Autoradiographic mapping of serotonin-1 receptors. Neuroscience 21:97–122.

Petersohn D, Thiel G (1996) Role of zinc-finger proteins Sp1 and zif268/egr-1 in transcriptional regulation of the human synaptobrevin II gene. Eur J Biochem 239:827–834.

Petersohn D, Schoch S, Brinkmann DR, Thiel G (1995) The human synapsin II gene promoter. Possible role for the transcription factor zif268/egr-1, polyoma enhancer activator 3, and AP2. J Biol Chem 270:24361–24369.

Phillipson OT, Kilpatrick IC, Jones MW (1987) Dopaminergic innervation of the primary visual cortex in the rat, and some correlations with human cortex. Brain Res Bull 18:621–633.

Pinaud R (2004) Experience-dependent immediate early gene expression in the adult central nervous system: evidence from enriched-environment studies. Int J Neurosci 114:321–333.

Pinaud R, Tremere LA, Penner MR (2000) Light-induced zif268 expression is dependent on noradrenergic input in rat visual cortex. Brain Res 882:251–255.

Pinaud R, Tremere LA, Penner MR, Hess FF, Robertson HA, Currie RW (2002) Complexity of sensory environment drives the expression of candidate-plasticity gene, nerve growth factor induced-A. Neuroscience 112:573–582.

Pinaud R, Vargas CD, Ribeiro S, Monteiro MV, Tremere LA, Vianney P, Delgado P, Mello CV, Rocha-Miranda CE, Volchan E (2003) Light-induced Egr-1 expression in the striate cortex of the opossum. Brain Res Bull 61:139–146.

Pisa M, Martin-Iverson MT, Fibiger HC (1988) On the role of the dorsal noradrenergic bundle in learning and habituation to novelty. Pharmacol Biochem Behav 30:835–845.

Pospelov VA, Pospelova TV, Julien JP (1994) AP-1 and Krox-24 transcription factors activate the neurofilament light gene promoter in P19 embryonal carcinoma cells. Cell Growth Differ 5:187–196.

Prusky G, Cynader M (1990) The distribution of M1 and M2 muscarinic acetylcholine receptor subtypes in the developing cat visual cortex. Brain Res Dev Brain Res 56:1–12.

Prusky GT, Shaw C, Cynader MS (1987) Nicotine receptors are located on lateral geniculate nucleus terminals in cat visual cortex. Brain Res 412:131–138.

Qu Y, Eysel UT, Vandesande F, Arckens L (2000) Effect of partial sensory deprivation on monoaminergic neuromodulators in striate cortex of adult cat. Neuroscience 101:863–868.

Reader TA (1978) The effects of dopamine, noradrenaline and serotonin in the visual cortex of the cat. Experientia 34:1586–1588.

Richfield EK, Young AB, Penney JB (1989) Comparative distributions of dopamine D-1 and D-2 receptors in the cerebral cortex of rats, cats, and monkeys. J Comp Neurol 286:409–426.

Rodriguez R, Kallenbach U, Singer W, Munk MH (2004) Short- and long-term effects of cholinergic modulation on gamma oscillations and response synchronization in the visual cortex. J Neurosci 24:10369–10378.

Roerig B, Katz LC (1997) Modulation of intrinsic circuits by serotonin 5-HT3 receptors in developing ferret visual cortex. J Neurosci 17:8324–8338.

Roerig B, Nelson DA, Katz LC (1997) Fast synaptic signaling by nicotinic acetylcholine and serotonin 5-HT3 receptors in developing visual cortex. J Neurosci 17:8353–8362.

Rossner S, Kumar A, Witzemann V, Schliebs R (1994) Development of laminar expression of the m2 muscarinic cholinergic receptor gene in rat visual cortex and the effect of monocular visual deprivation. Brain Res Dev Brain Res 77:55–61.

Rye DB, Wainer BH, Mesulam MM, Mufson EJ, Saper CB (1984) Cortical projections arising from the basal forebrain: a study of cholinergic and noncholinergic components employing combined retrograde tracing and immunohistochemical localization of choline acetyltransferase. Neuroscience 13:627–643.

Saper CB (1984) Organization of cerebral cortical afferent systems in the rat. II. Magnocellular basal nucleus. J Comp Neurol 222:313–342.

Sato H, Fox K, Daw NW (1989) Effect of electrical stimulation of locus coeruleus on the activity of neurons in the cat visual cortex. J Neurophysiol 62:946–958.

Sato H, Hata Y, Masui H, Tsumoto T (1987a) A functional role of cholinergic innervation to neurons in the cat visual cortex. J Neurophysiol 58:765–780.

Sato H, Hata Y, Hagihara K, Tsumoto T (1987b) Effects of cholinergic depletion on neuron activities in the cat visual cortex. J Neurophysiol 58:781–794.

Schliebs R, Walch C, Stewart MG (1989) Laminar pattern of cholinergic and adrenergic receptors in rat visual cortex using quantitative receptor autoradiography. J Hirnforsch 30:303–311.

Selden NR, Robbins TW, Everitt BJ (1990a) Enhanced behavioral conditioning to context and impaired behavioral and neuroendocrine responses to conditioned stimuli following ceruleocortical noradrenergic lesions: support for an attentional hypothesis of central noradrenergic function. J Neurosci 10:531–539.

Selden NR, Cole BJ, Everitt BJ, Robbins TW (1990b) Damage to ceruleo-cortical noradrenergic projections impairs locally cued but enhances spatially cued water maze acquisition. Behav Brain Res 39:29–51.

Shirokawa T, Kasamatsu T (1986) Concentration-dependent suppression by beta-adrenergic antagonists of the shift in ocular dominance following monocular deprivation in kitten visual cortex. Neuroscience 18:1035–1046.

Siciliano R, Fornai F, Bonaccorsi I, Domenici L, Bagnoli P (1999) Cholinergic and noradrenergic afferents influence the functional properties of the postnatal visual cortex in rats. Vis Neurosci 16:1015–1028.

Sillito AM, Kemp JA (1983) Cholinergic modulation of the functional organization of the cat visual cortex. Brain Res 289:143–155.

Smiley JF, Levey AI, Ciliax BJ, Goldman-Rakic PS (1994) D1 dopamine receptor immunoreactivity in human and monkey cerebral cortex: predominant and extrasynaptic localization in dendritic spines. Proc Natl Acad Sci U S A 91:5720–5724.

Sokoloff P, Schwartz JC (1995) Novel dopamine receptors half a decade later. Trends Pharmacol Sci 16:270–275.

Stichel CC, Singer W (1987a) Quantitative analysis of the choline acetyltransferase-immunoreactive axonal network in the cat primary visual cortex: I. Adult cats. J Comp Neurol 258:91–98.

Stichel CC, Singer W (1987b) Quantitative analysis of the choline acetyltransferase-immunoreactive axonal network in the cat primary visual cortex: II. Pre- and postnatal development. J Comp Neurol 258:99–111.

Stiles GL, Caron MG, Lefkowitz RJ (1984) Beta-adrenergic receptors: biochemical mechanisms of physiological regulation. Physiol Rev 64:661–743.

Takeuchi Y, Sano Y (1984) Serotonin nerve fibers in the primary visual cortex of the monkey. Quantitative and immunoelectronmicroscopical analysis. Anat Embryol (Berl) 169:1–8.

Tallon-Baudry C, Bertrand O, Peronnet F, Pernier J (1998) Induced gamma-band activity during the delay of a visual short-term memory task in humans. J Neurosci 18:4244–4254.

Thiel G, Schoch S, Petersohn D (1994) Regulation of synapsin I gene expression by the zinc finger transcription factor zif268/egr-1. J Biol Chem 269:15294–15301.

Tigges J, Tigges M, Cross NA, McBride RL, Letbetter WD, Anschel S (1982) Subcortical structures projecting to visual cortical areas in squirrel monkey. J Comp Neurol 209:29–40.

Tigges M, Tigges J, Rees H, Rye D, Levey AI (1997) Distribution of muscarinic cholinergic receptor proteins m1 to m4 in area 17 of normal and monocularly deprived rhesus monkeys. J Comp Neurol 388:130–145.

Videen TO, Daw NW, Rader RK (1984) The effect of norepinephrine on visual cortical neurons in kittens and adult cats. J Neurosci 4:1607–1617.

Voigt T, de Lima AD (1991a) Serotonergic innervation of the ferret cerebral cortex. I. Adult pattern. J Comp Neurol 314:403–414.

Voigt T, de Lima AD (1991b) Serotonergic innervation of the ferret cerebral cortex. II. Postnatal development. J Comp Neurol 314:415–428.

Volkmar FR, Greenough WT (1972) Rearing complexity affects branching of dendrites in the visual cortex of the rat. Science 176:1145–1147.

Wallace CS, Withers GS, Weiler IJ, George JM, Clayton DF, Greenough WT (1995) Correspondence between sites of NGFI-A induction and sites of morphological plasticity following exposure to environmental complexity. Brain Res Mol Brain Res 32:211–220.

Wang Y, Gu Q, Cynader MS (1997) Blockade of serotonin-2C receptors by mesulergine reduces ocular dominance plasticity in kitten visual cortex. Exp Brain Res 114:321–328.

Waterhouse BD, Azizi SA, Burne RA, Woodward DJ (1990) Modulation of rat cortical area 17 neuronal responses to moving visual stimuli during norepinephrine and serotonin microiontophoresis. Brain Res 514:276–292.

Wisden W, Errington ML, Williams S, Dunnett SB, Waters C, Hitchcock D, Evan G, Bliss TV, Hunt SP (1990) Differential expression of immediate early genes in the hippocampus and spinal cord. Neuron 4:603–614.

Wong WK, Ou XM, Chen K, Shih JC (2002) Activation of human monoamine oxidase B gene expression by a protein kinase C MAPK signal transduction pathway involves c-Jun and Egr-1. J Biol Chem 277:22222–22230.

Xiang Z, Prince DA (2003) Heterogeneous actions of serotonin on interneurons in rat visual cortex. J Neurophysiol 89:1278–1287.

Zilles K, Schroder H, Schroder U, Horvath E, Werner L, Luiten PG, Maelicke A, Strosberg AD (1989) Distribution of cholinergic receptors in the rat and human neocortex. Exs 57:212–228.

8

Critical Calcium-Regulated Biochemical and Gene Expression Programs involved in Experience-Dependent Plasticity

RAPHAEL PINAUD

Neurological Sciences Institue, OHSU, Portland, OR, USA

Introduction

In the previous chapter, we have discussed evidence that implicates neuro-modulators in the physiology of visual cortical neurons. Neuromodulators such as acetylcholine, noradrenaline and serotonin, and to a lesser extent dopamine, play clear roles in the regulation of firing patterns, RF structure and forms of experience-dependent plasticity in V1. It has become clearer over the years that alterations in firing patterns, including those that result from neuromodulatory input, during development of cortical synapses, critical period or adulthood trigger biochemical cascades that integrate alterations in membrane behavior to gene expression programs in the cell nucleus. These gene expression programs result in the synthesis of protein products that are involved in long-term alterations in cell physiology, a large number of them associated with synaptic and other forms of structural plasticity.

In this chapter I will focus on the biochemical cascades that have received the attention of a large fraction of neuroscientists, particularly those interested in investigating the molecular and cellular substrates of some forms of plasticity, including ocular dominance (OD) shifts, learning and memory. In addition, I will discuss research we and others have conducted aimed at identifying and characterizing activity-dependent genes associated with various forms of plasticity and possibly long-term modifications within visual cortical circuitry.

NMDA Receptors

Activation of the N-methyl-D-aspartate (NMDA)-type of glutamatergic receptors, and the biochemical pathways that are sensitive to, or get affected

R. Pinaud, L. A. Tremere, P. De Weerd(Eds.), Plasticity in the Visual System: From Genes to Circuits, 153-180, ©2006 Springer Science + Business Media, Inc.

by, this receptor have been intensely scrutinized over the past decade. NMDA-receptors have been on the forefront of investigations in the mechanisms involved in plasticity primarily because this subtype of receptors has been repeatedly implicated in enhanced synaptic efficacy (for reviews see Kleinschmidt et al., 1987; Cotman et al., 1988; Constantine-Paton et al., 1990; Daw et al., 1993; Hollmann and Heinemann, 1994). Unlike the two other families of glutamate receptors (Kainate and AMPA) NMDA-receptors allow for a calcium influx upon activation (Daw et al., 1993; Hollmann and Heinemann, 1994), which is known to be a key regulator of a number of intracellular events that range from protein phosphorylation to endocytosis and receptor trafficking, to name a few. In addition, NMDA-receptor gating not only requires ligand binding, but also concomitant neuronal depolarization for its activation (Collingridge and Lester, 1989; Daw et al., 1993; Hollmann and Heinemann, 1994). Depolarization has been shown to be required for receptor activation in order to relief the binding of magnesium cations that block receptor gating upon binding of glutamate in normal conditions (Daw et al., 1993; Hollmann and Heinemann, 1994; Mayer and Armstrong, 2004). These properties have been used by many investigators to propose that NMDA receptors are well positioned to act as coincidence detectors (of synaptic inputs), a mechanism that has also been forwarded to account for various forms of synaptic plasticity, including potentiation (Collingridge and Singer, 1990; Tang et al., 1999; Yuste et al., 1999; Karmarkar and Buonomano, 2002). Detailed reviews regarding the structure and function of NMDA receptors can be found elsewhere (Collingridge and Lester, 1989; Hollmann and Heinemann, 1994; Mayer and Armstrong, 2004).

Activity-sensitive Biochemical Pathways

As briefly stated above, NMDA receptor-activation not only allows for sodium and potassium flux, but also allows for an inward calcium conductance (Collingridge and Lester, 1989; Daw et al., 1993; Mayer and Armstrong, 2004). Neuronal activity also leads to the activation of voltage-sensitive calcium channels, which contribute to a significant proportion of activity-driven calcium influx (Berridge, 1998). A number of calcium-sensitive biochemical pathways have been detailed over the years (Bading et al., 1993; Gallin and Greenberg, 1995; Ginty, 1997). Here I will focus on those that have been more significantly involved in plasticity mechanisms.

Possibly the first cellular correlate of neuronal input is the activation of ionotropic and metabotropic channels. For the purposes of our discussion regarding the biochemical pathways associated with the induction of plastic changes, I start here on the metabotropic receptor-driven activation of phospholipases, which are a group of enzymes that play a role in the control of membrane phospholipids free fatty acid turnover (McMurray and Magee, 1972; Gatt and Barenholz, 1973; Majerus, 1992; Phillis and O'Regan, 2004). This process is tightly associated with membrane states such as transport and fluidity, directly impacting cell physiology. A number of phospholipases, including several members of the phospholipase A (PLA) and phospholipase C (PLC) families, are highly sensitive to variations in intracellular calcium levels, which directly regulate enzymatic activity (Majerus, 1992; Phillis and O'Regan, 2004). Enhanced catalytic activity of phospholipases, in particular PLC, has been repeatedly implicated in the early steps of the mechanisms associated with neuronal plasticity; PLC activation often occurs via the alpha-subunit of a number of G-proteins. The primary substrate for PLC is the membrane phosphatidylinositol 4,5-biphosphate (PIP_2) (Berridge, 1998). PLC catalyses the hydrolysis of PIP_2 into two products: diacylglycerol (DAG), a fatty acid that remains associated with the cell membrane, and inositol 1,4,5-triphosphate, that is apolar and therefore diffuses into the cell's cytoplasm (Fig. 8.1) (Berridge, 1998). These two molecules that result from PLC activation and its substrate hydrolysis carry out distinct, but possibly synergistic, pathways. DAG has been shown to effectively modulate protein kinase C (PKC) (Stahelin et al., 2005), which appears to tightly regulate the activity of adenylyl cyclase (AC), an enzyme that catalyses the conversion of ATP into cAMP, another powerful intracellular second messenger (Yoshimura and Cooper, 1993; Zimmermann and Taussig, 1996). Diffusible IP_3 has been shown to specifically bind to IP_3 receptors located in the endoplasmic reticulum (ER), which is a major source of intracellular calcium (Fig. 8.1) (Berridge, 1998). Upon binding, IP_3 triggers the release of intracellular calcium stores. This mechanism not only contributes to the activation of calcium-sensitive molecules intracellularly, but also to maintaining high cytoplasmic calcium levels; IP_3 receptors are sensitive to calcium and therefore exhibit calcium-induced calcium release from intracellular stores (Fig. 8.1) (Berridge, 1998). Two of such molecules that are affected by rising intracellular calcium are the adenylyl cyclases A1 and A8, both involved in the synthesis of cAMP (Ferguson and Storm, 2004). It is clear, therefore, that DAG and IP_3 pathways converge to the regulation of cAMP levels via AC activation (Fig. 8.1).

Protein Kinase A and Calmodulin

Enhanced cAMP levels exert powerful effects on the cAMP-dependent protein kinase (PKA); cAMP concentration rises contribute to unlocking of the catalytic domain of PKA and thus its activation (Figs. 8.1 and 8.2). Recent attention has been paid to PKA as increasing evidence implicates this kinase in visual cortical plasticity. For example, it has been shown that inhibition of PKA prevents monocular deprivation-induced OD shifts (Beaver et al., 2001). Interestingly, this same work showed that this form of plasticity is selective for the PKA pathway given that blockade of protein kinase G (PKG) had no effect on OD shifts (Beaver et al., 2001). In another set of experiments, it was demonstrated that inhibition of PKA activity decreased orientation selectivity in the visual cortex of cats that underwent monocular

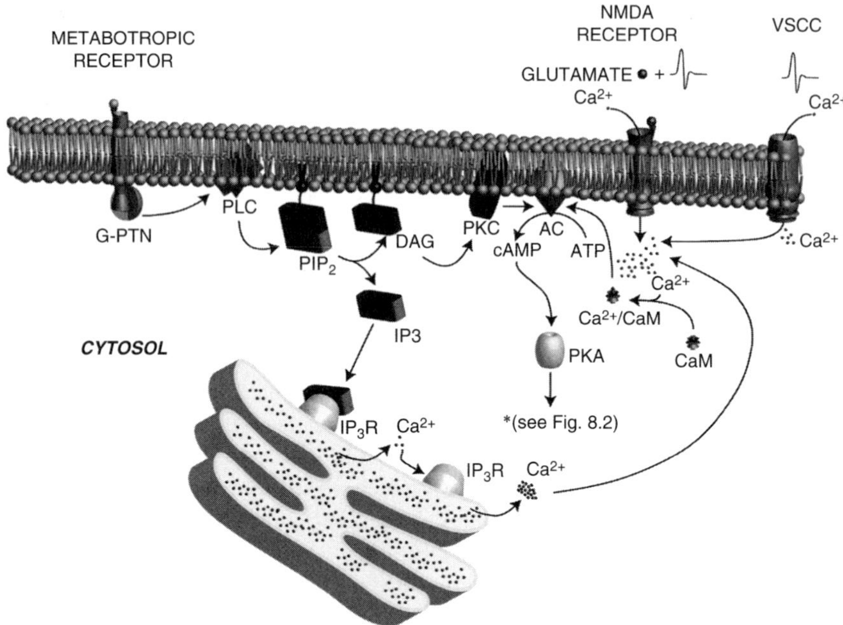

FIGURE 8.1. Convergence of activity-sensitive biochemical pathways associated with neuronal plasticity. Two pathways converge to PKA activation. The first one (left side of figure), mediated through metabotropic receptor activation, triggers the activation of a stimulatory g-protein, which in turn activates the PLC pathway. PLC activation leads to the hydrolysis of PIP_2 into DAG and IP_3. DAG regulates the activity of PKC, which in turn, modulates AC activity, thereby influencing intracellular cAMP levels. IP_3 diffuses into the cytoplasm and binds to specific IP_3 receptors in the ER, releasing Ca^{2+} from intracellular stores. Ca^{2+} potently triggers further release of calcium from the ER. The second pathway involves rises in intracellular Ca^{2+} levels either through activation of NMDA channels and VSCC, or through release of Ca^{2+} from intracellular stores (through the first pathway). Upon Ca^{2+} influx and binding, CaM undergoes a conformational change that affects its affinity for target proteins. For example, it has been shown that Ca^{2+}/CaM positively regulates AC leading to increased intracellular levels of cAMP. Both pathways trigger PKA activation through rises of cAMP levels. Downstream effects of PKA are indicated in Fig. 8.2. G-ptn, g-protein; PLC, phospholipase C; PIP_2, phosphatidylinositol 4,5-biphosphate; DAG diacylglycerol; IP_3, inositol 1,4,5-triphosphate; IP_3R, IP3 receptor Ca^{2+}, calcium; PKC, protein kinase C; AC, adenylyl cyclase; ATP, adenosine triphosphate; cAMP, cyclic adenosine monophosphate; Ca^{2+}/CaM, calcium/calmodulin protein complex; CaM, calmodulin; PKA, protein kinase A; ER, endoplasmic reticulum; NMDA, N-methyl-D-aspartate; VSCC, voltage-sensitive calcium channel.

deprivation, but not control animals (Beaver et al., 2002). Together these results suggest that PKA plays a critical role in some forms of experience-dependent plasticity in the visual cortex.

In addition to the mechanisms described above, calcium-binding protein calmodulin (CaM) also appears to play a critical role in mediating activity-dependent changes in cell physiology (Xia and Storm, 2005). Upon binding calcium, CaM undergoes a conformational change that exposes an occluded hydrophobic domain thereby enhancing its affinity for its target proteins (Fig. 8.2). Those targets include AC 1 and AC 8, calcium/calmodulin-dependent protein kinase (CaMK) I, II and IV, calcium channels, and a number of phosphodiesterases, to name a few (reviewed in Xia and Storm, 2005). Thus, calmodulin is also well positioned to orchestrate calcium-induced mechanisms that are involved in various forms of synaptic plasticity.

The MAPK/ERK Pathway

One of the key pathways that appear to integrate calcium signaling to the nuclear transcriptional regulation is the mitogen-activated protein kinase (MAPK)/extracellular signal-regulated kinase (ERK) pathway (Grewal et al., 1999; Sweatt, 2001). An intricate network of biochemical interactions consti-tutes this pathway; I will focus here on the main interactions leading to cal-cium-dependent transcriptional regulation (Fig. 8.2). More detailed reviews addressing these pathways can be found elsewhere (Grewal et al., 1999; Sweatt, 2001; Xia and Storm, 2005).

Calcium influx either through NMDA-receptor activation or voltage-sen-sitive calcium channels can directly participate in the activation of Ras, a small GTPase (Fig. 8.2). In its inactive state, Ras is bound to guanine diphos-phate (GDP). Upon activation, a number of guanine nucleotide exchange factors (GEFs), for example RasGRF (Farnsworth et al., 1995), mediate the exchange of GDP by guanine triphosphate (GTP). When GTP is bound, Ras becomes active. Activated Ras induces the recruitment of Raf (Raf proto-oncogene serine/threonine protein kinase), which contains in its structure a Ras binding domain as well as a kinase domain (Grewal et al., 1999; Carey et al., 2003). This event triggers a kinase cascade: Recruitment of Raf acti-vates its kinase domain that directly phosphorylates a cysteine rich domain within a second kinase, MAPK/ERK kinase (MEK) (Grewal et al., 1999; Kyosseva, 2004; Xia and Storm, 2005). When phosphorylated, MEK is acti-vated and proceeds to phosphorylate ERK either on tyrosine or threonine sites (TXY site). Upon phosphorylation, ERK dimerizes, dissociates from MEK and migrates to the cell nucleus, where it will act to regulate transcrip-tion (details below; Fig. 8.2) (Grewal et al., 1999). For the sake of space I have simplified the participants of the MAPK/ERK pathway, however, it should be noted that a number of different isoforms have been described for some molecules involved in this cascade. For example, three isoforms of Raf have been described (A-Raf, B-Raf and Raf-1 or c-Raf) (Hindley and Kolch,

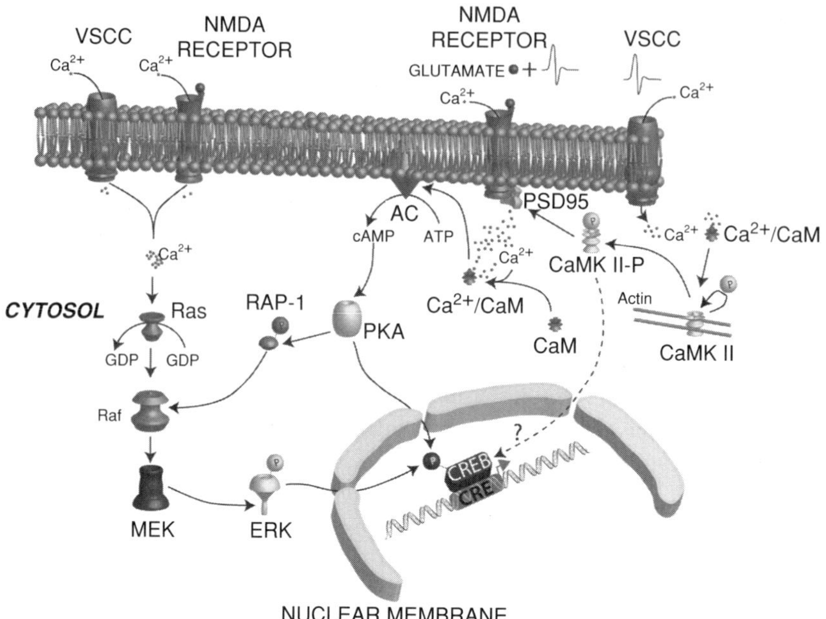

FIGURE 8.2. Calcium-sensitive biochemical cascades, involved in neural plasticity, integrate alterations in the cell membrane with the nuclear transcriptional machinery. Calcium influx either through VSCC or NMDA receptor activation activates the MAPK/ERK pathway (left of schematics). Calcium activates Ras by recruiting specific GEFs (not pictured), which exchange GDP for GTP. This process triggers a kinase cascade. Ras-GTP recruits a kinase known as Raf. Activated Raf phosphorylates MEK, which in turn phosphorylates, ERK. Phosphorylated ERK dimerizes and migrates to the nucleus where it will impact gene expression (below). Rises in cAMP through CaM modulation onto AC enhance kinase activity of PKA. PKA can phosphorylate RAP-1, which converges to the MAPK/ERK pathway by phosphoryating Raf. PKA can also directly impact transcriptional regulation by phosphorylating specific transcription factors, such as CREB. Ca^{2+}/CaM can also impact target proteins, such as CaMKII (right of schematics). Upon calcium-driven activation, CaMKII autophosphorylates and dissociates from actin filaments; it can then impact its target proteins such as PSD95. CaMKII might also contribute to phosphorylation of specific transcription factors, therefore directly influencing gene expression programs. These pathways integrate modifications in the cell surface to the cell's genome. Of particular interest to this review is the convergence of these biochemical pathways onto phosphorylation of the transcription factor CREB, which appears to integrate the activity of a population of inducible genes that contain the CRE element in their promoters (detailed in Fig. 8.3). Ca^{2+}, calcium; AC, adenylyl cyclase; ATP, adenosine triphosphate; cAMP, cyclic adenosine monophosphate; Ca^{2+}/CaM, calcium/calmodulin protein complex; CaM, calmodulin; PKA, cAMP-dependent protein kinase; NMDA, N-methyl-D-aspartate; VSCC, voltage-sensitive calcium channel; CaMKII, calcium/calmodulin-dependent protein kinase II; PSD95, 95 KDa post-synaptic density protein; GEF, guanine nucleotide exchange factors; GTP, guanine triphosphate; GDP, guanine diphosphate; Raf, Raf proto-oncogene serine/threonine protein kinase; MAPK, mitogen-activated protein kinase; ERK; extracellular signal-regulated kinase; MEK, MAPK/ERK kinase; RAP-1, Ras-related protein 1; CREB, cAMP responsive element binding protein; CRE, cAMP responsive element.

2002). In addition, two MEK isoforms (MEK 1 and 2) and two ERK isoforms (ERK1 and 2) have been described (Fukunaga and Miyamoto, 1998; English and Cobb, 2002).

Parallel pathways involving other key molecules also converge on to Raf activation. For example CaM, when bound to calcium, is also able to regulate the activity of AC, inducing increases in the intracellular levels of cAMP (Fig. 8.2) (Brostrom et al., 1975; Cheung et al., 1975; Brostrom et al., 1976). As mentioned above, enhanced cAMP levels activate PKA (Scott, 1991; Meinkoth et al., 1993; Brandon et al., 1997). It has been shown that PKA phosphorylates a Ras-related protein known as RAP1 (Ras-related protein 1) (Grewal et al., 2000a; Grewal et al., 2000b). RAP1 phosphorylation leads to activation of Raf. Thus Raf activation can proceed either through Ras or RAP1 activaton, providing a secondary mechanism leading to ERK activation (Fig. 8.2).

Visual stimulation has been shown to directly activate the MAPK/ERK pathway. For example, using an in-vitro essay, Kaminska and colleagues (1999) were able to demonstrate that MAPK/ERK is regulated by light in the rat visual cortex (Kaminska et al., 1999).

ERK activation has also been shown to be directly involved in plasticity paradigms in the rat visual system. In a slice preparation, pharmacological inhibition of the ERK pathway has been shown to interfere with LTP induced in the cortex (Di Cristo et al., 2001). In addition, local infusion of pharmacological antagonism directed at the ERK pathway in the visual cortex of rats has been shown to block OD shifts induced by monocular deprivation (Di Cristo et al., 2001). Together these results suggest that MAPK/ERK play critical roles in the experience-dependent plasticity during the critical period, as well as other forms of activity-dependent synaptic plasticity in the visual cortex.

Alpha-CaMKII

Another kinase that has attracted significant interest and spurred a plethora of queries regarding the molecular mechanisms underlying visual cortical plasticity is alpha-CaMKII. As described above, alpha-CaMKII is a target of CaM and, therefore, highly regulated by alterations in intracellular calcium levels (Xia and Storm, 2005). Furthermore, alpha-CaMKII can be directly activated by calcium (Fink and Meyer, 2002; Xia and Storm, 2005). Interestingly, once activated, alpha-CaMKII undergoes autophosphorylation, which maintains it constitutively active even in the absence of calcium (Fig. 8.2) (De Koninck and Schulman, 1998; Fink and Meyer, 2002; Lisman et al., 2002). Inactive CaMKII is usually associated with actin filaments; upon phosphorylation, it has been shown that this enzyme dissociates from actin, allowing it to freely disperse to the membrane where it acts on post-synaptic density (PSD) proteins, which associate to glutamatergic NMDA receptors (Fig. 8.2) (Shen and Meyer, 1999). Interestingly, upon interacting with NMDA receptors, alpha-CaMKII has been shown to remain in its activated conformation, allowing it to reverberate, and possibly amplify, calcium signals through time. Given that alpha-CaMKII activity is associated with calcium signals, that its activity is self-sustained independent of further variations in calcium levels, and that some of its targets involve plasticity-associated proteins such as the microtubule-associated protein 2 (MAP2) (Vaillant et al., 2002), it has been proposed that alpha-CaMKII might be involved in critical steps required for experience-dependent plasticity (Fink and Meyer, 2002; Lisman et al., 2002). In fact, earlier experiments conducted in hippocampal preparations have demonstrated that this enzyme is required for some forms of synaptic plasticity such as LTP, as well as memory formation (reviewed in Xia and Storm, 2005).

The contributions of alpha-CaMKII to experience-dependent plasticity in the visual cortex have been studied in a slice preparation obtained from alpha-CaMKII knock-out mice. It was found that both LTP and LTD induction was significantly affected in tissue obtained from knock-out mice. Interestingly, robust potentiation and depression could be elicited in 5-week old animals (Kirkwood et al., 1997). In the adult macaque visual cortex, alpha-CaMKII mRNA expression is regulated by monocular deprivation induced by intraocular tetrodotoxin (TTX) infusion: this procedure triggers a significant increase in alpha-CaMKII mRNA in OD columns associated with the deprived eye (Tighilet et al., 1998). This effect was highest for cortical layer IVc-beta, but undetectable for the remaining layers, suggesting that alpha-CaMKII expression in layer IVc-beta acts as a "plasticity filter" for activity-dependent plasticity in supragranular layers (Tighilet et al., 1998).

As indicated above, alpha-CaMKII undergoes autophosphorylation which sustains its activity even in the absence of calcium. A recent report by Taha and colleagues (2002) shows that mice carrying a mutated alpha-CaMKII that is unable to autophosphorylate exhibit significant decreases in monocular deprivation-induced OD plasticity (Taha et al., 2002). These findings clearly demonstrate a critical role for alpha-CaMKII activation to the normal shifts that occur as a result of monocular deprivation.

CRE-regulated Gene Expression

As indicated above, calcium influx either through activation of NMDA receptors or via voltage-sensitive calcium channels activates multiple signaling cascades that appear to be critical for various forms of activity-dependent synaptic plasticity in a number of systems, including visual cortical plasticity. These signaling cascades influence the activity of other messenger systems, forming an intricate second messenger network that controls a wide variety of cell functions that range from cell proliferation and cell death, to changes in neuronal excitability and synaptic transmission, to name a few. On the plasticity front, one of the transcriptional pathways that has been extensively investigated in relation to learning and formation of some forms of memory is the cAMP response element binding protein (CREB)-cAMP response element (CRE) (Figs. 8.2 and 8.3). Studies conducted in the hippocampus have shown that activation of the CREB-CRE pathway is associated with late-phase LTP, as well as long-term memory formation (Impey et al., 1996; Davis et al., 2000; Barco et al., 2002). Importantly, this pathway appears to be co-regulated by PKA and the ERK/MAPK pathways (and possibly the CaMKII pathway) (Figs. 8.2 and 8.3) (Impey et al., 1998; Davis et al., 2000), being well positioned to converge distinct biochemical signals and integrate those with the transcriptional machinery, thereby regulating a distinct set of genes.

CRE is a cis-element found in the promoter region of a number of genes that are expressed within the nervous system. It was originally described as a short domain (5'-TGACGTCA-3') within the somatostatin promoter where CREB was bound to (Montminy and Bilezikjian, 1987). Subsequent studies detected an almost identical sequence within the proto-oncogene c-fos promoter where CREB was also bound to (5'-TGACGTTT-3') (Sheng et al., 1988). CREB is a transcription factor whose activity is tightly regulated by intracellular calcium signals (Yin and Tully, 1996; Nguyen, 2001). Under resting conditions, dimerized CREB appears to be linked to the CRE site. Cell stimulation triggers a cascade of biochemical events that eventually activate CREB kinases, which are responsible for phosphorylating CREB at a specific serine residue (Ser-133) (reviewed in Shaywitz and Greenberg, 1999). CREB phosphorylation at Ser-133 triggers its binding to CREB-binding protein (CBP) (Fig. 8.3). CBP is then recruited and participates in the assembly of the transcriptional machinery in sites where phosphorylated CREB is found, thereby coordinating a wave of genomic activity that is integrated by the presence of the CRE-associated transcription (discussed below; Fig. 8.3) (Shaywitz and Greenberg, 1999).

In the visual system CRE-dependent transcription has been shown to play an important role in the regulation of plasticity. For example, Pham and co-workers (1999) utilized transgenic mice carrying a construct that consisted of multiple copies of CRE driving the expression of a lacZ reporter gene. These authors observed that monocular eyelid suture led to a prominent induction of CRE-driven lacZ expression within the cerebral cortex, indicating that this surgical procedure, which is a classical plasticity-inducing paradigm, activates CRE-mediated gene expression (Pham et al., 1999). In addition, these authors tested the effects of monocular deprivation in mature animals (P40-P44), after the critical period, and investigated CRE-lacZ transcription. It was found that CRE-lacZ transcription was markedly reduced in the visual cortex, suggesting that after the critical period, when plasticity mechanisms are presumably reduced, CRE-driven gene expression is greatly reduced as well (Pham et al., 1999).

A direct role for CREB in mediating OD plasticity was tested using viral vector approaches in the ferret V1 (Mower et al., 2002). In these experiments, authors intracortically infused herpes simplex viruses to infect V1 neurons with a double-negative insert for CREB (m-CREB) that blocks its activation

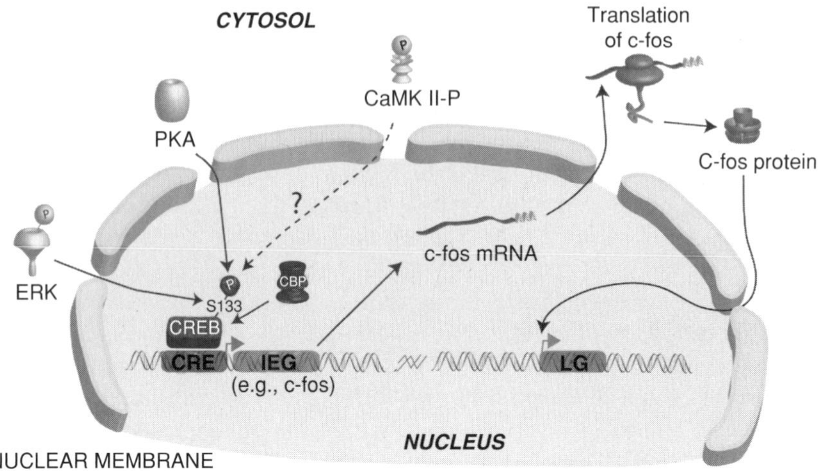

FIGURE 8.3. Waves of activity-dependent genomic responses via CREB phosphorylation. A number of kinases, including ERK and PKA are able to phosphorylate CREB at Ser133. At resting conditions, CREB is bound to the CRE. Upon its phosphorylation, CBP is recruited and the transcriptional machinery is activated thereby initiating the expression of genes that exhibit the CRE in their promoters. Among these genes are a number of IEGs, including c-fos. c-fos gene expression triggers the synthesis of a specific transcript that is transported to the cytoplasm where translation takes place. C-fos protein, a transcription factor, is then transported back to the nucleus where it will participate in the regulation of LG expression. ERK, extracellular signal-regulated kinase; PKA, cAMP-dependent protein kinase; CaMKII, calcium/calmodulin-dependent protein kinase II; CREB, cAMP responsive element binding protein; CRE, cAMP responsive element; CBP, CREB binding protein; IEG, immediate early gene; LG, late gene; mRNA, messenger ribonucleic acid; S133, serine 133.

due to a specific mutation. Coupling this approach with monocular deprivation, authors were able to observe that OD shifts were prevented in animals where CREB expression was blocked, but not in control animals that overexpressed CREB (Mower et al., 2002). In addition, Mower and colleagues showed that m-CREB effects on preventing OD plasticity were reversible according to decreased m-CREB activity. Subsequent studies utilizing the same approach have investigated whether CREB expression is required for the recovery of responses altered by monocular deprivation protocols (Liao et al., 2002). Using protocols of binocular vision recovery and reverse deprivation, it was shown that interference with CREB expression via expression of the mutant form of CREB does not prevent the recovery of responses associated with the originally deprived eye (Liao et al., 2002). Together these results provided strong evidence for the participation of CREB in the induction, but not recovery, of experience-dependent plastic changes within the developing visual cortex. Finally, as indicated above, it has been shown that

CRE-mediated gene expression decreases significantly by adulthood (Pham et al., 1999). However, expression of constitutively active forms of CREB in the mature visual system has been shown to sustain plastic states (Pham et al., 2004), suggesting that expression of this transcription factor generates conditions for enhanced plasticity, while its decrease, which is associated with normal aging, contributes to the decreased experience-dependent network malleability observed within the mature visual cortex.

Inducible-Immediate Early Gene Expression

As outlined in the previous section, phosphorylation of the transcription factor CREB appears to play an important role in the signaling pathway that interconnects modifications of activity in the cell surface, with the transcriptional machinery located in the cell nucleus, as part of a neural plastic response (Fig. 8.3). Transcription factors (TFs), which are proteins that often exhibit affinity for specific elements located in the promoter or enhancer regions of particular genes, and specific DNA binding properties, control (up- or down-regulation) transcription via trans-interactions with the transcription complex of RNA polymerase. The most studied TFs in sensory systems are often included into one of two classifications based on their expression properties: constitutive TFs (CTFs), such as CREB, are constitutively expressed and their activity is controlled by post-translational modifications, such as phosphorylation. Inducible TFs (ITFs) display expression regulated in response to cell stimulation, such as neuronal depolarization.

In the previous section I discussed calcium-dependent molecular events implicated in forms of synaptic plasticity in the visual cortex. In this section, I focus on activity-dependent regulation of ITFs and their relationship with experience-dependent plasticity. A detailed review on CTFs and ITFs in neural circuitry can be found elsewhere (Herdegen and Leah, 1998).

The regulation of CTFs such as CREB leads to waves of inducible gene expression programs in the cell nucleus as a result of activity/experience. One class of genes, commonly referred to as immediate early genes (IEGs; or primary response genes), in particular, has received significant attention from sensory neuroscientists for the past 15 years. IEGs are a class of genes that exhibit very rapid and transient expression in response to cell stimulation and the proteins encoded by these genes appear to play critical roles in cell physiology (reviewed in Hughes and Dragunow, 1995; Herdegen and Leah, 1998; Kaczmarek and Robertson, 2002). The expression of IEGs has been intensely scrutinized over the years for a number of reasons: 1) their expression has been shown to be highly associated with synaptic stimulation. Thus, IEG expression has been successfully used to probe neural activity in a variety of sensory systems and experimental conditions (Worley et al., 1991; Hughes and Dragunow, 1995; Kaczmarek and Chaudhuri, 1997; Herdegen and Leah, 1998; Kaczmarek and Robertson, 2002). 2) The expression of IEGs does not depend on new protein synthesis and, for this reason, it has been proposed that IEG expression represents the first inducible genomic response that occurs as a function of neuronal stimulation (Hughes and Dragunow, 1995; Kaczmarek and Chaudhuri, 1997; Herdegen and Leah, 1998; Kaczmarek and Robertson, 2002; Pinaud, 2004). 3) A wide range of IEGs encode ITFs that participate in the expression of a second, and delayed, subset of genes commonly referred to as "late response genes" (LGs), which are thought to execute more stable, long-term modifications in cellular architecture and function (Herdegen and Leah, 1998; Pinaud, 2004). ITFs therefore play a role in orchestrating appropriate and coordinated waves of LG expression (detailed below). 4) From a practical standpoint, identification of the products that result from IEG expression (either mRNA or protein) allows for the detection of cells that participate in any given experimental protocol with a single-cell resolution. In addition, double-labeling experiments that couple ITF expression with neurochemical markers (e.g., GAD65 expression for identification of GABAergic cells) (Pinaud et al., 2004) contributes to shedding light into the specific population of neurons that participate in a given experimental procedure. 5) Finally, the expression of several ITFs has been shown to depend on the activation of NMDA receptors and, therefore, have been proposed to participate in neural plastic responses in cortical circuits (Cole et al., 1989; Wisden et al., 1990; Steward and Worley, 2001b).

Given space constraints, I will not be able to thoroughly discuss research conducted in this area; however, it is my hope that the discussions that ensue will prompt the reader's interest in visiting the primary literature associated with these themes.

I have chosen to focus this section on three ITFs: the nerve growth factor induced gene-A (NGFI-A; also known as *zif-268*, *egr-1*, *krox-24* and *zenk*), the proto-oncogene *c-fos*, and *c-jun* (the AP-1 transcription factor), which have been on the forefront of investigations on ITFs in the visual system. In addition, I will discuss the expression of the an effector IEG, the activity-regulated cytoskeletal gene (*arc*), which exhibits interesting expression properties that potentially places it in a key position to mediate fast structural changes in neural circuitry as a function of experience. Importantly, the regulation of these genes has been shown to depend on NMDA-receptor activation (Cole et al., 1989; Wisden et al., 1990; Herdegen and Leah, 1998; Steward and Worley, 2001b).

NGFI-A

The IEG NGFI-A encodes a transcription factor that belongs to the zinc-finger class. Finger-like protuberances in its structure are stabilized by zinc and interact with a specific DNA motif found in the promoter of a wide variety of genes expressed within the nervous system, thereby potentially regulating their expression. NGFI-A has attracted significant attention over the years as the identity of LGs regulated by this protein becomes clearer (Pospelov et al., 1994; Thiel et al., 1994; Petersohn et al., 1995; Petersohn and Thiel, 1996; Carrasco-Serrano et al., 2000; Wong et al., 2002; Davis et al., 2003; Mello et al., 2004). Interestingly, NGFI-A appears to play significant roles on the expression of LGs regulatory that are related to the control of plasticity mechanisms (Fig. 8.3). For example, in-vitro studies have provided evidence for the regulation of NGFI-A over the synapsins I and II and synaptobrevin II genes, which are implicated in the control of the size of the readily releasable pool of neurotransmitters and neurotransmitter exocytosis (Thiel et al., 1994; Petersohn et al., 1995; Petersohn and Thiel, 1996). Other putative regulatory targets for NGFI-A include structural genes such as neurofilament (Pospelov et al., 1994), ligand-gated ion channels, such as the nicotinic acetylcholine receptor (Carrasco-Serrano et al., 2000) and metabolic enzymes, such as the monoamine oxidase B (Wong et al., 2002). It is therefore clear that NGFI-A is well positioned to integrate activity-dependent changes in the cell surface with the genomic responses thought to underlie the physical correlates of experience-dependent circuitry modifications (Fig. 8.3).

c-Fos and c-Jun: The AP-1

The proteins encoded by the Fos (C-Fos, FosB, Fra-1 and Fra-2) and Jun families (c-Jun, JunB and JunD) often hetero- or homo-dimerize through leucine zipper interactions located in motifs interspersed within their structure, contributing to the assembly of the transcription factor AP-1 (activator protein-1) (Morgan and Curran, 1991; Herdegen and Leah, 1998). The AP-1 TF recognizes and binds to a specific DNA motif thereby promoting the up- or down-regulation the activity of the target gene (Fig. 8.3) (Angel and Karin, 1991; Kobierski et al., 1991). Direction of the target gene expression regulated by AP-1 (up or down) is dictated by the specific combination of Fos and Jun components that participate on the dimerization, which is often a reflection of affinity for the DNA-binding motif. For example, Fos/JunD heterodimers exhibit high regulatory capabilities, while FosB/JunB bind to the DNA consensus but are inactive, therefore repressing transcriptional activity (Hughes and Dragunow, 1995); subsequent phosphorylation can further modify transcriptional capabilities of AP-1 (Herdegen and Leah, 1998).

Arc

Like NGFI-A and the AP-1 complex, the protein encoded by the IEG *arc* is activity-dependent. However, it encodes a growth factor and not a transcriptional regulator (Lyford et al., 1995). Arc has received significant attention over the past few years because of specific properties embedded in its expression program: *arc* expression results in the synthesis of an mRNA product that is not used for translation in the cell body, as occurs with the majority of transcripts, but rather the mRNA is transported by a specific set of proteins to dendrites, where it appears that the translation of Arc protein takes place locally in polyribosomal complex located in the post-synaptic cell, often times at the base of dendritic spines (Lyford et al., 1995; Steward et al., 1998;

Steward and Worley, 2001a, b). These properties, along with the fact that Arc associates with F-actin (Lyford et al., 1995) suggest that synthesis of Arc protein in the dendrites might contribute to relatively fast changes in synaptic architecture, particularly as it relates to dendritic configuration (e.g., retraction and elongation) in an activity-dependent fashion, as in the case of lesions or modifications in sensory input patterns (Grutzendler et al., 2002; Nimchinsky et al., 2002; Trachtenberg et al., 2002; Pinaud, 2004).

IEG Regulation by Visual Input in V1

Given the association of IEG expression with neuronal activity, the expression of this class of genes has been repeatedly and successfully used as a marker for neural activation across sensory systems (for reviews see Kaczmarek and Chaudhuri, 1997; Herdegen and Leah, 1998; Kaczmarek and Robertson, 2002). As briefly mentioned above, this methodology confers investigators a means to study large-scale patterns of activity with a single-cell resolution. In addition, the use of IEG expression as a mapping tool for neuronal activation allows for a high spatial resolution (e.g., being possible to study variations in activation patterns throughout multiple visual areas). Finally, experiments designed for this methodology very often employ awake-behaving animals, with minimal or without interference with the animal's "natural" behavior. The main drawback of such methodology, however, falls on the low temporal resolution of this approach. For an extensive review of IEG expression, methodological issues and applications, please refer to selected reviews (Herdegen and Leah, 1998; Kaczmarek and Robertson, 2002).

In the visual system, investigations on the effects of light deprivation on IEG expression have employed paradigms of permanent and reversible light deprivation, including monocular deprivation and retinal lesions, to name a few, across a number of experimental models; investigations on the impacts of visual experience on IEG regulation have used various experimental protocols that include light-stimulation and exposure of animals to complex visual environments.

IEG Regulation by Visual Deprivation and Stimulation

NGFI-A is expressed at relatively high basal levels in the rodent and cat visual cortices as a result of ongoing neuronal activity (Worley et al., 1991; Kaplan et al., 1996; Kaczmarek and Chaudhuri, 1997). Conversely, *c-fos* and *arc* are expressed at very low levels in the adult V1 (Beaver et al., 1993; Kaplan et al., 1996; Kaczmarek and Chaudhuri, 1997; Pinaud et al., 2001). For this reason, NGFI-A has often been used to probe the effects of visual deprivation in cortical circuits, while the low basal expression of *c-fos* and *arc* preclude their use for this purpose to a large extent. However, light-stimulation triggers significant increase in the products that result from both NGFI-A and *c-fos* induction in the mammalian visual cortex.

Work conducted by Worley and colleagues have shown that both NGFI-A mRNA and protein expression is markedly reduced as a result of TTX intraocular application in the rodent (Worley et al., 1991). In addition, dark-rearing protocols significantly reduced basal NGFI-A expression levels, while ensuing light exposure triggered a rapid and robust rescue of baseline gene expression levels (Worley et al., 1991; Kaczmarek and Chaudhuri, 1997; Pinaud et al., 2000). Subsequent eyelid suture and TTX intraocular infusion experiments have also shown significant NGFI-A expression downregulation in the rat visual cortex (Caleo et al., 1999).

Exposure to darkness followed by short light-stimulation events triggers a marked induction of NGFI-A in the rodent visual cortex (Worley et al., 1991; Kaczmarek and Chaudhuri, 1997; Pinaud et al., 2000). This light-induced NGFI-A expression spans throughout all cortical layers in both juvenile (during the critical period) and adult animals (Worley et al., 1991; Nedivi et al., 1996; Kaczmarek and Chaudhuri, 1997). Visual stimulation also triggers a robust C-fos expression in the rodent visual cortex. For example, it has been shown that pattern stimuli (gratings and dots) induce C-fos in the rat V1, with highest expression occurring in layers IV and VI and lower expression being detected in the supragranular layers (Montero and Jian, 1995).

In the cat, unilateral sectioning of the optic tract has been shown to significantly downregulate NGFI-A (*zif268*) mRNA in both supra and infragranular layers of V1 (Zhang et al., 1995). Long-term dark-rearing (1-week) of both young and adult cats also led to significant suppression of NGFI-A protein levels in V1 (Kaplan et al., 1996). Dark-adaptation followed by light-stimulation has been shown to robustly induce NGFI-A protein in an age-dependent, cortical layer-specific manner: while young cats (5-weeks old) exhibit significant NGFI-A expression across all cortical layers, adult animals undergo increased protein expression in supra- and infragranular layers, while immunolabeled cells in layer IV has been reported to be very low (Kaplan et al., 1996). These results differ from those obtained in the rodent visual cortex where visual deprivation affects more significantly the thalamo-recipient layer (Worley et al., 1991), possibly reflecting different cortical anatomical architecture and/or processing and computational properties across species. Finally, although *c-fos* is expressed at low basal levels, dark-induced downregulation of this gene has also been reported in both juvenile and adult cats (Kaplan et al., 1996). A marked *c-fos* and *junB* (components of the AP-1) mRNA expression has been shown in cats that have underwent dark-adaptation followed by a short light stimulation protocol (1 hr) (Rosen et al., 1992). For *c-fos*, highest expression following this protocol, however, occurs at 2 hrs after stimulus onset, being most prominent in layers II/III and VI, and decreases to levels similar to those found in control (non-stimulated) animals by 6 hrs (Beaver et al., 1993). This habituation of the expression levels occurs for both *c-fos* and *junB* (Rosen et al., 1992). A direct comparison between the cat expression levels of *c-fos* and NGFI-A during the critical period versus adulthood was conducted by Kaplan and colleagues (1996). It was found that dark-adaptation followed by light stimulation triggers a marked upregulation of both *c-fos* and NGFI-A protein across all cortical layers in 5-week old kittens (Kaplan et al., 1996). Interestingly, the same experimental protocol elicits a different expression profile in adult cats, whereby high expression levels were found in both supra- and infragranular layers, while very low immunoreactivity levels for both genes were detected for the granular layer, suggesting that the expression of these genes might reflect different degrees of plasticity across different cortical layers (discussed below).

In the primate visual cortex, NGFI-A is also expressed at high basal levels (Chaudhuri and Cynader, 1993; Silveira et al., 1996). Similarly to findings obtained in the cat V1, highest concentration of NGFI-A immunolabeled cells were detected in supragranular layers and layer VI, while a modest number of cells was detected in the thalamo-recipient layer IV and layer V (Chaudhuri et al., 1995; Okuno et al., 1997). Monocular deprivation induced by a variety of protocols such as enucleation, intraocular TTX infusion and eyelid suture have revealed a marked downregulation of NGFI-A in the ocular dominance columns associated with the deprived eye in the adult monkey (Chaudhuri and Cynader, 1993; Silveira et al., 1996). Interestingly, these changes in gene expression as a result of altered sensory input could be detected in a very fast time frame, as early as 1–3 hours after deprivation (Chaudhuri and Cynader, 1993; Chaudhuri et al., 1995; Silveira et al., 1996). Investigations on the expression profiles of NGFI-A throughout the critical period in the Cebus monkey have also revealed interesting results. As indicated above, monocular deprivation in the adult primate triggered a significant downregulation of NGFI-A in the deprived columns. Conversely, monocular deprivation failed to reveal ODCs at 3-month old animals, while at 6-months monkeys, ODCs were still scantily developed, suggesting that ODCs revealed by NGFI-A expression are formed late in the developmental process (Silveira et al., 1996).

Protocols of dark-adaptation followed by light-stimulation have also been conducted in the monkey V1. It has been shown that 24-hrs of dark-rearing followed by 2-hrs of monocular light stimulation triggers a marked NGFI-A induction in ODCs associated with the stimulated eye, while residual expression appears to be associated in the ODCs associated with the deprived eye (Chaudhuri et al., 1995). In this experiment, NGFI-A expression appeared to span throughout all cortical layers (Chaudhuri et al., 1995). *c-fos* expression at both the mRNA and protein levels also appears to be regulated in the vervet monkey, under a reverse occlusion paradigm. It has been reported that this procedure leads to *c-fos* expression in cortical layers II/II, IVc and VI irrespective of age (during the critical period or adulthood) (Kaczmarek and Chaudhuri, 1997). To my knowledge, *arc* expression profiles have not yet been investigated in the visual system of cats and monkeys.

It is clear that IEG expression reveals a pattern of neuronal activation in response to sensory input in the mammalian visual cortex. However, as neuronal activity does not absolutely predict the expression of IEGs (Sharp et al., 1993; Pinaud, 2004), it may be worthwhile to pose questions as to which specific qualities of stimulation actually drive IEG expression. As more data is generated on IEG expression, the more it becomes clear that they participate in various aspects of cell physiology such as cell death and trophic regulation, to name a few. However, it also appears that a subset of IEGs, including those discussed in this review, play significant roles to the induction and maintenance of plasticity in neuronal cells. It is precisely the fact that these IEGs are expressed as a reflection of neuronal activation patterns associated with the demonstration of plasticity that makes them "candidate-plasticity genes".

Plasticity in the Visual Cortex and IEG Expression

In an attempt to dissociate the effects of neuronal activation from gene mechanisms involved in the induction of plasticity, we and others have compared the expression profiles of IEGs in simple light-stimulation paradigms, with stimulation from complex visual environments, which are known to induce plastic changes in visual cortical circuitry (Volkmar and Greenough, 1972). Our paradigm of choice was the "enriched environment" (EE) protocol (reviewed in Rosenzweig et al., 1972; van Praag et al., 2000; Pinaud, 2004). This paradigm essentially consists of placing freely-ranging animals in a complex visual environment, and allows for a direct comparison of IEG expression patterns with animals that are also freely-ranging, but exposed to an impoverished visual environment. This paradigm has been shown to heavily and reliably induce plasticity in the visual cortex. For example, exposure of animals to enriched visual environments trigger massive alterations in cellular architecture, including enhanced dendritic arborizations and larger synaptic boutons, in the visual cortex (Rosenzweig et al., 1972; Volkmar and Greenough, 1972; van Praag et al., 2000; Pinaud, 2004).

If IEG expression is involved in the induction and/or regulation of plasticity in V1, one would predict that the expression profiles of genes of interest would differ in animals exposed to the EE, as opposed to those that received simple patterns of visual stimulation (and presumably did not undergo experience-dependent changes in cortical network organization). In order to test this hypothesis, we divided a population of young adult rats into three groups and housed them all in standard laboratory home cages. Animals in the first group (EE group) were removed from their home cages at the same time of the day and exposed to the EE setting for 1 hr daily, being subsequently returned to their lab cages. This procedure was repeated daily for a total of 21 days. The second group consisted of same age rats that were left undisturbed (UD group) in their lab home cages throughout the 21 days in which the experiment was carried out. Importantly, only animals that exhibited similar activity levels as the EE rats were included in this study. Finally, the third group of animals was composed of animals that were housed in the standard lab home cages, were handled in a similar fashion as those in the EE group, but returned to their lab home cages (handling-only; HO group). This last group of animals was included to control for any non-specific IEG expression related to the manipulation that EE animals underwent and stress (Pinaud et al., 2001; Pinaud et al., 2002; Pinaud, 2004).

We investigated the expression profiles of NGFI-A mRNA and protein for the experiment outlined above. As previously reported, NGFI-A mRNA expression, as revealed by in-situ hybridization, was observed at high basal levels in UD animals (Fig. 8.4). NGFI-A expression was detected across all cortical layers of the rat V1, with the exception of layer I (Wallace et al., 1995; Pinaud et al., 2002). Quantification of NGFI-A mRNA expression levels in V1 of HO animals revealed that these were not significantly different from those levels observed in animals belonging to the UD group (Pinaud et al., 2002). In EE animals, however, a very robust and significant upregulation of NGFI-A was observed in all cortical layers of V1, as compared to both UD and HO animals (Fig. 8.4) (Pinaud et al., 2002). Immunocytochemical analysis revealed that NGFI-A protein levels were also significantly increased in the V1 of EE animals compared to both UD and HO controls, while no differences were detected between these two latter groups. These findings suggest that it is the complexity of the visual environment that drove a differential regulation pattern for NGFI-A in the rat V1, rather than simple neuronal activity patterns. It is important to recall that both control groups (UD and HO) were fully awake, exhibited similar levels of activity as those in the EE group and, therefore, were being visually stimulated. However, NGFI-A expression remained at basal levels in V1 these groups.

We have also investigated the expression of the IEG *arc* in the adult rat V1 using the same paradigm as described above (Pinaud et al., 2001). One of the interesting features of *arc* is that the protein encoded by this gene is translated in dendrites in an activity-dependent fashion and, therefore, may play a direct role in dendritic reconfiguration (Steward et al., 1998; Pinaud et al., 2001; Steward and Worley, 2001a). We found that *arc*, unlike the transcription factor NGFI-A, is expressed at relatively low basal levels in the rodent V1 (Fig. 8.4). A direct comparison of *arc* mRNA expression levels in HO and UD animals failed to reveal any significant changes. Conversely, exposure of animals to an EE triggered a marked upregulation of *arc* mRNA across all cortical layers (Fig. 8.4) (Pinaud et al., 2001).

As detailed above, robust IEG induction just occurred in the EE group in a paradigm that triggers the widespread activation plasticity mechanisms; these results therefore suggested to us that both *arc* and NGFI-A induction may participate as part of the machinery involved in the early genomic response associated with plastic changes in the rat V1, rather than reflect simple neuronal activity. In fact, neuronal activity is necessary, but not sufficient for NGFI-A and *arc* induction in this paradigm.

Close inspection of the expression patterns of NGFI-A at both mRNA and protein levels revealed that although significant upregulation in the EE

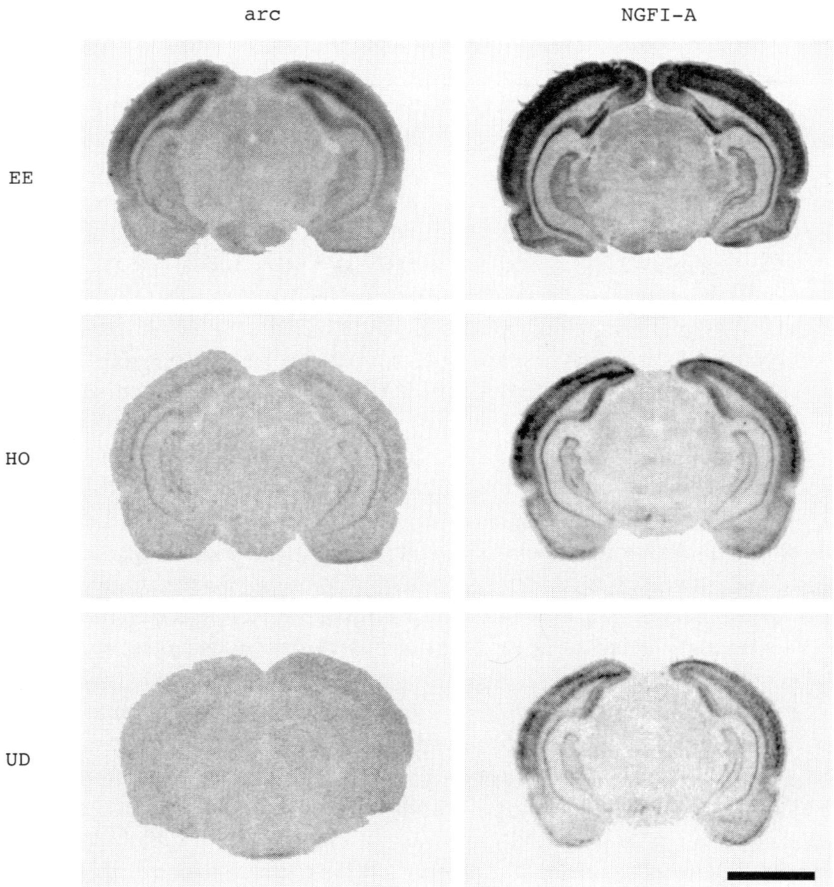

FIGURE 8.4. Experience-dependent expression of immediate early genes *arc* and NGFI-A in the visual cortex. In-situ hybridization autoradiograms depicting *arc* and NGFI-A expression in animals that were exposed to a complex visual environment (EE) for 1 hour/day for a total of 21 days, as compared to the expression levels of animals that were manipulated (HO) or left undisturbed in their home cages. Both NGFI-A and arc are markedly upregulated by exposure to the EE condition, as compared to both control groups. Expression levels for both IEGs are not different across control groups. Highest induction in response to the EE for both IEGs was detected in cortical layers III and V. Scale bar = 500μm. NGFI-A, nerve growth factor-induced gene A; *arc,* activity-regulated cytoskeletal gene; EE, enriched environment group; HO, handled-only group; UD, undisturbed group.

group occurred in all cortical layers, the highest experience-dependent induction was found in layers III and V (Fig. 8.4). In addition, the lowest expression levels for this IEG in the EE group were found in the thalamorecipient layer IV (Pinaud et al., 2002). Similarly, *arc* expression in the EE condition was not symmetric across layers: as was observed for NGFI-A, *arc* induction was more robust for layers III and V, while the lowest variation in transcript levels was detected for layer IV (Fig. 8.4) (Pinaud et al., 2001). These results are interesting given that they may be related to electrophysiological studies that suggest that layers III and V are potentially more plastic, as measured by their greater ease of inducing changes in synaptic strength such as LTP and LTD (Glazewski and Fox, 1996; Petersen and Sakmann, 2001). Conversely, it has been proposed that low plasticity levels in layer IV could be related to the stability in cortical map representation and possibly the maintenance of fidelity of information transfer at this level (Pinaud, 2004). Should this hypothesis prove to be correct, low expression levels of the candidate-plasticity genes *arc* and NGFI-A in layer IV might provide a direct visualization of reduced activity in components of the machinery involved in triggering experience-dependent reorganization of circuits in V1. By association, enhanced IEG expression in layers III and V may provide a direct indication of a high density of cells in V1 that are participating in the neural plastic response as a function of complex visual environment exposure.

The dramatic increase in gene expression in the EE condition may thus reflect enhanced levels, or specific patterns of activity, that result from exposure of animals to a complex visual environment. One possibility for the recruitment of gene expression programs associated with network rewiring in V1 is that mechanisms for detecting sub-optimal architecture and processing capabilities are in place in this cortical region. Upon exposure of animals to EEs, detection of sensory or information overload may trigger these programs in order to optimize cortical architecture and synaptic weights in a way to appropriately process information contained in this new, complex environment. This hypothesis remains to be tested experimentally, however.

IEG expression in EEs might also be related to exposure of animals to novel environments. Novelty has been shown to positively regulate the IEGs *c-fos* and *c-jun* in several brain areas, including the somatosensory and visual cortices (Papa et al., 1993; Zhu et al., 1995). In our experiments, however, novelty is not likely to play a key role in the regulation of IEGs given that expression levels were assessed in animals that were familiarized with the EE for 3 weeks.

Future Directions

In this chapter I discussed selected molecular events that appear to play critical roles in the induction and maintenance of plasticity in visual cortical neurons. Calcium influx, either via voltage-sensitive calcium channels or NMDA receptor activation, appears to initiate the cascade of biochemical events that lead to the integration of the cell membrane changes with genomic programs aimed at rewiring or re-weighting cortical networks.

I foresee significant interest in a number of areas associated with the topics discussed here, that will contribute to a better understanding of the events that lead to measurable plastic changes in the visual cortex as a function of experience. First, the development of molecular methodologies that allow for interference with the function of molecules of interest will shed light into their specific contributions to the process of plasticity in the visual cortex. Some of these methodologies, such as viral-mediated gene interference, have been used in visual system research and were discussed briefly here. However, it is my opinion that these approaches are significantly under-used in visual research, as compared to other research areas such as hippocampal physiology. The use of other powerful approaches such as RNA interference (RNAi), antisense technology and inducible and reversible gene expression (e.g., via reverse tetracycline-controlled transactivators (rtTA)) have not yet been routinely used in visual system research but provide the potential for exquisite control over gene expression mechanisms associated with experience-dependent plasticity. These and other approaches bring the promise of a true understanding of the individual contributions of plasticity-associated molecules. Furthermore successful control over these pathways can potentially enable scientists and physicians to ameliorate conditions resulting from the lack of, or abnormal plastic changes.

On the IEG front, the methodologies outlined above will allow for the study of their precise roles in orchestrating waves of gene expression aimed at activity-dependent reorganization of visual cortical circuits. In addition, particularly as it relates to transcriptional regulators, it will be critical to establish the identity of their regulatory targets in such a way to yield a more complete picture of the genetic and structural processes that represent plasticity in the visual cortex.

One of the areas that are, in my opinion, largely ignored in visual system research is activity-dependent receptor trafficking. In other models, such as the hippocampus, it has been shown that the trafficking of key molecules, for example AMPA and GABA receptors, to and from the cell membrane, contribute to critical aspects of activity-dependent facilitation and depression (Malinow and Malenka, 2002; Malenka, 2003; Stellwagen et al., 2005). Assuming that similar, if not identical, molecular and cellular mechanisms are in place to mediate forms of synaptic plasticity across brain regions, the study of receptor trafficking will not only contribute to understanding the physical substrates of plasticity in the adult visual system, but may also shed light into the mechanisms that regulate induction, maintenance and cessation of developmental plasticity during the critical period. These and other developments should promote exciting years to come in the study of visual system plasticity.

Acknowledgments: I thank Liisa Tremere for critical reading of this manuscript and Antonio Fortes for the generation of the figures included in this chapter.

References

Angel P, Karin M (1991) The role of Jun, Fos and the AP-1 complex in cell-proliferation and transformation. Biochim Biophys Acta 1072:129–157.

Bading H, Ginty DD, Greenberg ME (1993) Regulation of gene expression in hippocampal neurons by distinct calcium signaling pathways. Science 260:181–186.

Barco A, Alarcon JM, Kandel ER (2002) Expression of constitutively active CREB protein facilitates the late phase of long-term potentiation by enhancing synaptic capture. Cell 108:689–703.

Beaver CJ, Mitchell DE, Robertson HA (1993) Immunohistochemical study of the pattern of rapid expression of C-Fos protein in the visual cortex of dark-reared kittens following initial exposure to light. J Comp Neurol 333:469–484.

Beaver CJ, Ji Q, Fischer QS, Daw NW (2001) Cyclic AMP-dependent protein kinase mediates ocular dominance shifts in cat visual cortex. Nat Neurosci 4:159–163.

Beaver CJ, Fischer QS, Ji Q, Daw NW (2002) Orientation selectivity is reduced by monocular deprivation in combination with PKA inhibitors. J Neurophysiol 88: 1933–1940.

Berridge MJ (1998) Neuronal calcium signaling. Neuron 21:13–26.

Brandon EP, Idzerda RL, McKnight GS (1997) PKA isoforms, neural pathways, and behaviour: making the connection. Curr Opin Neurobiol 7:397–403.

Brostrom CO, Huang YC, Breckenridge BM, Wolff DJ (1975) Identification of a calcium-binding protein as a calcium-dependent regulator of brain adenylate cyclase. Proc Natl Acad Sci U S A 72:64–68.

Brostrom MA, Brostrom CO, Breckenridge BM, Wolff DJ (1976) Regulation of adenylate cyclase from glial tumor cells by calcium and a calcium-binding protein. J Biol Chem 251:4744–4750.

Caleo M, Lodovichi C, Pizzorusso T, Maffei L (1999) Expression of the transcription factor Zif268 in the visual cortex of monocularly deprived rats: effects of nerve growth factor. Neuroscience 91:1017–1026.

Carey KD, Watson RT, Pessin JE, Stork PJ (2003) The requirement of specific membrane domains for Raf-1 phosphorylation and activation. J Biol Chem 278: 3185–3196.

Carrasco-Serrano C, Viniegra S, Ballesta JJ, Criado M (2000) Phorbol ester activation of the neuronal nicotinic acetylcholine receptor alpha7 subunit gene: involvement of transcription factor Egr-1. J Neurochem 74:932–939.

Chaudhuri A, Cynader MS (1993) Activity-dependent expression of the transcription factor Zif268 reveals ocular dominance columns in monkey visual cortex. Brain Res 605:349–353.

Chaudhuri A, Matsubara JA, Cynader MS (1995) Neuronal activity in primate visual cortex assessed by immunostaining for the transcription factor Zif268. Vis Neurosci 12:35–50.

Cheung WY, Bradham LS, Lynch TJ, Lin YM, Tallant EA (1975) Protein activator of cyclic 3':5'-nucleotide phosphodiesterase of bovine or rat brain also activates its adenylate cyclase. Biochem Biophys Res Commun 66:1055–1062.

Cole AJ, Saffen DW, Baraban JM, Worley PF (1989) Rapid increase of an immediate early gene messenger RNA in hippocampal neurons by synaptic NMDA receptor activation. Nature 340:474–476.

Collingridge GL, Lester RA (1989) Excitatory amino acid receptors in the vertebrate central nervous system. Pharmacol Rev 41:143–210.

Collingridge GL, Singer W (1990) Excitatory amino acid receptors and synaptic plasticity. Trends Pharmacol Sci 11:290–296.

Constantine-Paton M, Cline HT, Debski E (1990) Patterned activity, synaptic convergence, and the NMDA receptor in developing visual pathways. Annu Rev Neurosci 13:129–154.

Cotman CW, Monaghan DT, Ganong AH (1988) Excitatory amino acid neurotransmission: NMDA receptors and Hebb-type synaptic plasticity. Annu Rev Neurosci 11:61–80.

Davis S, Vanhoutte P, Pages C, Caboche J, Laroche S (2000) The MAPK/ERK cascade targets both Elk-1 and cAMP response element-binding protein to control long-term potentiation-dependent gene expression in the dentate gyrus in vivo. J Neurosci 20:4563–4572.

Davis W, Jr., Chen ZJ, Ile KE, Tew KD (2003) Reciprocal regulation of expression of the human adenosine 5′-triphosphate binding cassette, sub-family A, transporter 2 (ABCA2) promoter by the early growth response-1 (EGR-1) and Sp-family transcription factors. Nucleic Acids Res 31:1097–1107.

Daw NW, Stein PS, Fox K (1993) The role of NMDA receptors in information processing. Annu Rev Neurosci 16:207–222.

De Koninck P, Schulman H (1998) Sensitivity of CaM kinase II to the frequency of Ca2+ oscillations. Science 279:227–230.

Di Cristo G, Berardi N, Cancedda L, Pizzorusso T, Putignano E, Ratto GM, Maffei L (2001) Requirement of ERK activation for visual cortical plasticity. Science 292:2337–2340.

English JM, Cobb MH (2002) Pharmacological inhibitors of MAPK pathways. Trends Pharmacol Sci 23:40–45.

Farnsworth CL, Freshney NW, Rosen LB, Ghosh A, Greenberg ME, Feig LA (1995) Calcium activation of Ras mediated by neuronal exchange factor Ras-GRF. Nature 376:524–527.

Ferguson GD, Storm DR (2004) Why calcium-stimulated adenylyl cyclases? Physiology (Bethesda) 19:271–276.

Fink CC, Meyer T (2002) Molecular mechanisms of CaMKII activation in neuronal plasticity. Curr Opin Neurobiol 12:293–299.

Fukunaga K, Miyamoto E (1998) Role of MAP kinase in neurons. Mol Neurobiol 16:79–95.

Gallin WJ, Greenberg ME (1995) Calcium regulation of gene expression in neurons: the mode of entry matters. Curr Opin Neurobiol 5:367–374.

Gatt S, Barenholz Y (1973) Enzymes of complex lipid metabolism. Annu Rev Biochem 42:61–90.

Ginty DD (1997) Calcium regulation of gene expression: isn't that spatial? Neuron 18:183–186.

Glazewski S, Fox K (1996) Time course of experience-dependent synaptic potentiation and depression in barrel cortex of adolescent rats. J Neurophysiol 75:1714–1729.

Grewal SS, York RD, Stork PJ (1999) Extracellular-signal-regulated kinase signalling in neurons. Curr Opin Neurobiol 9:544–553.

Grewal SS, Fass DM, Yao H, Ellig CL, Goodman RH, Stork PJ (2000a) Calcium and cAMP signals differentially regulate cAMP-responsive element-binding protein function via a Rap1-extracellular signal-regulated kinase pathway. J Biol Chem 275: 34433–34441.

Grewal SS, Horgan AM, York RD, Withers GS, Banker GA, Stork PJ (2000b) Neuronal calcium activates a Rap1 and B-Raf signaling pathway via the cyclic adenosine monophosphate-dependent protein kinase. J Biol Chem 275:3722–3728.

Grutzendler J, Kasthuri N, Gan WB (2002) Long-term dendritic spine stability in the adult cortex. Nature 420:812–816.

Herdegen T, Leah JD (1998) Inducible and constitutive transcription factors in the mammalian nervous system: control of gene expression by Jun, Fos and Krox, and CREB/ATF proteins. Brain Res Brain Res Rev 28:370–490.

Hindley A, Kolch W (2002) Extracellular signal regulated kinase (ERK)/mitogen activated protein kinase (MAPK)-independent functions of Raf kinases. J Cell Sci 115: 1575–1581.

Hollmann M, Heinemann S (1994) Cloned glutamate receptors. Annu Rev Neurosci 17:31–108.

Hughes P, Dragunow M (1995) Induction of immediate-early genes and the control of neurotransmitter-regulated gene expression within the nervous system. Pharmacol Rev 47:133–178.

Impey S, Mark M, Villacres EC, Poser S, Chavkin C, Storm DR (1996) Induction of CRE-mediated gene expression by stimuli that generate long-lasting LTP in area CA1 of the hippocampus. Neuron 16:973–982.

Impey S, Obrietan K, Wong ST, Poser S, Yano S, Wayman G, Deloulme JC, Chan G, Storm DR (1998) Cross talk between ERK and PKA is required for Ca2+ stimulation of CREB-dependent transcription and ERK nuclear translocation. Neuron 21:869–883.

Kaczmarek L, Chaudhuri A (1997) Sensory regulation of immediate-early gene expression in mammalian visual cortex: implications for functional mapping and neural plasticity. Brain Res Brain Res Rev 23:237–256.

Kaczmarek L, Robertson HA (2002) Immediate early genes and inducible transcription factors in mapping of the central nervous system function and dysfunction. Amsterdam: Elsevier Science B.V.

Kaminska B, Kaczmarek L, Zangenehpour S, Chaudhuri A (1999) Rapid phosphorylation of Elk-1 transcription factor and activation of MAP kinase signal transduction pathways in response to visual stimulation. Mol Cell Neurosci 13:405–414.

Kaplan IV, Guo Y, Mower GD (1996) Immediate early gene expression in cat visual cortex during and after the critical period: differences between EGR-1 and Fos proteins. Brain Res Mol Brain Res 36:12–22.

Karmarkar UR, Buonomano DV (2002) A model of spike-timing dependent plasticity: one or two coincidence detectors? J Neurophysiol 88:507–513.

Kirkwood A, Silva A, Bear MF (1997) Age-dependent decrease of synaptic plasticity in the neocortex of alphaCaMKII mutant mice. Proc Natl Acad Sci U S A 94: 3380–3383.

Kleinschmidt A, Bear MF, Singer W (1987) Blockade of "NMDA" receptors disrupts experience-dependent plasticity of kitten striate cortex. Science 238:355–358.

Kobierski LA, Chu HM, Tan Y, Comb MJ (1991) cAMP-dependent regulation of proenkephalin by JunD and JunB: positive and negative effects of AP-1 proteins. Proc Natl Acad Sci U S A 88:10222–10226.

Kyosseva SV (2004) Mitogen-activated protein kinase signaling. Int Rev Neurobiol 59:201–220.

Liao DS, Mower AF, Neve RL, Sato-Bigbee C, Ramoa AS (2002) Different mechanisms for loss and recovery of binocularity in the visual cortex. J Neurosci 22:9015–9023.

Lisman J, Schulman H, Cline H (2002) The molecular basis of CaMKII function in synaptic and behavioural memory. Nat Rev Neurosci 3:175–190.

Lyford GL, Yamagata K, Kaufmann WE, Barnes CA, Sanders LK, Copeland NG, Gilbert DJ, Jenkins NA, Lanahan AA, Worley PF (1995) Arc, a growth factor and activity-regulated gene, encodes a novel cytoskeleton-associated protein that is enriched in neuronal dendrites. Neuron 14:433–445.

Majerus PW (1992) Inositol phosphate biochemistry. Annu Rev Biochem 61:225–250.

Malenka RC (2003) Synaptic plasticity and AMPA receptor trafficking. Ann N Y Acad Sci 1003:1–11.

Malinow R, Malenka RC (2002) AMPA receptor trafficking and synaptic plasticity. Annu Rev Neurosci 25:103–126.

Mayer ML, Armstrong N (2004) Structure and function of glutamate receptor ion channels. Annu Rev Physiol 66:161–181.

McMurray WC, Magee WL (1972) Phospholipid metabolism. Annu Rev Biochem 41:129–160.

Meinkoth JL, Alberts AS, Went W, Fantozzi D, Taylor SS, Hagiwara M, Montminy M, Feramisco JR (1993) Signal transduction through the cAMP-dependent protein kinase. Mol Cell Biochem 127–128:179–186.

Mello CV, Velho TA, Pinaud R (2004) Song-induced gene expression: a window on song auditory processing and perception. Ann N Y Acad Sci 1016:263–281.

Montero VM, Jian S (1995) Induction of c-fos protein by patterned visual stimulation in central visual pathways of the rat. Brain Res 690:189–199.

Montminy MR, Bilezikjian LM (1987) Binding of a nuclear protein to the cyclic-AMP response element of the somatostatin gene. Nature 328:175–178.

Morgan JI, Curran T (1991) Stimulus-transcription coupling in the nervous system: involvement of the inducible proto-oncogenes fos and jun. Annu Rev Neurosci 14:421–451.

Mower AF, Liao DS, Nestler EJ, Neve RL, Ramoa AS (2002) cAMP/Ca2+ response element-binding protein function is essential for ocular dominance plasticity. J Neurosci 22:2237–2245.

Nedivi E, Fieldust S, Theill LE, Hevron D (1996) A set of genes expressed in response to light in the adult cerebral cortex and regulated during development. Proc Natl Acad Sci U S A 93:2048–2053.

Nguyen PV (2001) CREB and the enhancement of long-term memory. Trends Neurosci 24:314.

Nimchinsky EA, Sabatini BL, Svoboda K (2002) Structure and function of dendritic spines. Annu Rev Physiol 64:313–353.

Okuno H, Kanou S, Tokuyama W, Li YX, Miyashita Y (1997) Layer-specific differential regulation of transcription factors Zif268 and Jun-D in visual cortex V1 and V2 of macaque monkeys. Neuroscience 81:653–666.

Papa M, Pellicano MP, Welzl H, Sadile AG (1993) Distributed changes in c-Fos and c-Jun immunoreactivity in the rat brain associated with arousal and habituation to novelty. Brain Res Bull 32:509–515.

Petersen CC, Sakmann B (2001) Functionally independent columns of rat somatosensory barrel cortex revealed with voltage-sensitive dye imaging. J Neurosci 21:8435–8446.

Petersohn D, Thiel G (1996) Role of zinc-finger proteins Sp1 and zif268/egr-1 in transcriptional regulation of the human synaptobrevin II gene. Eur J Biochem 239: 827–834.

Petersohn D, Schoch S, Brinkmann DR, Thiel G (1995) The human synapsin II gene promoter. Possible role for the transcription factor zif268/egr-1, polyoma enhancer activator 3, and AP2. J Biol Chem 270:24361–24369.

Pham TA, Impey S, Storm DR, Stryker MP (1999) CRE-mediated gene transcription in neocortical neuronal plasticity during the developmental critical period. Neuron 22:63–72.

Pham TA, Graham SJ, Suzuki S, Barco A, Kandel ER, Gordon B, Lickey ME (2004) A semi-persistent adult ocular dominance plasticity in visual cortex is stabilized by activated CREB. Learn Mem 11:738–747.

Phillis JW, O'Regan MH (2004) A potentially critical role of phospholipases in central nervous system ischemic, traumatic, and neurodegenerative disorders. Brain Res Brain Res Rev 44:13–47.

Pinaud R (2004) Experience-dependent immediate early gene expression in the adult central nervous system: evidence from enriched-environment studies. Int J Neurosci 114:321–333.

Pinaud R, Tremere LA, Penner MR (2000) Light-induced zif268 expression is dependent on noradrenergic input in rat visual cortex. Brain Res 882:251–255.

Pinaud R, Penner MR, Robertson HA, Currie RW (2001) Upregulation of the immediate early gene arc in the brains of rats exposed to environmental enrichment: implications for molecular plasticity. Brain Res Mol Brain Res 91:50–56.

Pinaud R, Tremere LA, Penner MR, Hess FF, Robertson HA, Currie RW (2002) Complexity of sensory environment drives the expression of candidate-plasticity gene, nerve growth factor induced-A. Neuroscience 112:573–582.

Pinaud R, Velho TA, Jeong JK, Tremere LA, Leao RM, von Gersdorff H, Mello CV (2004) GABAergic neurons participate in the brain's response to birdsong auditory stimulation. Eur J Neurosci 20:1318–1330.

Pospelov VA, Pospelova TV, Julien JP (1994) AP-1 and Krox-24 transcription factors activate the neurofilament light gene promoter in P19 embryonal carcinoma cells. Cell Growth Differ 5:187–196.

Rosen KM, McCormack MA, Villa-Komaroff L, Mower GD (1992) Brief visual experience induces immediate early gene expression in the cat visual cortex. Proc Natl Acad Sci U S A 89:5437–5441.

Rosenzweig MR, Bennett EL, Diamond MC (1972) Brain changes in relation to experience. Sci Am 226:22–29.

Scott JD (1991) Cyclic nucleotide-dependent protein kinases. Pharmacol Ther 50:123–145.

Sharp FR, Sagar SM, Swanson RA (1993) Metabolic mapping with cellular resolution: c-fos vs. 2-deoxyglucose. Crit Rev Neurobiol 7:205–228.

Shaywitz AJ, Greenberg ME (1999) CREB: a stimulus-induced transcription factor activated by a diverse array of extracellular signals. Annu Rev Biochem 68: 821–861.

Shen K, Meyer T (1999) Dynamic control of CaMKII translocation and localization in hippocampal neurons by NMDA receptor stimulation. Science 284:162–166.

Sheng M, Dougan ST, McFadden G, Greenberg ME (1988) Calcium and growth factor pathways of c-fos transcriptional activation require distinct upstream regulatory sequences. Mol Cell Biol 8:2787–2796.

Silveira LC, de Matos FM, Pontes-Arruda A, Picanco-Diniz CW, Muniz JA (1996) Late development of Zif268 ocular dominance columns in primary visual cortex of primates. Brain Res 732:237–241.

Stahelin RV, Digman MA, Medkova M, Ananthanarayanan B, Melowic HR, Rafter JD, Cho W (2005) Diacylglycerol-induced membrane targeting and activation of protein kinase Cepsilon: Mechanistic differences between PKCdelta and epsilon. J Biol Chem.

Stellwagen D, Beattie EC, Seo JY, Malenka RC (2005) Differential regulation of AMPA receptor and GABA receptor trafficking by tumor necrosis factor-alpha. J Neurosci 25:3219–3228.

Steward O, Worley PF (2001a) A cellular mechanism for targeting newly synthesized mRNAs to synaptic sites on dendrites. Proc Natl Acad Sci U S A 98:7062–7068.

Steward O, Worley PF (2001b) Selective targeting of newly synthesized Arc mRNA to active synapses requires NMDA receptor activation. Neuron 30:227–240.

Steward O, Wallace CS, Lyford GL, Worley PF (1998) Synaptic activation causes the mRNA for the IEG Arc to localize selectively near activated postsynaptic sites on dendrites. Neuron 21:741–751.

Sweatt JD (2001) The neuronal MAP kinase cascade: a biochemical signal integration system subserving synaptic plasticity and memory. J Neurochem 76:1–10.

Taha S, Hanover JL, Silva AJ, Stryker MP (2002) Autophosphorylation of alphaCaMKII is required for ocular dominance plasticity. Neuron 36:483–491.

Tang YP, Shimizu E, Dube GR, Rampon C, Kerchner GA, Zhuo M, Liu G, Tsien JZ (1999) Genetic enhancement of learning and memory in mice. Nature 401: 63–69.

Thiel G, Schoch S, Petersohn D (1994) Regulation of synapsin I gene expression by the zinc finger transcription factor zif268/egr-1. J Biol Chem 269:15294–15301.

Tighilet B, Hashikawa T, Jones EG (1998) Cell- and lamina-specific expression and activity-dependent regulation of type II calcium/calmodulin-dependent protein kinase isoforms in monkey visual cortex. J Neurosci 18:2129–2146.

Trachtenberg JT, Chen BE, Knott GW, Feng G, Sanes JR, Welker E, Svoboda K (2002) Long-term in vivo imaging of experience-dependent synaptic plasticity in adult cortex. Nature 420:788–794.

Vaillant AR, Zanassi P, Walsh GS, Aumont A, Alonso A, Miller FD (2002) Signaling mechanisms underlying reversible, activity-dependent dendrite formation. Neuron 34:985–998.

van Praag H, Kempermann G, Gage FH (2000) Neural consequences of environmental enrichment. Nat Rev Neurosci 1:191–198.

Volkmar FR, Greenough WT (1972) Rearing complexity affects branching of dendrites in the visual cortex of the rat. Science 176:1145–1147.

Wallace CS, Withers GS, Weiler IJ, George JM, Clayton DF, Greenough WT (1995) Correspondence between sites of NGFI-A induction and sites of morphological plasticity following exposure to environmental complexity. Brain Res Mol Brain Res 32:211–220.

Wisden W, Errington ML, Williams S, Dunnett SB, Waters C, Hitchcock D, Evan G, Bliss TV, Hunt SP (1990) Differential expression of immediate early genes in the hippocampus and spinal cord. Neuron 4:603–614.

Wong WK, Ou XM, Chen K, Shih JC (2002) Activation of human monoamine oxidase B gene expression by a protein kinase C MAPK signal transduction pathway involves c-Jun and Egr-1. J Biol Chem 277:22222–22230.

Worley PF, Christy BA, Nakabeppu Y, Bhat RV, Cole AJ, Baraban JM (1991) Constitutive expression of zif268 in neocortex is regulated by synaptic activity. Proc Natl Acad Sci U S A 88:5106–5110.

Xia Z, Storm DR (2005) The role of calmodulin as a signal integrator for synaptic plasticity. Nat Rev Neurosci 6:267–276.

Yin JC, Tully T (1996) CREB and the formation of long-term memory. Curr Opin Neurobiol 6:264–268.

Yoshimura M, Cooper DM (1993) Type-specific stimulation of adenylylcyclase by protein kinase C. J Biol Chem 268:4604–4607.

Yuste R, Majewska A, Cash SS, Denk W (1999) Mechanisms of calcium influx into hippocampal spines: heterogeneity among spines, coincidence detection by NMDA receptors, and optical quantal analysis. J Neurosci 19:1976–1987.

Zhang F, Vanduffel W, Schiffmann SN, Mailleux P, Arckens L, Vandesande F, Orban GA, Vanderhaeghen JJ (1995) Decrease of zif-268 and c-fos and increase of c-jun mRNA in the cat areas 17, 18 and 19 following complete visual deafferentation. Eur J Neurosci 7:1292–1296.

Zhu XO, Brown MW, McCabe BJ, Aggleton JP (1995) Effects of the novelty or familiarity of visual stimuli on the expression of the immediate early gene c-fos in rat brain. Neuroscience 69:821–829.

Zimmermann G, Taussig R (1996) Protein kinase C alters the responsiveness of adenylyl cyclases to G protein alpha and betagamma subunits. J Biol Chem 271: 27161–27166.

9

The Molecular Biology
of Sensory Map Plasticity
in Adult Mammals

LUTGARDE ARCKENS

Laboratory of Neuroplasticity and Neuroproteomics, Katholieke Universiteit Leuven,
Naamsestraat 59, B-3000 Leuven, Belgium

1. Introduction

The visual cortex of mammals is immature at birth, both anatomically and physiologically, and develops gradually in the first weeks and months of postnatal life. During this period, visual stimulation induces specific patterns of neuronal activity in the central visual system that contribute to the establishment of visual perception and visually guided behaviour. Visual neurons will eventually be tuned to adequately interpret the information encoded in environmental stimulation patterns. To achieve this, a substantial anatomical organization takes place. Strengthening, remodelling and eliminating synapses help in creating the adult-specific neuronal circuitry.

The visual cortex shares a common feature with the other sensory cortical areas devoted to touch and hearing in that they all present their respective sensory epithelial surfaces in a topographic manner. Visual cortex is retinotopically organized, that is neighbouring cortical regions respond to neighbouring points in visual space. Similarly, neighbouring skin sites are somatotopically presented in somatosensory cortex, whereas auditory cortex is organized according to tonotopic coordinates. Two decades of research have clearly shown that these orderly maps in neocortex are not static entities but instead remain malleable throughout an animal's life. Evidence for such plasticity in the mature sensory systems comes from sensory deprivation or specific sensory stimulation of these systems. These manipulations force the cortical neurons to deal with a specifically new sensory environment and cause changes of their molecular composition and, consequently, their signalling properties (Kaas, 1991; Donoghue, 1995; Buanomano and Merzenich, 1998).

Bilateral retinal lesions initially lead to functional deficits. Immediately after lesioning the retina, neurons in primary visual cortex (striate cortex, area 17) with a receptive field situated in the center of the damaged region in

R. Pinaud, L. A. Tremere, P. De Weerd (Eds.), Plasticity in the Visual System:
From Genes to Circuits, 181-203, ©2006 Springer Science + Business Media, Inc.

the retina become silenced, i.e. they no longer respond to stimulation within their original receptive field (Kaas, 1991; Kaas et al., 1990; Gilbert & Wiesel, 1992; Chino et al., 1992, 1995; Eysel, 1992; Schmid et al., 1996; Dreher et al., 2001). Striate neurons at the edges of the lesion projection zone (LPZ) however, remain responsive but display enlarged receptive fields that are slightly shifted towards the intact retina surrounding the lesion (Gilbert & Wiesel, 1992). In ensuing weeks and months, recovery of the lost functions occurs. Much larger receptive field shifts take place, which can ultimately result in the complete filling-in of the LPZ (Gilbert & Wiesel, 1992; Chino et al., 1992; 1995). This extensive remodelling of cortical topography thus involves an enlargement of the representation of perilesion retina at the expense of the cortical area previously dedicated to the lesioned retina (Gilbert & Wiesel, 1992). This functional reorganization of the cortical retinotopic maps in primary visual area 17 is accompanied over time by biochemical and morphological modifications, involving the LPZ, surrounding and remote cortical zones. The cortical long-range horizontal connections of pyramidal cells are commonly considered the most likely structural mediators of topographic map reorganization in visual cortex, first through the 'unmasking' of existing sub-threshold connections, in a second phase by experience-dependent strengthening of their synapses and finally through sprouting of new collaterals (Darian-Smith & Gilbert, 1994, 1995; Das & Gilbert, 1995a,b; Chino et al., 1995; Chino, 1999; Calford, 2002; Calford et al., 2003). How these rearrangements in strength and anatomy of cortical connections depend on changes in the repertoire of genes and proteins, expressed by the relevant brain cells, has been the subject of intensive investigations and the first molecular models of cortical plasticity are now emerging.

2. Overview of known molecular events

During topographic map reorganization several molecular actions take place. Many genes are switched on or off and a number of protein products undergo posttranslational modifications. Tables 9.1 and 9.2 summarize today's knowledge on the temporal and spatial character of the changes in molecular expression patterns as induced by retinal lesions in cat primary visual cortex. The currently identified genes and proteins are involved in different processes like neurite growth, synapse maturation, signalling pathways, active membrane processes, neurotransmitter release or uptake, metabolic pathways and enzymatic changes. Two main approaches have been used to gain insight in the molecular basis of cortical plasticity. Table 9.1 summarizes the data gathered with a first powerful and extensively used approach where the problem was tackled by trying to identify selected gene families based on the existing literature. Table 9.2 deals with the results from an mRNA differential display approach, a complementary method that allowed a blind screening of the total genome.

TABLE 9.1. Summary of the temporal and spatial changes in molecular expression in the LPZ and remote visual cortex of adult cats with retinal lesions.

Time post-lesion	3 days LPZ	R	1 week LPZ	R	2 weeks LPZ	R	4 weeks LPZ	R	2-3 months LPZ	R	1 year LPZ	R	>2 years LPZ	R	Source	Methods used
Neurotransmitters, neuromodulators, their transporters and receptors																
Glutamate					↓	↑	↓	↑	/	↑					Arckens et al, 2000	ICC
							/								Qu et al., 2003	WB
															Massie et al., 2003	Microdialysis
																ISH
Glial glutamate transporters GLAST & GLT-1	/				↑	↓	↑	↓							Massie et al., 2003	WB
																Microdialysis
Glutamate R GluR1	↓				↓	/	↓	/							Van Damme et al., 2002	ISH
															Van Damme, 2003	ICC
																WB
GluR2	↓				↓	/	↓	/								
NR1	↓				↓	/	↓	/								
NR2A	/				↑	/	/	/								
NR2B	/				↓	/	↑	/								
GABA							/	↑	/	↑					Massie et al., 2003	microdialysis
GAD 65 and 67					↓				/						Rosier et al., 1995	ICC
															Arckens et al., 1998	ISH
GABA –A and -B R					/										Rosier et al., 1995	RA
Noradrenaline						↑									Qu et al., 2000	HPLC
Dopamine						↑										
DOPAC and HVA					↓											
Serotonin					↓											
5-HIAA					↓	↑										

(Continued)

TABLE 9.1. Summary of the temporal and spatial changes in molecular expression in the LPZ and remote visual cortex of adult cats with retinal lesions.—(*cont'd*)

Time post-lesion	3 days		1 week		2 weeks		4 weeks		2-3 months		1 year		>2 years		Source	Methods used
	LPZ	R	LPZ	R	LPZ	R	LPZ	R	LPZ	R	LPZ	R	LPZ	R		
Immediate early genes																
Fos	↓		↓		↓		↓		↓		↓				Arckens et al., 2000 (but see Obata et al., 1999)	ICC ISH cPCR
Zif268	↓		↓		↓		↓		↓		↓				Van der Gucht et al., 2003	
CREB	↑		↑		↑		↑		↑		↑		↑		Obata et al., 1999	ICC
p-CREB													↑			
MEF2	↑				↑	↓	↑	↓							Leysen et al., 2004	DD-RTPCR, ICC, WB, sqPCR
Growth factors and their receptors																
BDNF	↑		↑		↑		↑		↑		↑		↑		Obata et al., 1999	ICC
NT-3	↑		↑		↑		↑		↑		↑		↑			
NGF	↑		↑		↑		↑		↑		↑		↑			
IGF1	↑		↑		↑		↑		↑		↑		↑			
P75	↑		↑		↑		↑		↑		↑		↑			
TrKA	/		/		/		/		/		/		/			
TrKB	↑		↑		↑		↑		↑		↑		↑			
TrKC	↑		↑		↑		↑		↑		↑		↑			
Proteins related to neurite development and synapse maturation																
Synapsin 1,2			↑		↑		↑		↑		↑		↑		Obata et al., 1999	ICC
Synaptophysin	/		/		/		/		/		/		/			
MAP-2	↑		↑		↑		↑		↑		↑		↑			

TABLE 9.1. Summary of the temporal and spatial changes in molecular expression in the LPZ and remote visual cortex of adult cats with retinal lesions.—(*cont'd*)

Time post-lesion	3 days		1 week		2 weeks		4 weeks		2-3 months		1 year		>2 years		Source	Methods used
	LPZ	R	LPZ	R	LPZ	R	LPZ	R	LPZ	R	LPZ	R	LPZ	R		
GAP-43	/		/		/		/		/		/		/		Obata et al., 1999, Baekelandt et al., 1996	ICC, ISH (personal observation)
Enzymes																
CaMkinase II alfa			↑								↑				Obata et al., 1999 but see Van den Bergh et al., 2003	ISH WB ICC
CaMkinase II alfa autophos-phorylation	/				↑		(↑)								Van den Bergh et al., 2003	WB
Protein kinase C beta 1 and beta 2	/		/		/		/		/		/				Personal observation	ISH
Cu/Zn superoxide dismutase	↑														Leysen et al., 2002	DD-RTPCR, sqPCR
Chaperones																
Cyclophilin A	↓		↓		↓		↓		↓		↓				Arckens et al., 2003	DD-RTPCR, ISH, ICC, WB, cPCR
GAPDH	/				/		/								Van der Gucht et al., 2003; Arckens et al., 2003	cPCR, WB

blank cells: conditions not investigated; / no significant effect of the lesion, ↑ significantly increased expression compared to normal cat, ↓ significantly decreased expression compared to normal cat, LPZ: lesion projection zone in area 17, R: Remote, peripheral portion of area 17, arrows in italics: post-lesion survival time not clearly discussed; ICC: immunocytochemistry, ISH: in situ hybridization, WB: Western blotting, cPCR: competitive PCR, sqPCR: semi-quantitative PCR, DD-RTPCR: mRNA differential display

TABLE 9.2. Summary of differentially expressed clones identified by DDRT-PCR 3 and 14 days after the induction of retinal lesions in adult cat.

Condition	Identification	Accession n°	% ID	Species
LPZ 3d	Glucose-6 phosphate dehydrogenase	M24470	72 % (216/296)	Human
	Cu/ZnSOD	K00065	74 % (180/241)	Human
	MEF2A	NM005587	79 % (780/987)	Human
	Retinoic acid receptor protein	HS05227	85% (141/166)	Human
	EAAT-2 (GLT1)	U03505	98 % (225/230)	Human
LPZ 14d	Protocadherin 42	L11370	90% (205/226)	Human
	Sodium iodide symporter	D87920	91% (153/168)	Human
	Sop2p-like protein	Y08999	93% (113/121)	Human
	Lipoamide dehydrogenase 2	J03620	88% (210/236)	Human
	EAAT-2 (GLT1)	D85884	87% (114/130)	Human
R 3d	Neurofibromin	M89914	88 % (152/173)	Human
	SCIP	M72711	89 % (100/112)	Rat
	Thymosin beta-4	X16053	87 % (141/161)	Mouse
	Beta-adaptin	M34175	75 % (169/226)	Human
	RET II	X15786	85 % (207/241)	Human
	GST1-GTP binding protein	X17644	82,5 % (80/97)	Human
	Cyclophilin A	AF023860	87 % (311/355)	Macaque monkey
R 14d	PI3-kinase p85-α subunit	M61906	94% (173/184)	Human
	KIAA0534 protein	AB011106	92% (167/181)	Human
	Latrophilin 3 splice variant	AF111090	91% (166/181)	Bovine
	N-ethylmaleimide-sensitive factor	U03985	93% (292/313)	Human
	Eukaryotic initiation factor 4A-II (eIF4A2)	D30655	97% (313/321)	Human
	Diacylglycerol kinase gamma	D26135	91% (244/268)	Human
	Dystonin	U31850	96% (335/348)	Human
	Beta spectrin	M74773	92% (155/168)	Mouse
	Transportin 2	AF019039	86% (145/169)	Human

% ID, the identity obtained by nucleotide database search; Species, the species with which sequence similarity was highest; LPZ: higher mRNA expression in the lesion projection zone; R: higher mRNA expression in remote peripheral visual cortex; numbers in parenthesis: number of identical nucleotides over total length of the cDNA sequence

2a. Involvement of Several Transcription Factors in Cortical Plasticity

Various transcription factors have already been implicated in retinotopic map plasticity and brain plasticity in general (Kaczmarek & Nikolajew, 1990; Kaplan et al., 1996, Kaczmarek & Chaudhuri, 1997, Obata et al., 1999, Arckens et al., 2000b, Pinaud, 2004). Members of four distinct transcription factor families have been unambiguously correlated to cortical plasticity. With *c-fos* belonging to the AP-1 family, *zif268* to the zinc finger family, *creb* to the ATF/CREB family and *mef2* to the MADS box family of transcription factors, the extent by which the regulation of gene expression is significantly altered after sensory deprivation as induced by retinal lesions became apparent. Some of these transcription factors involved in map plasticity also participate in other protein expression-dependent forms of brain plasticity like long-term potentiation and long-term memory (Hughes and Dragunow, 1995; Herdegen and Leah, 1998). Next to being markers for cortical neuronal activity, transcription factors emerge as components in a cascade of events that leads to long-term changes in cortical circuitry. Evidently, the complex and overlapping expression patterns of the different transcription factors further suggest that they do not function individually in neurons. Rather they will work in concert with a network of transcription factors to exert specific functions (for review see Hughes and Dragunow, 1995; Herdegen and Leah, 1998).

CREB (Obata et al., 1999) and the MEF2 transcription factors (Leysen et al., 2004) showed augmented expression in the LPZ of primary visual area 17. In contrast, the immediate early genes *c-fos* and *zif268* displayed a clearly decreased expression within the LPZ (Arckens et al., 2000b; Van der Gucht et al., 2003). Without doubt, charting the expression of the specific target genes of each of these transcription factors, the so-called 'late-response genes', will contribute to our understanding of the molecular mechanisms of adult brain plasticity. As an example, several neuron-specific genes, like synapsin 2, neurogranin, and VIP, contain putative MEF2 sites in their upstream regulatory regions and some have indeed been shown to be activated by MEF2, e.g. the NR1 subunit of NMDA type glutamate receptors and the cytoskeleton neurofilament protein (Chin et al., 1994; Sato et al., 1995; Hahm and Eiden, 1998, Krainc et al., 1998; Skerjanc and Wilton, 2000). Recently, Western blotting experiments already provided evidence for plasticity-related changes in NR1 subunit expression in the LPZ of area 17 in adult cat (Van Damme et al., 2002). Obata et al. (1999) also showed a clear effect of retinal lesions on synapsin 1 and 2 immunoreactivity within 3 days. Here *zif268* could also come into play since Thiel and colleagues (1994) showed that *zif268* too could regulate synapsin 1 gene expression.

Remarkable about the modulation of the transcription factor expression in the LPZ of area 17 of animals with retinal lesions is the sustained deviation of *c-fos, zif268* and CREB expression levels and the emergence of changes in the phosphorylation state of CREB as late as two years after trauma (Obata et al., 1999; Arckens et al., 2000b). These long-term effects on the regulation of gene transcription may indicate that the related brain regions remain in demand of dynamic adjustments for a long post-traumatic time, an observation also made for the auditory system upon acoustic trauma (Michler and Illing, 2003).

2b. Altered Neurotransmitter Systems – Changes in the Excitation/inhibition Balance

A variety of studies have suggested that the balance between excitation and inhibition modulates plastic responses in the central nervous system (CNS) after peripheral or central injuries (Garraghty et al., 1991; Jones, 1993; Feldman et al., 1999; Arckens et al., 1998, 2000a, Myers et al., 2000; Boroojerdi et al., 2001; Tighilet & Lacour, 2001; Tremere et al., 2001; Zepeda et al., 2004). Reductions of cortical inhibition through a decrease in the availability of GABA or an increase in glutamate transmission, or a combinatorial effect, have been suggested as possible mechanisms underlying functional reorganization. Changes at the amino acid level could reflect the activation of adaptive pre-synaptic mechanisms, whereas modification in the receptor subunits could be related to post-synaptic responses that lead to changes in the excitation-inhibition neurotransmission balance. Only few studies correlate time-dependent changes in neurotransmitter systems with functional recovery, although they may reflect biochemical adjustments involved in shaping neuronal receptive fields during functional reorganization.

Following retinal lesions, the release and uptake of inhibitory and excitatory neurotransmitters in cat primary visual cortex are re-regulated in a complex and multifaceted manner. Microdialysis revealed reduced glutamate levels within the LPZ of area 17 (Qu et al., 2003, Massie et al., 2003a,b). This decrease of the excitatory amino acid appeared specific because contents of non-neurotransmitter amino acids, as measured in parallel, did not change in a similar magnitude, suggesting that retinal lesions do not affect all amino acids in the same way, as would be expected if changes in general metabolism were responsible. The reduced neurotransmitter levels were the result of a decreased synthesis/release of the amino acid glutamate in the first weeks post-lesion. Indeed, an effect on neurotransmitter uptake was only observed in animals that survived longer (Massie et al., 2003a), as revealed through the investigation of the expression of glial glutamate transporter molecules, and uptake inhibition experiments in which these transporters were blocked with L-trans-pyrrolidine-3,4-dicarboxylic acid (PDC), a transportable glutamate analogue that produces the most selective and potent inhibition of glutamate uptake (Bridges et al., 1991; Thomsen et al., 1994).

Within the first weeks, altered glutamate receptor subunit expression was also observed in the LPZ. The number of AMPA receptors appeared reduced, since both the GluR1 and GluR2 subunit levels decreased shortly after making the retinal lesions, with a partial restoration of normal expression levels at 30 days post-lesion (Van Damme et al., 2002). Alternatively, the GluR2 down-regulation may be interpreted as an increase in Ca^{2+}-permeable AMPA subunit receptors in the LPZ (Hollmann et al., 1991). Increased Ca^{2+} influx through GluR2-negative AMPA receptors could then control the phosphorylation of AMPA subunits, thereby triggering long-lasting changes in synaptic efficacy or connectivity (Jonas et al., 1994).

Different observations were made for the NMDA type glutamate receptors in cat area 17 after the induction of retinal lesions. The subunit composition of the NMDA receptors changed drastically with post-lesion survival time. The total number of NMDA receptors looked lower in the LPZ due to a general down-regulation of NR1, the mandatory subunit of the NMDA receptor. We observed the lowest level for NR1 at 14 days post-lesion, which coincides in time with the presence of a substantial functionally silent LPZ. In contrast, the NR2A subunit expression peaked at 14 days post-lesion in the LPZ, whereas the NR2B subunit did at 30 days corresponding to the phase in the reorganization in which glutamate levels had already substantially recovered (Arckens et al., 2000a; Massie et al., 2003a; Qu et al., 2003) This parallel increase in glutamate probably reflects excitatory adjustments at the pre- and postsynaptic level, pointing to an increased excitability within the LPZ from 2 to 4 weeks post-lesion on. The restoration of approximately normal NR1 levels 30 days post-lesion also correlates well in time with such a progression of the functional reorganization.

Thus, the excitatory neurotransmitter system, including the amino acid glutamate, the glial glutamate transporters and the NMDA and AMPA receptors, participates intimately in the synaptic changes that contribute to functional reorganization after peripheral lesions.

Despite intensive investigation of the inhibitory neurotransmitter system in animals with retinal lesions, the findings for GABA in the LPZ are much less striking as compared to glutamate. GABA immunocytochemistry, $GABA_A$ and $GABA_B$ receptor autoradiography or analysis of extracellular fluid GABA concentrations could not reveal any substantial GABA-related changes within the LPZ (Rosier et al., 1995; Arckens et al., 2000a, Massie et al., 2003b). Only GAD, the enzyme that synthesizes GABA out of glutamate, displayed a lesion-dependent immunoreactivity pattern. The decrease in detectable GAD-immunoreactive terminals was interpreted as a decreased synaptic inhibition, possibly through a hampered axonal transport of the enzyme, since the number of GAD-immunoreactive cell bodies simultaneously increased. With GAD_{65} being responsible for the GABA synthesis-on-demand at synaptic sites, such a restrain on the transport of GAD_{65} could indeed lead to local, subtle changes in pre-synaptic, vesicular GABA supply and thus inhibition (Soghomonian and Martin, 1998). In accordance, Engel and colleagues (2001) have recently shown that GABA metabolism can determine inhibitory synaptic strength. Therefore pre-synaptic GABA content might be a regulated mechanism for synaptic plasticity.

The critical role of the central neuromodulatory systems in maintaining and shaping the neuronal network of the cerebral cortex is not yet clearly understood. The catecholamine and serotonin content of the LPZ in area 17 of animals with retinal lesions has been investigated by reverse phase high performance liquid chromatography and electrochemical detection (Qu et al., 2000). The dopamine metabolites homovallinic acid (HVA) and dihydroxyphenilacetic acid (DOPAC) appeared reduced 14 days post-trauma, which could point to a change in dopamine metabolism in the LPZ. Levels of serotonin and its metabolite 5-HIAA were also decreased in the LPZ. Lack of knowledge on the temporal aspect of the neuromodulator changes and on how the specific receptors react to retinal lesions makes it difficult to speculate on the specific mechanisms by which dopamine and serotonin might modulate cortical reorganization. Since in normal animals their respective receptors are distributed throughout supra- and infragranular layers in sensory cortices (Dyck and Cynader, 1993; Rakic and Lidow, 1995), serotonin and dopamine might modulate the strength of cortico-cortical projections. Serotonin modulates LTP/LTD-like processes in developing visual cortex (Kojic et al., 2000, Edagawa et al., 2001). Likewise, transient changes in the serotonin levels in the LPZ could thus facilitate LTP/LTD-like adaptations of synaptic contacts upon sensory deprivation. Indeed, in primary auditory cortex dopamine-induced remodelling of the sound-frequency representations has also been described (Stark and Scheich, 1997, Bao et al., 2001).

2c. Growth Factors

Lesion-induced changes in cortical topography are furthermore accompanied by a rapid increase in the expression of neurotrophins, including BDNF and NT-3, and relevant receptors (Obata et al., 1999). Neurotrophins are important regulators of synaptic development and plasticity in both the central and peripheral nervous system. Neurotrophins can modulate synaptic transmission at the pre- and postsynaptic level in a target-specific fashion. In the visual cortex, BDNF can influence GABAergic intracortical inhibitory neurons and serotonergic afferents from the Raphe nucleus (Berardi et al., 2003). The observation that the elevation of the neurotrophin levels is sustained for up to two years after induction of retinal lesions may reflect the fact that even though visually driven activity restores in the cortical scotoma, the level of activity never fully returns to that of the surrounding cortex and some imbalance of activity persists (Das and Gilbert, 1995b).

2d. Synapse Efficacy, Axon Growth, Dendritic Sprouting

The sculpting, maintenance and remodelling of axonal and dendritic arbors happens under influence of an amalgam of intracellular and extracellular factors. Synapsin I has been implicated to play a role in the establishment of synaptic contacts during the development of the mammalian CNS as well as during the process of re-innervation within the adult brain after partial deafferentation (De Camilli et al., 1983, Brock and Callaghan, 1987, Moore and Bernstein, 1989, Melloni et al., 1994). The detection of changes in synapsin I levels has been used as parameter for changes in the number of nerve terminals within a certain brain region. Moreover it has been reported that the appearance of detectable synapsin I levels in developing and sprouting synapses would coincide with the acquisition of function by those synapses. The persistent upregulation of synapsin immunoreactivity in the LPZ thus seems to correlate with ongoing changes in synaptic density and activity (Obata et al., 1999; Walaas et al., 1988, Moore and Bernstein, 1989, Melloni et al., 1994).

Dendritic and axonal sprouting has also been associated with MAP2 and GAP-43 expression respectively. MAP2 immunoreactivity has been considered a good marker for dendrites based on experimental evidence that correlates dendritic growth with an increase in MAP2 immunoreactivity (Philpot et al., 1997; Sanchez et al., 2000; Bury and Jones, 2002). At every time point studied, the LPZ in area 17 of retinal lesion cats displayed a modified MAP2 immunolabeling (Obata et al., 1999). The MAP2 upregulation therefore suggests dendritic sprouting as a possible mechanism by which cortical reorganization occurs (Zepeda et al., 2004). GAP43 is mainly present in growth cones. Its expression after injury has been correlated with axonal regeneration (Baekelandt et al., 1994; Stroemer et al., 1995). Nevertheless axonal sprouting does not solely depend on GAP43 levels (Szele et al., 1995). The absence of modulation of GAP43 immunoreactivity cannot rule out that axonal sprouting takes place at some point in the recovery process (Baekelandt et al., 1996; Obata et al., 1999). Indeed axonal remodelling of horizontal connections has been shown to occur in the LPZ of area 17 albeit from 8 months post-lesion on (Darian-Smith and Gilbert, 1994).

2e. Systematic Screening for Plasticity-related Molecules

Differential display is an mRNA fingerprinting technique allowing the systematic screening for differences in gene expression between experimental conditions (Liang and Pardee, 1992). The possibility to detect up- and down-regulated genes makes it an attractive screening method. We have used differential display to compare bi-directionally the repertoire of genes expressed in the LPZ and remote, non-deprived visual cortex of adult cats with small homonymous central retinal lesions (Arckens et al., 2003, Leysen et al., 2004). Differential display qualified as the method of choice since no *a priori* assumptions had to be made as to the possible identity of the genes of interest. The twenty-seven genes isolated (listed in Table 9.2) execute diverse functions in mammalian brain. Whereas some showed a higher expression level in the LPZ of area 17, even more displayed a lower expression in the LPZ compared to remote visual cortex 3 to 14 days post-lesion. It is tempting to categorize these genes as plasticity activator and suppressor genes, like the memory activator and suppressor genes that either support or put inhibitory constraints on the storage of long-term memory (Abel et al., 1998). For 4 out of 27 genes (Table 9.2), the glial glutamate transporter EAAT2, cyclophilin A, MEF2A and Cu/Zn super oxide dismutase (Cu/Zn SOD), we have gathered detailed experimental evidence on the temporal and spatial regulation of the expression of these genes and their protein products in relation to cortical plasticity (Leysen et al., 2000, Massie et al. 2003a, Arckens et al., 2003, Leysen et al. 2004).

The plasticity-related character of the expression of the other 23 genes reported in Table 9.2 should be interpreted with caution, but based on their cellular functions, several of these genes could act in concert with molecules listed in Table 9.1. SCIP (suppressed camp-inducible POU) or Oct-6 is a POU domain transcription factor with a temporally and spatially regulated expression during CNS development, and appears highly expressed in cortical layer V neurons, that project to subcortical brain regions (Frantz et al., 1994). Retinoic acid receptors contribute to LTP and LTD, at least in hippocampus, implicating that vitamin A and its derivatives, the retinoids, might contribute to synaptic plasticity (Chiang et al., 1998). The fine regulation of retinoid-mediated gene expression is also important for optimal brain functioning and higher cognition (Etchamendy et al., 2003). Thymosin beta-4 is thought to participate in neurite outgrowth by sequestration of G-actin necessary for growth cone extension (Vartiainen et al., 1996; Carpintero et al., 1999). Beta-adaptin plays a role in receptor-mediated endocytosis (Fergusson, 2001). N-ethylmaleimide-sensitive factor (NSF) interacts directly and selectively with the intracellular C-domain of the GluR2 subunit of the AMPA receptor, and this interaction can be enhanced by BDNF. Blocking NSF results in reduced AMPA receptor-mediated synaptic transmission (Song et al., 1998; Narisawa-Saito et al., 2002). Based on these functions, the genes listed in Table 9.2 have potential as regulators of brain plasticity and thus clearly deserve attention in future research.

3. Changes inside and outside the lesion projection zone as a function of post-lesion survival time

Today's knowledge on the temporally and spatially specific modulations of plasticity-associated genes and proteins (Table 9.1 and 9.2) implies two important aspects for a better understanding of the mechanism of adult plasticity. Firstly, the molecular response of the LPZ at 14 days post-lesion is different, as compared to earlier or later time points. Ca^{++}/Calmodulin dependent protein kinase II alpha only showed significantly elevated levels of autophoshorylation 14 days after the induction of retinal lesions (Van den Bergh et al., 2003a). Similarly, the impact of retinal lesions on AMPA-type glutamate receptor subunit expression was also most intense after a 14 day-recuperation period, but nevertheless present form 3 to 30 days post-lesion (Van Damme et al., 2002). Related data on extracellular fluid glutamate levels and the expression of glial high-affinity Na^+/K^+-dependent glutamate transporters suggested a shift in the mechanism that controls the activation of glutamate receptors, from a differential synthesis/release of glutamate to a differential re-uptake (Massie et al., 2003a). Likewise, Obata and colleagues (1999) reported an effect of retinal lesions on neurotrophins in two waves, one starting a few days after lesioning and ending at 1 month, and a second one from 3 months to more than two years post-lesion. What mediates this wave-like feature of the molecular effects of retinal lesions is not yet clear and calls for more applied research to define the exact molecular mechanisms of cortical plasticity. The understanding of the exact sequence of different mechanisms could prove useful for the development of novel approaches for rehabilitation after central or peripheral lesions of the nervous system based on goal-directed pharmacological interventions to boost the recovery capacity intrinsically present in sensory cortex.

Secondly, we established that the cortical reorganization after retinal lesioning is certainly also driven by molecular changes occurring outside the LPZ, i.e. in remote non-deprived cortex (Qu et al., 2000, 2003, Massie et al., 2003a,b). Every neurotransmitter and neuromodulator that was investigated showed clearly increased levels in regions of area 17 devoted to peripheral vision as compared to control regions in normal animals (Table 9.1). The effect on extracellular fluid glutamate and GABA levels in remote area 17 turned out to increase with post-lesion survival time. The pronounced increase in glutamate could point to hyperexcitability and hyperactivity in remote visual cortex. The observation that also all the neuromodulatory systems were upregulated hints to a similar conclusion. Noradrenaline levels were specifically augmented, and as discussed by Pinaud (2000, 2004), noradrenergic input to the visual cortex regulates light-induced *zif268* expression. Therefore a quantitative investigation of the temporal and spatial modulations of *zif268* and *c-fos* expression in the visual cortex should deliver final and conclusive evidence for the reaction of the complete visual cortex to limited retinal lesions. An ongoing modification of the physiological state and the function of non-deprived visual cortex appear necessary to allow functional reorganization of the cortical circuitry. Ample evidence indicates that also after brain damage a profound reorganization occurs in the spared cerebral cortex, as a contribution to recovery. Reorganizational effects take place at perilesional sites but also in remote cortical regions and even in the contralateral hemisphere, and may be driven by reciprocal intracortical connections. Such remote effects on the brain as a whole following restricted cortical lesions suggest a different contribution of distinct perilesional areas to the process of recovery and compensation for lost functions (Eysel et al., 1999; Keyvani et al., 2002; Frost et al., 2003; Reinecke et al., 2003).

4. Involvement of primary and higher order sensory areas, supra- and infragranular layers, inputs from the dLGN and the superior colliculus

So far electrophysiological investigations elegantly revealed the time window in which the first faze of cortical reorganization takes place in area 17, with an immediate effect on receptive field (RF) size minutes after the lesion and a more persistent effect on RF position in ensuing weeks (Kaas et al., 1990; Gilbert and Wiesel, 1992; Chino et al., 1992, 1995; Eysel, 1992; Dreher et al., 2001). Moreover the distance in millimeter over which restoration of neuronal activity could be measured in visual cortex, as determined by the maximal shift of RFs towards the center of the LPZ, identified the horizontal connections as a valid structural basis for map plasticity, as was confirmed with tracer studies (Darian-

Smith and Gilbert, 1994; Das and Gilbert, 1995a; Calford et al., 2003). Nevertheless, during the investigation of the effect of retinal lesions on immediate early gene expression in cat visual cortex it became painstakingly clear that not only primary visual cortex is traumatized by the lesions. In no less than 10 visual areas containing a projection of the lesioned retinal zone, as predicted by the retinotopic maps of Rosenquist (1984), and expressing an appreciable basal *c-fos* or *zif268* level, a LPZ was detected as characterized by clearly decreased IEG expression levels (Arckens et al., 2000b). These LPZs, like the one in area 17, became smaller with post-lesion survival time, since they restored normal IEG levels with time from the border of the LPZ inwards. Furthermore, it became evident that the lesion affects not only the supragranular layers but also the infragranular layers. The fact that these different cortical layers seemed to achieve this in an independent fashion (Obata et al., 1999; Arckens et al., 2000a,b), could point to a laminar stratification of synaptic modifiability in cortex resulting in a layer-specific contribution to brain plasticity (Kaplan et al., 1996, Kaczmarek and Chaudhuri, 1997). These observations urge contemporary scientists to broaden the current view on how map plasticity might be brought about in neocortex. Future research should reveal whether higher order sensory areas react in concert with primary areas to retinal lesions. Do higher order areas rely on the restoration of function within primary areas or do higher order areas, at least partially, instruct primary visual cortex via feedback connections, in other words is map plasticity driven in a bottom-up or a top-down fashion. Additionally, there is the need for a concise and detailed analysis of the role of all six cortical layers in map plasticity (Diamond et al., 1994; Waleszcyk et al., 2003). Previous investigations almost exclusively focused on the role of the supragranular layers leaving us with the definite contribution of the horizontal connections in these layers but with a void on knowledge outside cortical layers II and III. Also, although the dLGN has been accurately excluded as an essential source for the guidance of the plastic phenomena that occur at the cortical level (Eysel et al., 1981; Eysel, 1982), the superior colliculus (SC) received little attention (Ortega et al., 1995). If higher order areas would be true players in cortical plasticity, the SC might turn out an important source of remaining and necessary visual input to drive cortical reorganization.

5. Scope and future of the field

To date, neuronal plasticity in the visual system has been studied essentially in cats, ferrets and primates, but the mouse visual cortex has been largely neglected. Although the mouse is becoming the animal of choice in many disciplines, it has not been used that often for the study of sensory processing in the central nervous system. Mice are nocturnal animals and are believed to rely more on tactile information and olfactory cues and merely use their visual system as an event detector. However, the interest in mouse visual cortex has increased over the past few years and the study of plasticity in the mouse visual cortex is becoming of particular interest because it might be advantageous to examine the involvement of specific proteins in cortical plasticity with the use of transgenic animals. Only recently, Hübener and colleagues (2003) started to perform retinal lesion experiments in mice. Optical imaging of intrinsic signals indicated that retinal lesions lead to a reorganization of the cortical retinotopic map in mice too (Schuett et al., 2002; Mrsric-Flogel et al., 2003). The retinotopic map in the mouse visual cortex thus offers a unique opportunity to study the structural and molecular mechanisms of neocortical reorganization, as a valuable complement to more traditional animal models. Investigation of knockout and transgenic mice should enable to discriminate a true causal relationship between specific molecular events and lesion-induced map reorganization from the simultaneous occurrence of otherwise independent phenomena. Genetic manipulation of protein expression using lentiviral vector technology, possibly in combination with RNA interference also holds great promise for future investigations concerning the molecular basis of brain plasticity (Van den Haute et al, 2003).

The last decade, comparison of mRNA expression levels between distinct physiological or developmental conditions, using innovative genomics approaches, e.g. mRNA differential display (Pazman et al., 2000, Yang et al., 2002), cDNA microarrays (Lomax et al., 2000; Pasinetti, 2001, Kobori et al. 2002; Lachance and Chaudhuri, 2004) or subtractive hybridization (Prasad and Cynader, 1994; Feng et al., 2004), has resulted in exciting advances in our understanding of the molecular mechanisms underlying the functioning of the mammalian brain. However, quantitative variations in messenger molecules do not necessarily reflect corresponding changes in the concentration of specific functional proteins, especially when the final step for the synthesis of a functional protein encompasses posttranslational modifications. Since proteins are the ultimate effector-molecules of the cell, our knowledge on brain plasticity could be truly advanced by orienting future research towards the protein level, taking advantage of methods in which no *a priori* assumptions have to be made as to the possible identity of the proteins of interest (Lubec et al., 2003; Kim et al., 2004; Choudhary and Grant, 2004). Fortunately, genome sequencing projects have led to the publication of an increasing number of genome sequence databases in recent years, creating the opportunity for automated identification of proteins by mass spectrometry, leading to an enormous boost in the area of protein identification, protein-protein interaction and protein characterization sciences.

Until recently, differential protein expression analysis comprised two-dimensional electrophoresis (2-DE) for quantitative protein display and mass spectrometric (MS) identification of the gene products of interest (for a recent review, see Pandey and Mann, 2000). In a neuroscience context, these techniques have thus far primarily been utilized for constructing protein expression maps of brain regions, like the hippocampus (Edgar et al., 1999). Our recent investigation on the identification of age-related visual cortex proteins (Van den Bergh et al., 2003b) nicely revealed the exciting potentials of an innovative protein visualization procedure, two-dimensional difference gel electrophoresis (2D-DIGE). The characteristic fluorescent labelling of the different protein samples and their separation within one gel make of 2D-DIGE a true differential protein display method (Van den Bergh and Arckens, 2004). A comparative 2D-DIGE analysis between the LPZ and the peripheral regions of primary and higher order visual areas in function of post-lesion survival time should lead to the identification of new critical players in cortical plasticity.

In conclusion, in past years, the use of state of the art technology has led to the identification of several proteins involved in different forms of brain plasticity. Nevertheless, many questions remain about the identity and the exact nature of the effector-molecules critical to brain circuitry rearrangements in an injury response. Only the identification of more positive and negative mediators of adult brain plasticity can eventually lead to the elucidation of the underlying molecular cascade which will boost the development of new pharmaceuticals useful in the treatments for sensory loss and brain damage and for goal-directed improvement of post-lesional recovery of brain function.

Acknowledgements: I wish to acknowledge all past and present members of the laboratory of Neuroplasticity and Neuroproteomics. Our work at the Katholieke Universiteit Leuven was supported by grants of the Queen Elisabeth Medical Foundation, the Fund for Scientific Research—Flanders (F.W.O.), and the Research Fund (OT 01/22) of the Katholieke Universiteit Leuven, Belgium.

References

T. Abel, K.C. Martin, D. Bartsch, and E.R. Kandel, Memory suppressor genes: inhibitory constraints on the storage of long-term memory, *Science* **279**, 338–341 (1998)

L. Arckens, U.T. Eysel, J.-J. Vanderhaeghen, G.A. Orban, and F. Vandensande, Effect of sensory deafferentation on the GABAergic circuitry of the adult cat visual system, *Neurosci.* 83, 381–391 (1998)

L. Arckens, G. Schweigart, Y. Qu, G. Wouters, D.V. Pow, F. Vandesande, U.T. Eysel, and G.A. Orban, Cooperative changes in GABA, glutamate and activity levels: the missing link in cortical plasticity, *Eur. J. Neurosci.* 12, 4222–4232 (2000a)

L. Arckens, E. Van der Gucht, U.T. Eysel, G.A. Orban, and F. Vandesande, Investigation of cortical reorganization in area 17 and nine extrastriate visual areas through the detection of changes in immediate early gene expression as induced by retinal lesions, J. *Comp. Neurol.* 425, 531–544 (2000b)

L. Arckens, E. Van der Gucht, G. Van den Bergh, A. Massie, I. Leysen, E. Vandenbussche, U.T. Eysel, R. Huybrechts, and F. Vandesande, Differential display implicates cyclophilin A in adult cortical plasticity, *Eur. J. Neurosci.* 18, 1–15 (2003)

V. Baekelandt, L. Arckens, W. Annaert, U.T. Eysel, G.A. Orban, and F. Vandesande, Alterations in GAP-43 and synapsin provide evidence for synaptic reorganization in adult cat dorsal lateral geniculate nucleus following retinal lesions, *Eur. J. Neurosci.* 6,754–765 (1994)

V. Baekelandt, U.T. Eysel, G.A. Orban, and F. Vandesande, Long-term effects of retinal lesions on growth-associated protein 43 (GAP-43) expression in the visual system of adult cats. *Neurosci Lett.* 208,113–6 (1996)

S. Bao, V. T. Chan, and M. M. Merzenich, Cortical remodelling induced by activity of ventral tegmental dopamine neurons, *Nature* 412, 79–83 (2001)

N. Berardi, T. Pizzorusso, G.M. Ratto, and L. Maffei, Molecular basis of plasticity in the visual cortex, *Trends in Neurosci.* 26, 369–378 (2003)

B. Boroojerdi, F. Battaglia, W. Muellbacher, and L.G. Cohen, Mechanisms underlying rapid experience-dependent plasticity in the human visual cortex, *Proc. Natl Acad. Sci. USA* 98, 14698–14701 (2001)

R.J. Bridges, M.S. Stanley, M.W. Anderson, C.W. Cotman, and A.R. Camberlin, Conformationally defined neurotransmitter analogues. Selective inhibition of glutamate uptake by one pyrrolidine-2,4-dicarboxylate diasteromer. *J. Med. Chem.* 34, 717–725 (1991)

T.O. Brock and P.O. Callaghan, Quantitative changes in the synaptic vesicle protein synapsin I and p38 and the astrocyte-specific protein glial fibrillary acidic protein are associated with chemical-induced injury to the rat central nervous system, *J. Neurosci.* 7, 931–942 (1987)

D.V. Buonomano and M.M. Merzenich, Cortical plasticity: from synapses to maps, *Annu. Rev. Neurosci.* 21, 149–186 (1998)

S.D. Bury and T.A. Jones, Unilateral sensorimotor cortex lesions in adult rats facilitate motor skill learning with the "unaffected" forelimb and training-induced dendritic structural plasticity in the motor cortex, *J. Neurosci.* 22, 8597–606 (2002)

M.B. Calford, Mechanisms for acute changes in sensory maps, *Adv. Exp. Med. Biol.* 508, 451–460 (2002)

M.B. Calford, L.L. Wright, A.B. Metha, and V. Taglianetti, Topographic plasticity in primary visual cortex is mediated by local corticocortical connections, *J. Neurosci.* 23, 6434–6442 (2003)

P. Carpintero, R. Anadon, S. Diaz-Regueira, and J. Gomez-Marquez, Expression of thymosin beta4 messenger RNA in normal and kainite-induced rat forebrain, *Neurosci.* 90, 1433–1444 (1999)

M.Y. Chiang, D. Misner, G. Kempermann, T. Schikorski, V. Giguere, H.M. Sucov, F.H. Gage, C.F. Stevens, and R.M. Evans, An essential role for retinoid receptors RARbeta and RXRgamma in long-term potentiation and depression, *Neuron* 21, 1353–1361 (1998)

L.-S. Chin, L. Li, and P. Greengard, Neuron-specific expression of the synapsin II gene is directed by a specific core promoter and upstream regulatory elements, *J. Biol. Chem.* 269, 18507–18513 (1994)

Y.M. Chino, The role of visual experience in the cortical topographic map reorganization following retinal lesions, *Restor. Neurol. Neurosci.* 15, 165–176 (1999)

Y.M. Chino, J.H. Kaas, E.L. Smith III, A.L. Langston, and H. Cheng, Rapid reorganization of cortical maps in adult cats following restricted deafferentation in retina, *Vision Res.* 32, 789–796 (1992)

Y.M. Chino, E.L. Smith III, J.H. Kaas, Y. Sasaki, and H. Cheng, Receptive-field properties of deafferented visual cortical neurones after topographic map reorganization in adult cats, *J. Neurosci.* 15, 2417–2433 (1995)

J. Choudhari and S.G.N. Grant, Proteomics in postgenomic neuroscience: the end of the beginning, *Nature Neurosci.* 7, 440–445 (2004)

C. Darian-Smith and C.D. Gilbert, Axonal sprouting accompanies functional reorganization in adult cat striate cortex, *Nature* 368, 737–740 (1994)

C. Darian-Smith and C.D. Gilbert, Topographic reorganization in the striate cortex of the adult cat and monkey is cortically mediated, *J. Neurosci.* 15, 1631–1647 (1995)

A. Das and C.D. Gilbert, Long-range horizontal connections and their role in cortical reorganization revealed by optical recording of cat primary visual cortex, *Nature* 375, 780–784 (1995a)

A. Das and C.D. Gilbert, Receptive field expansion in adult visual cortex is linked to dynamic changes in strength of cortical connections, *J. Neurophysiol.* 74, 779–792 (1995b)

P. De Camilli, R. Cameron, and P. Greengard, Synapsin I (protein I), a nerve terminal-specific phosphoprotein. I Its general distribution in synapses of the central and peripheral nervous system demonstrated by immunofluorescence in frozen and plastic sections, *J. Cell Biol.* 96:1337–1354 (1983)

M.E. Diamond, W. Huang, and F.F. Ebner, Laminar comparison of somatosensory cortical plasticity, *Science* 265, 1885–1888 (1994)

J.P. Donoghue, Plasticity of adult sensorymotor representations, *Curr. Opin. Biol.* 5, 749–754 (1995)

B. Dreher, W. Burke, and M.B. Calford, Cortical plasticity revealed by circumscribed retinal lesions or artificial scotomas, *Prog. Brain Res.* 13, 217–246 (2001)

R. H. Dyck and M. S. Cynader, Autoradiographic localization of serotonin receptor subtypes in cat visual cortex: transient regional, laminar, and columnar distributions during postnatal development, *J. Neurosci.* 13, 4316–4338 (1993)

Y. Edagawa, H. Saito, and K. Abe, Endogenous serotonin contributes to a developmental decrease in long-term potentiation in the rat visual cortex, *J. Neurosci.* 21(5), 1532–1537 (2001)

P.F. Edgar, J.E. Douglas, C. Knight, G.J.S. Cooper, R.L.M. Faull, and R. Kydd, Proteome map of the human hippocampus, *Hippocampus* 9, 644–650 (1999)

D. Engel, I. Pahner, K. Schulze, C. Frahm, H. Jarry, G. Ahnert-Hilger, and A. Draguhn, Plasticity of rat central inhibitory synapses through GABA metabolism, *J. Physiol.* 535.2, 473–482 (2001)

N. Etchamendy, V. Enderlin, A. Marighetto, V. Pallet, P. Higueret, and R. Jaffard, Vitamin A deficiency and relational memory deficit in adult mice: relationships with changes in brain retinoid signalling, *Behav. Brain Res.* 145, 37–49 (2003)

U.T. Eysel, Functional reconnections without new axonal growth in a partially denervated visual relay nucleus, *Nature* 299, 442–444 (1982)

U.T. Eysel, Remodelling receptive fields in sensory cortices, *Curr. Biol.* 2, 389–391 (1992)

U.T. Eysel, F. Gonzalez-Aguilar, and U. Mayer, Time-dependent decrease in the extent of visual deafferentation in the lateral geniculate nucleus of adult cats with small retinal lesions. *Exp. Brain Res.* 41,256–263 (1981)

U.T. Eysel, G. Schweigart, T. Mittmann, D. Eyding, Y. Qu, F. Vandesande, G. Orban, and L. Arckens, Reorganization in the visual cortex after retinal and cortical damage, *Res. Neurol. Neurosci.* 15, 153–164 (1999)

D.E. Feldman, R.A. Nicoll, and R.C. Malenka, Synaptic plasticity at thalamocortical synapses in developing rat somatosensory cortex: LTP, LTD, and silent synapses, *J. Neurobiol.* 41, 92–101 (1999)

Y. Feng, H. Ling Liang, and M. Wong-Riley, Differential gene expressions in the visual cortex of postnatal day 1 versus day 21 rats revealed by suppression subtractive hybridization, *Gene* 329, 93–101 (2004)

S.S. Ferguson, Evolving concepts in G protein-coupled receptor endocytosis: the role in receptor desensitization and signalling, *Pharmacol. Rev.* 53, 1–24 (2001)

G.D. Frantz, A.P. Bohner, R.M. Akers, and S.K. McConnell, Regulation of the POU domain gene SCIP during cerebral cortical development, *J. Neurosci.* 14, 472–485 (1994)

S.B. Frost, S. Barbay, K.M. Friel, E.J. Plautz, and R.J. Nudo, Reorganization of remote cortical regions after ischemic brain injury: a potential substrate for stroke recovery, J. *Neurophysiol.* 89, 3205–3214 (2003)

P. Garraghty, E. LaChica, and J. Kaas, Injury-induced reorganization of somatosensory cortex is accompanied by reduction in GABA staining, *Somatosens. Mot. Res.* 8, 347–354 (1991)

C.D. Gilbert and T.N. Wiesel, Receptive field dynamics in adult primary visual cortex, *Nature* 356, 150–152 (1992)

S.H. Hahm and L.E. Eiden, Five discrete cis-active domains direct cell type-specific transcription of the vasoactive intestinal peptide (VIP) gene. *J. Biol. Chem.* 273, 17086–17904 (1998)

T. Herdegen and J.D. Leah, Inducible and constitutive transcription factors in the mammalian nervous system: control of gene expression by Jun, Fos and Krox, and CREB/ATF proteins, *Brain Res. Rev.* 28, 370–490 (1998)

M. Hollmann, M. Hartley, and S. Heinemann, Ca^{2+} permeability of KA-AMPA-gated glutamate receptor channels depends on subunit composition, *Science* 252, 851–853 (1991)

M. Hübener, Mouse visual cortex, *Curr. Opin. Neurobiol.* 13, 413–420 (2003)

P. Hughes and M. Dragunow, Induction of immediate-early genes and the control of neurotransmitter-regulated gene expression within the nervous system, *Pharmacol. Rev.* 47, 133–78 (1995)

P. Jonas, C. Racca, B. Sakmann, P.H. Seeburg, and H. Monyer, Differences in Ca^{2+} permeability of AMPA-type glutamate receptor channels in neocortical neurons caused by differential GluR-B subunit expression, *Neuron* 12, 1281–1289 (1994)

E.G. Jones, GABAergic neurons and their role in cortical plasticity in primates, *Cereb. Cortex* 3, 361–372 (1993)

J.H. Kaas, Plasticity of sensory and motor maps in adult animals, *Annu. Rev. Neurosci.* 14, 137–167 (1991)

J.H. Kaas, L.A. Krubitzer, Y.M. Chino, A.L. Langston, E.H. Polley, and N. Blair, Reorganization of retinotopic cortical maps in adult mammals after lesions of the retina, *Science* 248, 229–231 (1990)

L. Kaczmarek and E. Nikolajew, c-fos protooncogene expression and neuronal plasticity, *Acta Neurobiol. Exp.* 50, 173–179 (1990)

L. Kaczmarek and A. Chaudhuri, Sensory regulation of immediate-early gene expression in mammalian visual cortex: implication for functional mapping and neuronal plasticity, *Brain Res. Rev.* 23, 237–256 (1997)

I.V. Kaplan, Y.H. Guo, and G.D. Mower, Immediate early gene expression in cat visual cortex during and after the critical period - differences between EGR-1 and fos proteins, *Mol. Brain Res.* 36, 12–22 (1996)

K. Keyvani, O.W. Witte, and W. Paulus, Gene expression profiling in perilesional and contralateral areas after ischemia in rat brain, J. *Cereb. Blood Flow & Metab.* 22, 153–160 (2002)

S.I. Kim, H. Voshol, J. van Oostrum, T.G. Hastings, M. Cascio, and M.J. Glucksman, Neuroproteomics: expression profiling of the brain's proteomes in health and disease, *Neurochem Res.* 29, 1317–1331 (2004)

N. Kobori, G.L. Clifton, and P. Dash, Altered expression of novel genes in the cerebral cortex following experimental brain injury, *Mol Brain Res.* 15, 148–158 (2002)

L. Kojic, R. H. Dyck, Q. Gu, R. M. Douglas, J. Matsubara, and M. S. Cynader, Columnar distribution of serotonin-dependent plasticity within kitten striate cortex, *Proc. Natl Acad. Sci. USA* 97, 1841–1844 (2000)

D. Krainc, G. Bai, S. Okamoto, M. Carles, J.W. Kusiak, R.N. Brent, and S.A. Lipton, Synergistic activation of the *N*-Methyl-d-aspartate receptor subunit 1 promoter by myocyte enhancer factor 2C and Sp1, *J. Biol. Chem.* 273, 26218–26224 (1998)

P.E. Lachance and A. Chaudhuri, Microarray analysis of developmental plasticity in monkey primary visual cortex, *J. Neurochem.* 88, 1455–1469 (2004)

I. Leysen, R. Huybrechts, U.T. Eysel, G.A. Orban, F. Vandesande, and L. Arckens, Differential display reveals altered SOD1 gene expression in sensory-deprived cat visual cortex, *Soc. Neurosci. Abstr.* 26 2194, 825.7 (2000)

I. Leysen, E. Van der Gucht, U.T. Eysel, R. Huybrechts, and L. Arckens, Time-dependent changes in the expression of the MEF2 transcription factor family during topographic map reorganization in mammalian visual cortex, *Eur. J. Neurosci.* in press (2004)

P. Liang and A.B. Pardee, Differential display of eukaryotic messenger RNA by means of the polymerase chain reaction *Science* 257, 967–970 (1992)

M.L. Lomax, L. Huang, Y. Cho, and T.-W.L. Gong, R.A. Altschuler, Differential display and gene array to examine auditory plasticity, *Hearing Res.* 47, 293–302 (2000)

G. Lubec, K. Krapfenbauer, and M. Fountoulakis, Proteomics in brain research: potential and limitations, *Prog. Neurobiol.* 69, 193–211.

A. Massie, L. Cnops, K. Van Damme, F. Vandesande, and L. Arckens, Glutamate levels and transport in cat (*Felis catus*) area 17 during cortical reorganization following binocular retinal lesions, *J. Neurochem.* 84, 1387–1397 (2003a)

A. Massie, L. Cnops, I. Smolders, K. Van Damme, E. Vandenbussche, F. Vandesande, U.T. Eysel, and L. Arckens, Extracellular GABA concentrations in area 17 of cat visual cortex during topographic map reorganization following binocular central retinal lesioning, *Brain Res.*, 976, 100–108 (2003b)

R.H. Melloni, P.J. Apostolides, J.E. Hamos, and L.J. Degennaro, Dynamics of synapsin I gene expression during the establishment and restoration of functional synapses in the rat hippocampus, *Neurosci.* 4, 683–703 (1994)

S.A. Michler and R.-B. Illing, Molecular plasticity in the rat auditory brain stem: modulation of expression and distribution of phosphoserine, phospho-CREB and TrkB after noise trauma, *Audiol. Neurootol.* 8, 190–206 (2003)

R.Y. Moore and M.E. Bernstein, Synaptogenesis in the rat suprachiasmatic nucleus demonstrated by electron microscopy and synapsin I immunoreactivity, *J. Neurosci.* 9, 2151–2162 (1989)

T.D. Mrsic-Flogel, M. Vaz Afonso, U.T. Eysel, T. Bonhoeffer, and M. Hübener, Reorganization of mouse visual cortex after retinal lesions, *Soc. Neurosci. Abstr.*,29, 567.4 (2003)

W.A. Myers, J.D. Churchill, N. Muja, and P.E. Garraghty, Role of NMDA receptors in adult primate cortical somatosensory plasticity, *J. Comp. Neurol.*, 418, 373–382 (2000)

M. Narisawa-Saito, Y. Iwakura, M. Kawamura, K.Araki, S. Kozaki, N. Takei, and H. Nawa, Brain-derived neurotrophic factor regulates surface expression of amino

3—hydroxy-5-methyl-4-isoaxoleproprionic acid receptors by enhancing the N-ethylmaleimide-sensitive factor/GluR2 interaction in developing neocortical neurons, *J. Biol. Chem.* 277, 40901–40910 (2002)

S. Obata, J. Obata, A. Das, and C.D. Gilbert, Molecular correlates of topographic reorganization in primary visual cortex following retinal lesions, *Cereb. Cortex* 9, 238–248 (1999)

F. Ortega, L. Hennequet, R. Sarria, P. Streit, and P. Grandes, Changes in the pattern of glutamate-like immunoreactivity in rat superior colliculus following retinal and visual cortical lesions, *Neurosci.* 67, 125–134 (1995)

A. Pandey and M. Mann, Proteomics to study genes and genomes, *Nature* 405, 837–846 (2000)

G.M. Pasinetti, Use of cDNA microarray in the search for molecular markers involved in the onset of Alzheimer's disease dementia, *J. Neurosci. Res.* 65, 471–476 (2001)

C. Pazman, J. Castelli, X. Wen, and R. Somogyi, Large-scale identification of differentially expressed genes during neurogenisis, *Molec. Neurosci.* 11, 1–6 (2000)

B.D. Philpot, J.H.Lim, S.Halpain, and P.C. Brunjes, Experience-dependent modifications in MAP2 phosphorylation in rat olfactory bulb, *J Neurosci.* 17, 9596–604 (1997)

R. Pinaud, L. A. Tremere, and M. R. Penner, Light-induced zif268 expression is dependent on noradrenergic input in rat visual cortex, *Brain Res.* 882, 251–255 (2000)

R. Pinaud, Experience-dependent immediate early gene expression in the adult central nervous system: evidence from enriched-environment studies. *Int. J. Neurosci.* 114, 321–33 (2004)

S.S. Prasad and M.S. Cynader, Identification of cDNA clones expressed selectively during the critical period for visual cortex development by subtractive hybridisation, *Brain Res.* 639, 73–84 (1994)

Y. Qu, U.T. Eysel, F. Vandesande, and L. Arckens, Effect of sensory deprivation on monoaminergic neuromodulators in striate cortex of adult cat, *Neurosci.* 101, 863–868 (2000b)

Y. Qu, A. Massie, E. Van der Gucht, E. Vandenbussche, F. Vandesande, and L. Arckens, Retinal lesions affect extracellular aspartate and glutamate levels in sensory-deprived and remote non-deprived area 17 regions as revealed by *in vivo* microdialysis in awake adult cat, *Brain Res.* 962, 199–206 (2003)

P. Rakic and M.S. Lidow, Distribution and density of monoamine receptors in the primate visual cortex devoid of retinal input from early embryonic stages, *J. Neurosci.* 15, 2561–74 (1995)

S. Reinecke, H.R. Dinse, H. Reinke, and O.W. Witte, Induction of bilateral plasticity in sensory cortical maps by small unilateral cortical infarcts in rats, *Eur. J. Neurosci.* 17, 623–627 (2003)

A.C. Rosenquist, Connections of visual cortical areas in the cat. In: Peters A, Jones EG, editors: Cerebral cortex. volume 3 New York: Plenum, pp. 81–117 (1984)

A.M. Rosier, L. Arckens, H. Demeulemeester, G.A. Orban, U.T. Eysel, Y.-J. Wu, and F. Vandesande, Effect of sensory deafferentation on immunoreactivity of GABAergic cells and on GABA receptors in the adult cat visual cortex, *J. Comp. Neurol.* 359, 476–489 (1995)

C. Sanchez, J. Diaz-Nido, and J. Avila, Phosphorylation of microtubule-associated protein 2 (MAP2) and its relevance for the regulation of the neuronal cytoskeleton function, *Prog. Neurobiol.* 61,133–68 (2000)

T. Sato, D.-M. Xiao, H. Li, F.L. Huang, and K.-P. Huang, Structure and regulation of the gene encoding the neuron-specific protein kinase C substrate neurogranin (RC3 protein), *J. Biol. Chem.* 270, 10314–10322 (1995)

L.M. Schmid, M.G.P. Rosa, M.B. Calford, and J.S. Ambler, Visuotopic reorganization in the primary visual cortex of adult cats following monocular and binocular retinal lesions, *Cereb. Cortex* 6, 388–405 (1996)

S. Schuett, T. Bonhoeffer, and M. Hübener, Mapping retinotopic structure in mouse visual cortex with optical imaging, *J. Neurosci.* 22, 6549–6559 (2002)

I.S. Skerjanc and S. Wilton, Myocyte enhancer factor 2C upregulates MASH-1 expression and induces neurogenesis in P19 cells, *FEBS Lett.* 472, 53–56 (2000)

J.J. Soghomonian and D.L. Martin DL.Two isoforms of glutamate decarboxylase: why? *Trends Pharmacol. Sci.* 19, 500–505 (1998)

I. Song, S. Kamboj, J. Xia, H. Dong, D. Liao, and R.L. Huganir, Interaction of the N-Ethylmaleimide-sensitive factor with the AMPA receptors, *Neuron* 21, 393–400 (1998)

H. Stark and H. Scheich, Dopaminergic and serotonergic neurotransmission systems are differentially involved in auditory cortex learning: a long-term microdialysis study of metabolites, *J. Neurochem.* 68, 691–697 (1997)

R.P. Stroemer, T.A. Kent, and C.E. Hulsebosch, Neocortical neural sprouting, synaptogenesis, and behavioral recovery after neocortical infarction in rats, *Stroke.* 26, 2135–44 (1995)

F.G. Szele, C. Alexander, and M.F. Chesselet, Expression of molecules associated with neuronal plasticity in the striatum after aspiration and thermocoagulatory lesions of the cerebral cortex in adult rats, *J. Neurosci.* 15, 4429–4448 (1995)

G. Thiel, S. Schoch, and D. Petersohn, Regulation of synapsin 1 gene expression by the zinc finger transcription factor zif-268/egr-1, *J. Biol. Chem.* 269, 15294–15301 (1994)

B. Tighilet and M. Lacour, Gamma amino butyric acid (GABA) immunoreactivity in the vestibular nuclei of normal and unilateral vestibular neurectomized cats, Eur. *J. Neurosci.* 13, 2255–2267 (2001)

C.Thomsen, L. Hansen, and P.D. Suzdak, L-Glutamate uptake inhibitors may stimulate phosphoinositide hydrolysis in baby hamster kidney cells expressing mGluR1a via heteroexchange with L-glutamate without direct activation of mGluR1a, *J. Neurochem.* 63, 2038–2047 (1994)

L. Tremere, T.P. Hicks, and D.D. Rasmusson, Role of inhibition in cortical reorganization of the adult raccoon revealed by microiontophoretic blockade of GABA (A) receptors, *J. Neurophysiol.* 86, 94–103 (2001)

K. Van Damme, U.T. Eysel, F. Vandesande, and L. Arckens, NMDA receptor subunit NR2A is involved in adult cortical plasticity, *Soc. Neurosci. Abstr.* 28, 760.7 (2002)

G. Van den Bergh, U.T. Eysel, E. Vandenbussche, F. Vandesande, and L. Arckens, Retinotopic map plasticity in adult cat visual cortex is accompanied by changes in Ca^{2+} / calmodulin-dependent protein kinase II alpha autophosphorylation, *Neurosci.* 120, 133–42 (2003a)

G. Van den Bergh, S. Clerens, L. Cnops, F. Vandesande, and L. Arckens, Fluorescent two-dimensional difference gel electrophoresis (2D-DIGE) and mass spectrometry identify age-related protein expression differences for the primary visual cortex of kitten and adult cat, *J. Neurochem.* 85, 193–205 (2003b)

G. Van den Bergh and L. Arckens Fluorescent two-dimensional difference gel electrophoresis unveils the potential of gel-based proteomics. *Curr. Opin. Biotechnol.* 15, 38–43 (2004)

C. Van den Haute, K. Eggermont, B. Nuttin, Z. Debyser, and V. Baekelandt, Lentiviral vector-mediated delivery of short hairpin RNA results in persistent knockdown of gene expression in mouse brain, *Hum. Gene Ther.* 14, 1799–1807 (2003)

E. Van der Gucht, A. Massie, B. De Klerck, K. Peeters, K. Winters, H. Gerets, S. Clerens, F. Vandesande, and L. Arckens, Molecular cloning and differential expression of the cat immediate early gene c-fos, *Mol. Brain Res.* 111, 198–210 (2003)

N. Vartiainen, I. Pykkonen, T. Hokfelt, and *J.* Koistinaho, Induction of thymosin beta(4) mRNA following focal brain ischemia, *Neuroreport* 7, 1613–1616 (1996)

S.I. Walaas, R. Jahn, and P. Greengard, Quantitiation of nerve terminal populations: synaptic vesicle-associated proteins as markers for synaptic density in the rat neostriatum, *Synapse* 2, 516–520 (1988)

W.J. Waleszcyk, C. Wang, J.M. Young, W. Burke, M.B. Calford, and B. Dreher, Laminar differences in plasticity in area 17 following retinal lesions in kitten or adult cat, *Eur. J. Neurosci.* 17, 2351–2368 (2003)

C.B. Yang, Y.T. Zheng, G.Y. Li, and G.D. Mower, Identification of Munc13-3 as a candidate gene for critical period plasticity in visual cortex, *J. Neurosci.* 22, 8614–8618 (2002)

A. Zepeda, F. Sengpiel, M.A.Guagnelli, L. Vaca, and C. Arias, Functional reorganization of visual cortex maps after ischemic lesions is accompanied by changes in expression of cytoskeletal proteins and NMDA and GABA(A) receptor subunits. *J. Neurosci.* 24,1812–21 (2004)

10

Plasticity of Retinotopic Maps in Visual Cortex of Cats and Monkeys After Lesions of the Retina or Primary Visual Cortex

JON H. KAAS, CHRISTINE E. COLLINS
AND YUZO M. CHINO

[1]Department of Psychology, Vanderbilt University, Nashville, TN, USA and
[2]University of Houston, Houston, TX, USA

Introduction

The visual systems of mammals, like the somatosensory and auditory systems, consist of a hierarchy of nuclei and cortical areas that represent receptor surfaces in topographic patterns (Felleman and Van Essen, 1991; Bullier, 2004; Rosa and Tweedale, 2004). The question raised here is what happens to the pattern of representation when part of the activating inputs to one of these levels has been removed. The basic answer for all of these sensory-perceptual systems is that the deprived zones of neurons usually do not remain unresponsive to sensory stimuli. Instead, they recover responsiveness to remaining inputs that always activated other pools of neurons, but were previously ineffective sources of activation for the pool of neurons that became deprived (Kaas and Florence, 2001). The visual system is especially interesting in this regard, because neurons across binocular positions of visual cortex are responsive to two receptor surfaces, the visually matched positions of the retina of each eye. Thus, it is possible to totally deprive parts of the cortical representation of normal sources of visual activation, or deprive these parts of the sources of activation from only one eye. Both types of deprivation, as well as deprivations of higher visual centers by small lesions of primary visual cortex, reveal cryptic features of visual system organization, as well as the capacity of deprived neurons to acquire new sources of activation. This capacity of the visual system to reorganize appears to extend throughout life, and suggests that the apparent stability of this system under normal conditions is the consequence of a counter-balancing of factors that could cause change.

R.Pinaud, L.A. Tremere, P. De Weerd(Eds.), Plasticity in the Visual System:
From Genes to Circuits, 205-227, ©2006 Springer Science + Business Media, Inc.

Features of the Mammalian Visual System that Provide a Substrate for Plasticity: A Brief Review

Several well-known features of visual system organization are important to keep in mind when considering how retinotopic or visuotopic patterns in visual structures reorganize. Perhaps the most important feature is that a given cortical neuron in V1 receives multiple connections from various sources and a fine balance among these input signals, mostly binocular, ultimately determines how the cell will fire action potentials and signal the location of visual objects in space. These include feed-forward thalamic inputs from each eye, signals from the local excitatory and inhibitory neurons that are known to adjust neuronal drive, signals arriving via the intrinsic long-range horizontal connections that can integrate information over large cortical distances, and feedback inputs from the higher order visual areas that are also involved in long-range signal integrations. First, we need to consider the parallel and then converging pathways from each eye to visual cortex. The retina of each eye has a line of decussation separating a nasal portion where neurons (ganglion cells) project to layers in the contralateral (dorsal) lateral geniculate nucleus (LGN) to form representations of the contralateral visual hemifield via the contralateral eye. In a complimentary manner, the smaller temporal portion of the retina has ganglion cells that project to layers in the ipsilateral LGN to form representations of the smaller binocular portion of the contralateral visual hemifield via the ipsilateral eye. The relays of these two sets of maps based on the ipsilateral or contralateral eye may converge in layer 4 of primary visual cortex (V1) of most non-primate mammals, or remain separate in alternating narrow bands of cells or clusters of cells in layer 4 of most primates (see Casagrande and Kaas, 1994). If the convergence does not occur in layer 4, it occurs as processing proceeds from layer 4 to more superficial and deeper layers. As these binocular layers are the sources of activating inputs to other cortical visual areas, all higher-order visual areas are binocular. Thus, a focal lesion of one eye would totally deprive groups of neurons of visual activation in the layers of the LGN that are subserved by that eye, and possibly neurons in bands or clusters in V1, but not in other layers of V1 or in other cortical areas. The exception, of course, is in the small monocular portion of V1 subserved by the nasal extreme of the contralateral eye.

The second feature of the visual system that is relevant to the issue of plasticity is that the system is serially organized, although parallel connections are an important feature as well (see Felleman and Van Essen, 1991; Bullier, 2004). If the retina is the first stage or level of processing, the LGN is the second stage, V1 is a third stage, and V2 is a fourth stage (although V2 and other higher order areas may have parallel inputs from the LGN). In primates, several areas may be considered to collectively constitute a fifth stage (V3, MT, DM; see Kaas and Lyon, 2001). At each stage, there is some divergence of the topographic pattern of projections from one level to the next, so the retinotopic (visuotopic) pattern of representation becomes less precise (Rosa and Tweedale, 2004). Neurons are responsive to larger arrays of retinal receptors, and the receptive fields for these neurons are larger. However, the full impact of this divergence is not apparent in normal mammals because each relay activates both relay neurons and intrinsic inhibitory neurons at the next level, and the inhibitory neurons that drive some of the excitatory inputs below threshold. Thus, the full extent of the excitatory receptive fields of cortical neurons is only apparent when inhibition is locally blocked (see Dykes, 1997; Chowdhury and Rasmusson, 2003). These inhibitory neurons can be responsive to feed-forward activation, recurrent axons of relay neurons, or feedback from higher visual areas. Neurons in the LGN also receive inhibitory inputs from the visual segment of the reticular nucleus. Since inhibitory neurons generally have larger or different receptive fields than adjacent excitatory neurons, and so provide lateral inhibition, any focal lesion of the retina or a retinotopically organized nucleus or cortical area immediately alters the balance between excitatory and inhibitory effects on neurons at several levels of the visual system. Thus, some of the deprived neurons at several levels become responsive to previously subthreshold sources of activation. In addition, lesions of higher level visual representations may remove feedback to inhibitory neurons in lower level representations, thus changing receptive fields (Ergenzinger et al., 1998). Thus, some neurons in the visual system would have altered or larger receptive fields, common measures of plasticity. However, as this change reflects only a new balance in a dynamic system (Chapman and Stone, 1996; Cavanaugh et al., 2002a, 2002b) the immediate changes have also been called pseudo plasticity or the "iceberg" effect as much of the excitatory receptive field is normally hidden below "the water line" (see Kaas and Collins, 2003).

The network of inhibitory neurons also has an important delayed effect on the visual system after injury. Both the expression of the inhibitory neurotransmitter GABA by inhibitory intrinsic neurons and the receptors for this transmitter on excitatory relay neurons are highly regulated by ongoing levels of activity in these neurons. Thus, the immediate consequence of damage to the retina, LGN, or any area of visual cortex is that large populations of neurons that depend on the damaged and lost neurons for activation become unresponsive to visual stimuli and are only "spontaneously" active. This decreased level of activity down-regulates the expression of GABA and GABA receptors, and the down-regulated populations of neurons become less inhibited and more responsive to remaining sources of activation that were previously subthreshold. The down-regulation of inhibition can be extremely rapid. Early studies showed marked effects days to weeks after a sensory loss (Garraghty et al., 1991; Jones, 1993; Rosier et al., 1995), but considerable reductions also occur within minutes to hours after a nerve block that reduces activation (Ziemann et al., 1998). Thus, studies of the plasticity that occurs within hours after lesions of the visual system may be largely studies of the iceberg effect and effects of reducing inhibition.

Another relevant aspect of inhibition in the visual system is that inhibitory neurons tend to play an antagonistic role in local circuits by suppressing responses for higher contrast stimuli beyond the "classic" receptive field. Responses may be most suppressed for stimuli in and outside the minimal receptive field that are mismatched in orientation or direction of movement. Thus, inhibition sharpens the contrast in neuron responsiveness between stimuli differing in location, orientation, movement, and so on. This normal role for inhibition also may be expressed in the damaged visual system to prevent reactivated neurons from acquiring mismatched receptive field properties or receptive fields in more than one location in the visual field.

A third feature of visual system organization that appears to be important in studies of plasticity is the array of horizontal connections (Fig. 10.1) in visual areas that communicate between different parts of the map (Gilbert and Wiesel, 1989; Casagrande and Kaas, 1994; Das and Gilbert, 1995; Toth et al., 1996; Stettler et al., 2002; Tucker and Fitzpatrick, 2003). Such horizontal connections normally exert subthreshold excitatory or inhibitory effects on the feedforward responses, depending on stimulus contrast and configuration that may help entrain the responses of excited neurons in different parts of the map. Deprived neurons may become responsive to such weak excitatory effects of horizontal connections, perhaps as a result of a reduction in inhibition. If the weak excitations become effective in initiating action potentials, the weak connections may be strengthened, according to some proposals (Hess and Donoghue, 1994), via the mechanism of long-term potentiation (LTP). Thus strengthened, the effectiveness of the connections may persist after activity levels rise and inhibition returns. Horizontal connections would expand the representations of parts of the visual field within a cortical area, as normal locations of activation would be relayed to deprived parts of the representation. In a related matter, recent studies in monkey visual cortex indicate that signal integration over large space or distance in individual V1 neurons depends heavily on the feedback connections from the extrastriate visual areas V2, V3, and MT (Hupe et al., 1998; Hupe et al., 2001a; Hupe et al., 2001b; Sceniak et al., 2001; Albright and Stoner, 2002; Angelucci et al., 2002; Cavanaugh et al., 2002a, 2002b; Stettler et al., 2002; Bair et al., 2003; Rockland, 2004). Contrary to traditional views, these feedback influences have a much faster time course than that of the intrinsic long-range connections in V1 (Bringuier et al., 1999; Hupe et al., 2001a; Hupe et al., 2001b). The functional relationship between the long-range horizontal connections and the feedback connections in the role of signal integrations over large areas is not clearly understood, however, the extent of signal interactions over space appears to be greater for the feedback connections (Angelucci et al., 2002; Cavanaugh et al., 2002a, 2002b). Also, signal interactions via the horizontal connections could be influenced by inputs transmitted by these feedback connections.

A fourth feature of visual systems that is relevant in studies of plasticity is that the axons and dendrites of neurons are in a balanced state of growth

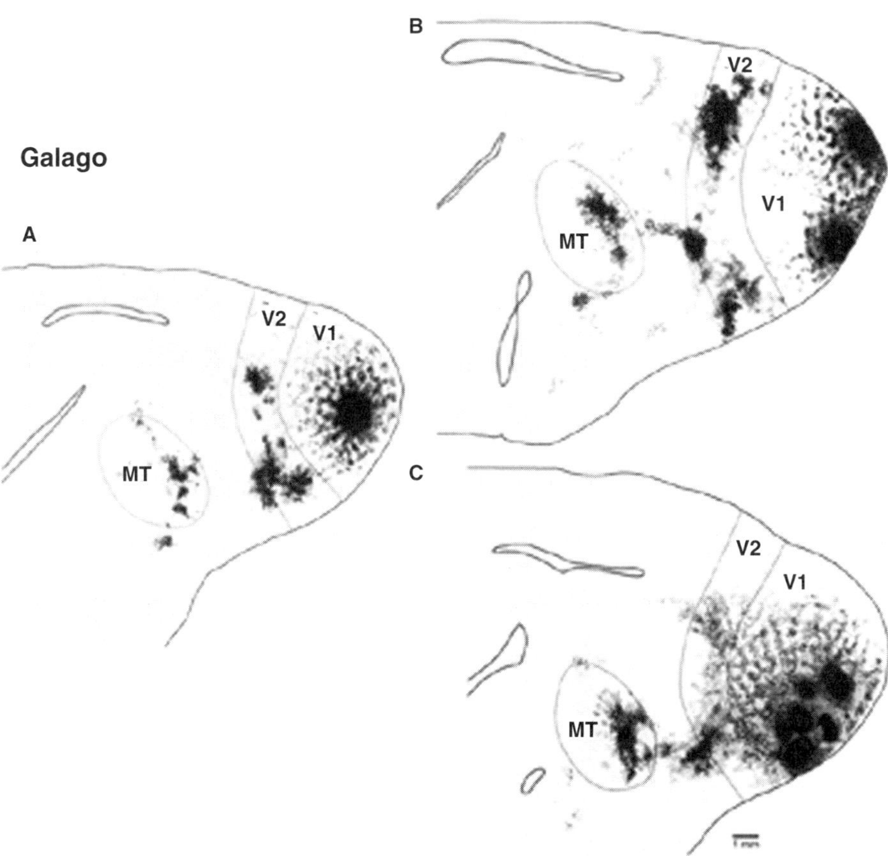

FIGURE 10.1. Three examples of intrinsic and extrastriate connections of V1 (area 17) of prosimian galagos. The illustrations are of the dorsolateral portion of the left cerebral hemisphere that has been flattened and cut parallel to the surface to reveal a surface view of the pattern of connections. A. An injection of WGA-HRP was placed in dorsolateral V1 where it densely labeled a core of neurons (black oval). Neurons in the core projected to locations in surrounding V1 for distances of 2–3 mm, and neurons in this projection zone projected back to the core. The patchy pattern of fine and course dots indicates these labeled terminals and neurons. Note that both the labeled terminals and neurons reflect a network of intrinsic horizontal connections that are densely distributed over short distances and are sparse or absent over distances larger than 3-4 mm. These horizontal connections are also unevenly distributed. In primates, they are dense over longer distances in the cytochrome oxidase (CO) blobs, and the patchy pattern in the figure superimposes on the CO blob pattern. Other foci of labeled neurons and axon terminals are in the second visual area, V2, and the middle temporal visual area, MT. The illustration is based on label in several adjacent brain sections cut parallel to the surface of V1 and adjoining cortex in the study of Cusick and Kaas (1988). B. Results from another case with two injections in V1. C. Results from a third case with six closely spaced injections in V1.

and retraction. Synapses are being lost and regained (Cotman and Nieto-Sampedro, 1982). Axon and dendritic branches grow and retract (Wolff and Missler, 1992). Neurons respond to a number of factors that promote or inhibit growth. Many or most of these factors are regulated by levels of neuronal activity or are responsive to neuron injury. Damage to the visual system and depressed levels of activity in pools of neurons seem to alter the balance between growth and regressive factors, thus, deprived neurons accept invasions of axons of active neurons. Most notably, the growth of somatosensory inputs into deprived portions of the spinal cord and brainstem has been described (Florence et al., 1989; Jain et al., 2000; Wu and Kaas, 2002), but the growth of new horizontal connections in deprived portions of somatosensory (Florence et al., 1998) and visual cortex (Darian-Smith and Gilbert, 1994) has also been reported. Thus, the growth of new connections, even in the mature brain, is likely to be an important factor in visual system plasticity. However, the aforementioned evidence for growth in visual cortex is very preliminary and thus, the role of new growth in reorganizing the damaged visual system is largely unknown. Because small, but significant changes in local connections are difficult to detect, a more comprehensive approach is required to resolve this critical issue of the role of new growth in visual system plasticity in adults.

The Effect of Focal Lesions of the Retina on Retinotopic Representations in the LGN of Cats and Monkeys

The results of a series of experiments on the effects of small, focal (5-10°) lesions of the retina in cats and monkeys lead to the following conclusions. First, monocular lesions totally deprive portions of the layers of the LGN receiving input from the lesioned retina of visual activation (Eysel et al., 1980; Eysel, 1982). These lesions leave a long-lasting or permanent core of unresponsive neurons in the deafferented retinotopic zones of the affected layers. Nevertheless, a fringe of neurons along the outer border of the deafferented zones recovered responsiveness to visual stimuli within a few weeks following the lesion. The changes in receptive field centers for these neurons were less than 1°. Most likely, the changes resulted from the potentiation of synapses on the dendrites of these bordering neurons, as some of the dendrites would likely extend into portions of the LGN with presumed retinal inputs. Possibly, new synapses were generated, but the limited change in functional connections did not appear to depend on any notable growth of axon arbors into the deafferented zone (Stelzner and Keating, 1977; Baisden et al., 1980; Eysel, 1982). In contrast, during the course of 6–8 months of recovery, axons of somatosensory afferents grow into the deafferented parts of the dorsal column nuclei in the somatosensory system of monkeys (Jain et al., 2000). These conclusions, based on the effects of monocular lesions of the retina, are supported by results from cats and monkeys with small, binocularly matched lesions of the retina (Darian-Smith and Gilbert, 1995). Weeks after the lesions, zones persisted in the LGN where the neurons were unresponsive to visual stimuli, while recordings from the cortex at the same time, revealed no such unresponsive zones. The persistence of unresponsive zones of neurons in the layers of the LGN at times when deprived cortical neurons have recovered responses (see below) indicates that the recoveries in cortex are not simply relayed from the LGN, or from a "healed" retina (see Horton and Hocking, 1998) to the LGN. Thus, mechanisms that operate at the cortical level must be responsible for most of the cortical reorganization. Nevertheless, the limited recovery of responses in the outer fringe of deprived LGN neurons would contribute to the much more extensive reactivation in cortex.

The Effects of Binocular Lesions of the Retina on the Retinotopy and Responsiveness of V1 in Cats and Monkeys

Our early experiments on the effects of retinal lesions on the retinotopic organization and responsiveness of primary visual cortex (Kaas et al., 1990) were motivated by the evidence that removing sensory input from part of the hand of a monkey altered the map of that hand in contralateral somatosensory cortex (Merzenich et al., 1983a; Merzenich et al., 1983b). Cutting the median nerve ablated inputs from cutaneous receptors from the thumb half of the glabrous skin of the hand, and thereby removed the peripheral source of sensory activation of about half the topographic representation of the hand in area 3b (S1). Over a period of weeks, the neurons throughout the deprived portion of area 3b became responsive to remaining inputs from the rest of the hand. As most of S1 represents only the contralateral body surface, while V1 represents the contralateral visual hemifield via overlapping inputs relayed from each eye, the analogous experiment for the visual system seemed to require visuotopically matched lesions of the retina of both eyes. Thus, our experimental plan was to totally deprive neurons in a restricted part of V1 of visual activation, allow a period of recovery so that the system could reorganize, and then record from neurons throughout the deprived zone of cortex. As it can be difficult to precisely match lesions in the two eyes, a small 5–10° lesion that extended across all retinal neuronal layers was placed in paracentral vision of one eye and all inputs were removed from the other eye of adult cats. Thus, visual cortex was activated monocularly, and the visual system was "blind", not only to visual stimuli falling on the optic disc (the blind spot), but also to visual stimuli falling on the missing part of the retina (which was lesioned with a laser). The deprived zone of V1 was the portion of the retinotopic map that received inputs from the neuron column in the LGN that no longer received inputs from the retina. From the position of the lesion relative to the area centralis and the optic disc, and the published maps of the retinotopy of V1 (Tusa et al., 1978), the location of the deprived zone of V1 could be closely approximated. Microelectrode recordings throughout and around this zone allowed the outer boundary of the deprived zone to be defined more precisely. Neurons in rows of recording sites across normally activated cortex always corresponded to predictable progressions of receptive field locations, as expected from published maps of V1. However, neurons in the deprived zones do not contribute to these normal progressions, as this would require receptive fields based on the lesioned part of the retina. Thus, the deprived zone could be identified either by sites with neurons that were unresponsive to visual stimuli, or neurons that had a receptive field in locations that did not correspond to those expected from normal retinotopic progressions. The deprived zones could also be identified histochemically, since activity levels in these zones are unlikely to be completely normal (see Horton and Hocking, 1998). However, after months of recovery, neurons throughout the deprived zones in V1 responded vigorously to visual stimuli, (Kaas et al., 1990). Receptive fields for neurons in the deprived zones were abnormally located in positions just outside the lesion of the retina. For rows of recording sites that crossed the deprived zone, receptive fields would progress toward the retinal lesion, pile up at the margin of the lesion, and then suddenly jump to the other side of the lesion and pile up before progressing away from the lesion (Fig. 10.2). Some neurons had double receptive fields, with one on either side of the retinal lesion. Thus, neurons, across a 3-6 mm zone of V1 that normally represented five or more degrees of visual field, were fully responsive to visual stimuli and had receptive fields that were displaced to the intact retina around the margin of the lesion (Kaas et al., 1990; Chino et al., 1992; Schmid et al., 1996). Very similar results were obtained in more difficult experiments where small lesions were closely matched in size and visuotopic position for the two eyes (Gilbert and Wiesel, 1992; Darian-Smith and Gilbert, 1994; Chino et al., 1995; Darian-Smith and Gilbert, 1995). Thus, the ability of deprived neurons in primary visual cortex of cats to recover responsiveness to visual stimuli and contribute to an altered visuotopic representation became well established.

The experiments on cats also demonstrated that the full recovery of responsiveness took time. Immediately after retinal lesions, neurons centered in the deprived zone of V1 were unresponsive to visual stimuli (Chino et al., 1992; Gilbert and Wiesel, 1992; Darian-Smith and Gilbert, 1995). However, neurons along the outer fringe of the deprived zone were responsive, and they had abnormally large receptive fields that were outside of the retinal lesion. This "unmasking" (Wall, 1977) of previously silent inputs most likely reflects the immediate reduction in lateral inhibition following reduced activation of GABAergic neurons in cortex, and to a lesser extent in the LGN. The reactivation of neurons in the central core of the deprived cortex takes weeks to months to occur, and it may not be complete after large retinal lesions (Darian-Smith and Gilbert, 1995).

The effects of small lesions of the retina on the responsiveness of neurons in V1 appear to be similar in macaque monkeys to those in cats (Darian-Smith and Gilbert, 1995). Matched retinal lesions of about 5° in diameter deprived zones of 10-12 mm in diameter in V1. As in cats, neurons in the deprived zone of V1 gained responsiveness to visual stimuli over long recovery times and had receptive fields that were displaced to parts of the retina sur-

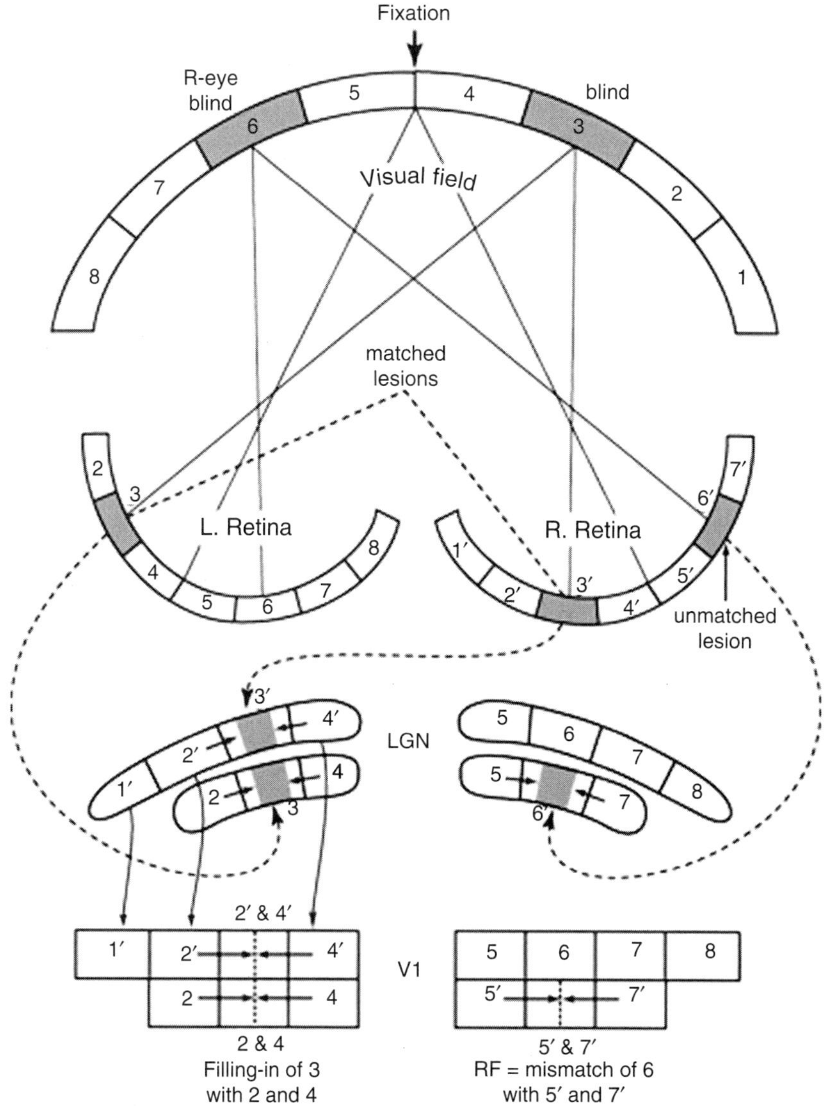

FIGURE 10.2. The effects of visuotopically matched and mismatched lesions of the retina on visuotopic organization in the LGN and primary visual cortex (V1). When matching parts of the retina of the right (3′) and left (3) eyes are lesioned, parts that view the same portion of the visual field (3), the lesions create a blind spot (scotoma), and adjacent sectors of LGN layers (3 & 3′) are deprived of retinal projections. Neurons in the central cores of these LGN sectors remain unresponsive to visual stimuli even after months of recovery, but neurons on the outer fringe of these sectors may become responsive to previously subthreshold inputs from parts of the retina around the lesions (2 & 4; 2′ & 4′). Neurons in a region of cortex with inputs from LGN positions 3 and 3′ no longer can be responsive to the blind visual field position 3. However,

rounding the lesion. However, even after a year of recovery, a small central core of neurons could remain unresponsive. Thus, there are limits on the extent of cortex that is reactivated even after 1 year of recovery. In an earlier study where even larger regions of cortex were deprived by bilateral lesions of the foveal retina in macaque monkeys, the reactivation of deprived cortex appeared to be incomplete after 75 days, with many neurons unresponsive or only weakly responsive to visual stimuli (Heinen and Skavenski, 1991).

The effects of restricted monocular lesions of the retina on the responses of neurons in V1

The results of early experiments on the effects of binocular and monocular lesions of the retina in cats and monkeys appeared to indicate that after monocular lesions, neurons remained responsive to inputs from the intact eye, and they did not acquire new, displaced receptive fields based on inputs from the lesioned eye (Kaas et al., 1990; Chino et al., 1992; Gilbert and Wiesel, 1992). The lack of responses to stimuli in the lesioned eye suggested that the inputs from the normal eye provided enough activation to suppress the expression and strengthening of latent inputs relayed from the lesioned eye. Such suppression was demonstrated when Chino et al. (1992) discovered that after a monocular lesion, unresponsive deprived neurons became responsive to visual stimuli within hours of enucleating the intact eye. However, Schmid et al. (1995) subsequently reported that monocularly deprived neurons in V1 of cats were often responsive to visual stimuli, having large receptive fields just outside the deactivated zone of retina in these experiments. As the deactivated zone in the retina was produced by a bubble of retinal detachment following a laser lesion, there might be some question about the extent of the deactivated portion of the retina. In addition, responsiveness to visual stimuli was clearly depressed within the deprived zone of cortex. Nevertheless, many deprived neurons apparently did respond to visual stimuli, and had new, large, displaced receptive fields. After recovery times of weeks or longer, the responses of neurons in deprived cortex to visual stimuli

---

neurons in position 3 become responsive to visual stimuli in positions 2 & 4 via the right (2′ & 4′) and left (2 & 4) eyes. Binocular neurons in V1 are not in conflict, even in the reactivated zone of deprived neurons, and thus, suppression of inputs from one eye by the other is not expected. When a lesion (6′) in one eye is not matched by a lesion in the other eye, neurons in the monocularly deprived segment of V1 remain responsive to their normally active inputs from the unlesioned eye (6) and responsiveness to the lesioned eye (6′) appears to be absent during many recording and testing procedures. However, when such responsiveness is detected, the receptive fields for the lesioned eye are displaced to adjoining parts of the visual field (5 & 7) so that the binocular neurons have receptive fields in different locations for each eye. This receptive field mismatch would be expected to generate inhibition that would suppress activity based on the lesioned eye.

in the lesioned eye were enhanced, and the displaced receptive fields were smaller (Schmid et al., 1995; Calford et al., 2000; Chino et al., 2001; Calford et al., 2003). The response properties of neurons were similar when tested with either eye (Matsuura et al., 2002; Calford et al., 2003), even though the displaced and normal receptive fields for these binocular neurons were no longer in register.

It is not clear why neurons in the deprived zone of cortex were found to be unresponsive to the stimuli in the lesioned eye in some experiments (Kaas et al., 1990; Chino et al., 1992; Gilbert and Wiesel, 1992; Schmid et al., 1996) and responsive in others (Schmid et al., 1995; Schmid et al., 1996; Calford et al., 2000; Chino et al., 2001). In the study of Chino et al. (2001), the recovery times after monocular retinal lesions extended over years, and the lesions were made in juvenile cats. These factors may be important, but not essential. In retrospect, one might expect some level of responsiveness via the short-range network of horizontal connections that emerges between nearby groups of neurons with similar response properties during normal development. Thus, clusters of neurons outside the deprived zone of cortex would be activated in a normal fashion, and horizontal connections from these clusters would excite clusters in the deprived zone. In support of this possibility, Calford et al. (2003) found that lesions of visual cortex that are outside the deprived zone could abolish responses of neurons in the deprived zone. The common finding has been that neurons in the deprived zone are not immediately responsive to visual stimuli in the lesioned eye after a monocular lesion (Calford et al., 1999), or in either eye after matched or overlapping lesions in both eyes (Heinen and Skavenski, 1991; Gilbert and Wiesel, 1992). Only neurons in the outer fringe of the deprived zone typically are responsive, and they are unlikely to be totally deprived. This difference in results suggests that the normally present network of horizontal connections is usually not effective in driving deprived neurons in mammals with monocular lesions, but under some conditions, it can be.

One possibility is that testing conditions alter the dynamics of the cortical networks to potentiate the previously weak excitatory connections. Over just a few minutes of visual stimulation presented just outside the receptive fields of V1 neurons in normal cats (Pettet and Gilbert, 1992) and monkeys (De Weerd et al., 1995), the receptive fields sometimes expanded to include the stimulated zone. Pettet and Gilbert (1992) suggest that the visual responses of some of the neurons in the deprived cortex immediately after retinal lesions reflect the normal dynamics of cortical neurons that alter receptive field size. They suggest that repeated stimulation may adapt inhibition, tipping the balance toward excitation, or somehow potentiate excitatory horizontal connections.

The more pronounced recoveries of visually evoked responses in monocularly and binocularly deprived cortical zones is likely to be the result of mechanisms that increase the effectiveness of these horizontal connections, and maintain the higher levels of effectiveness. The lesions are followed by lower levels of activity in the deprived zones, and by the down-regulation of the inhibitory neurotransmitter GABA, and GABA receptors (Rosier et al., 1995; Arckens et al., 1998), that allow excitatory inputs to be more effective (however, see Pernberg et al., 1998; Eyding et al., 2002). Visual experience may have a conditioning effect that strengthens some synapses (Eysel et al., 1998; McLean and Palmer, 1998). The deprived zones express higher levels of neurotropin and growth-associated proteins as early as two days after the retinal lesions (Obata et al., 1999), and this appears to lead to the growth of new horizontal connections (Darian-Smith and Gilbert, 1994; Florence et al., 1998).

It is important to keep in mind, however, that arguments against the idea of "potentiating" excitatory connections have been made for cat visual cortex (De Angelis et al., 1995; Chapman and Stone, 1996) and monkey V1 (Cavanaugh et al., 2002a) an alternative, simpler explanation that does not involve elaborate plastic changes was offered instead. This issue must be resolved by future experiments involving a more comprehensive approach.

Alterations in human perception after focal retinal lesions

In humans with retinal lesions, the scotoma produced by the lesion fills in perceptually, much as the natural blind spot fills in (see Komatsu et al., 2000 for cortical mechanisms of filling-in of the blind spot representation in V1 of monkeys), so that the scotoma is not noticed (Gerrits and Timmerman, 1969; Sergent, 1988). One would expect this result from the recovery of visual responsiveness of neurons in the deprived zones in V1 of cats and monkeys. Thus, these reactivated neurons appear to contribute to perception in much the same way as in normal individuals, with the exception that neurons are signaling the presence of visual stimuli that are located elsewhere (Davis et al., 1998).

Plasticity in V1 after lesions of V1

After a small lesion in V1, part of the representation of the contralateral visual hemifield is lost (Fig. 10.3). However, such lesions produce less of an impairment in visual perception in monkeys than equivalent lesions of the retina (Weiskrantz and Cowey, 1967) suggesting that perilesion regions of cortex recover some of the information normally represented in the site of the lesion. In humans, the scotoma produced by a cortical lesion may become smaller in size over time with training (Zihl and von Cramon, 1985; Kasten et al., 1998) providing further evidence for reorganization of visual cortex to recover inputs that formerly activated only the lesion site. In cats, with a small lesion of V1, large increases in the sizes of receptive fields were observed for neurons in the perilesion zone after two months of recovery (Eysel and Schweigart, 1999). However, after only two days of recovery neurons in the perilesion zone had only slightly larger than normal receptive fields, although these receptive fields could be further enlarged with conditioning stimuli

(Schweigart and Eysel, 2002). Similar results were obtained in optical imaging experiments after lesions of V1 (Zepeda et al., 2004). Immediately after the lesion, optical imaging revealed an area of cortex that was devoid of visual activation, corresponding to the ischemic cortex. However, the region of visual space represented in the perilesional tissue expanded over five weeks of recovery, compensating in part for the loss of visual information.

The Reactivation of Neurons in Higher-order Visual Areas After Restricted Lesions of Primary Visual Cortex

In primates, most of the projections of the LGN are to primary visual cortex (Bullier et al., 1994; Stepniewska et al., 1999). The small number of neurons that project to extrastriate areas such as V2, V3, DM, and MT do not appear to be independently capable of activating cortical neurons. Thus, lesions or cooling of striate cortex render large regions of extrastriate visual cortex unresponsive to visual stimuli (e.g., Schiller and Malpeli, 1977; Girard et al., 1991). This dependence on a single source of visual activation allows the potential of extrastriate visual areas to reorganize in adult primates to be evaluated by depriving visual areas of some or all of the direct and indirect inputs from V1. To date, this potential has been most fully studied in visual area MT (V5), with different results from different laboratories.

The middle temporal area, MT, is a visual area in the middle of the dorsal end of the temporal lobe. MT contains a visuotopic representation of the contralateral visual hemifield (Allman and Kaas, 1971; Gattass and Gross, 1981; Van Essen et al., 1981) and receives topographically organized feedforward inputs from V1 and V2 (e.g., Weller and Kaas, 1983; Stepniewska and Kaas, 1996). The other notable input is from a subdivision of the inferior pulvinar that in turn receives inputs from visual cortex, but not from the superior colliculus (see Stepniewska et al., 2003). The direct inputs to MT from the LGN are sparse (Stepniewska et al., 1999). This anatomy suggests that the driving input to MT comes from V1 and V2, rather than the pulvinar and LGN, as visually evoked activity in V2 depends on V1 (Schiller and Malpeli, 1977). All visually evoked activity in MT would appear to depend on V1. However, MT is only about 1/10th the size of V1 (Frahm et al., 1998), so projections from V1 and V2 converge in MT, and neurons in MT have receptive fields that are considerably larger than those in V1 and V2. Thus, small focal lesions of V1 would be expected to have limited impact on neurons in MT, while large or complete lesions of V1 would be expected to have the immediate consequences of rendering most or all of MT neurons unresponsive to visual stimuli. However, responsiveness to visual stimuli would persist if MT has a significant alternative source of visual activation that depends on the LGN or superior colliculus rather than V1. In addition, the responsiveness to visual stimuli would return if previously weak and subthreshold sources of activation become potentiated. Preserved parts of V1, for example, could expand their territories of effective activation in MT, as the retinotopic projections of V1 to MT are somewhat divergent, and V2 projections to MT could contribute to this expansion. In addition, the missing part of the visual hemifield in V1 could be partially recovered due to the potentiation of the subthreshold fringes of damaged arbors of geniculostriate axons (see Jenkins and Merzenich, 1987). As V1 neurons acquire new receptive fields, this information would be relayed to MT. Also, the dense network of horizontal connections in MT (Krubitzer and Kaas, 1990) provides a mechanism for the spread of activity from portions of MT with effective inputs to other parts of MT without such inputs. As callosal connections are widespread in MT (Cusick et al., 1984; Krubitzer and Kaas, 1990), an intact V1 in the opposite cerebral hemisphere could indirectly activate deprived MT. Finally, the sparse direct inputs to MT from the LGN could become potentiated and effective or more effective, and superior colliculus inputs to the pulvinar could indirectly access MT via other cortical areas and become an effective or more effective source of activation. Thus, the anatomy of the visual system suggests a number of possibilities.

Experimentally, it is not clear what happens or why, as results from different laboratories on different primates differ. In early studies on macaque monkeys, large lesions of V1 did not completely inactivate MT when tested after weeks of recovery (Rodman et al., 1989). Instead, many neurons responded to visual stimuli and were selective for direction of stimulus motion as neurons are in normal MT. After subsequent lesions of the superior colliculus in the same monkeys, MT neurons no longer responded to visual stimuli (Rodman et al., 1990). As the lesions were large and possibly complete, the responses of neurons in MT did not appear to depend on preserved parts of V1. Thus, the results suggested that inputs from the superior colliculus to the pulvinar somehow became (or remained) an effective source of visual activation. In subsequent experiments in which the activity of neurons in part of V1 of macaques was blocked with a cooling probe (Girard et al., 1992), many neurons that appeared to be in the deprived portion of MT remained responsive to visual inputs, and had the direction selective response properties that characterize MT neurons. These results suggested that alternative inputs to MT were effective sources of activation, or rapidly became effective. Thus, there was no evidence of the potentiation of weak connections after a recovery period. However, the cooling procedure may not have blocked all outputs from part of V1 to MT.

At least a short-term dependence of MT on V1 inputs was suggested by the finding of Maunsell et al. (1990) when a block of LGN layers totally deactivated neurons in MT. Yet, the sample size was small, so it is possible that responsive neurons were not detected. In addition, the total loss of responsiveness could have reflected the inactivation of both axons projecting to V1, and those few projecting to MT.

In New World monkeys, and prosimian galagos, results have been different, but not consistently so. Lesions of part of V1 in owl monkeys produced zones in MT where neurons were completely unresponsive to visual stimuli in recordings obtained over several hours immediately after the lesions (Kaas and Krubitzer, 1992). However, some neurons along the outer fringe of the unresponsive zone appeared to have receptive fields that were displaced from within, to just outside, the cortical scotoma. Thus, it was possible that a reduction of lateral inhibition in MT allowed preserved subthreshold inputs from V1 to reactivate some neurons. However, there was no evidence for an effective, alternative source of activation based on the totally preserved visual hemifield representations in the superior colliculus. Very similar results were obtained in other adult owl monkeys and marmosets studied two or more months after partial lesion of V1 (Collins et al., 2003). A small lesion that removed about 5° of representation in V1 failed to produce any zone of deactivation in MT, but MT neurons often have receptive fields of 5° or more in diameter, so responsiveness could have been based on preserved, and possibly potentiated, inputs from V1. Larger lesions, removing parts of V1 representing 20° or more of visual hemifield consistently produced zones in MT where neurons were unresponsive to visual stimuli. Other parts of MT were normally responsive to visual stimuli, and all responsiveness could be related to intact parts of V1. There was no evidence for an alternative source of activation based on the superior colliculus. However, neurons along the margin of the deprived zone in MT may have had displaced receptive fields, and the responses of some of these neurons to visual stimuli were abnormal in that they rapidly habituated. Thus, even after months of recovery, a major reactivation of MT did not occur. Any recovery of responsiveness that may have occurred was limited to a fringe of neurons that were probably not completely deprived of V1 inputs.

More recently, activity patterns evoked by visual stimuli in MT of galagos and one owl monkey were revealed by optical imaging just before and after the activity of part of V1 was blocked with muscimol (Collins et al., 2005). The deactivation of part of V1 produced a "hole" in the activation pattern in MT, providing further evidence that MT depends on V1 for activation.

Somewhat different results were reported for MT neurons after partial lesions of V1 in small New World marmoset monkeys (Rosa et al., 2000). In the deprived zone of MT immediately after the lesions, about 20% of the neurons remained responsive to visual stimuli in part of the visual hemifield corresponding to lesioned portions of V1. Other responsive neurons appeared to have new receptive fields that were displaced outside the scotoma. Thus, the results in this study suggest that about 1/5th of the neurons in MT remain or rapidly acquire responsiveness to alternative sources of visual activation after V1 lesions, apparently based on the preserved visual hemifield representation in the superior colliculus, while a number of other deprived neurons rapidly acquire new receptive fields based on preserved inputs from V1.

Interpretations of these differing results from macaques, owl monkeys, marmosets, and galagos remain uncertain, as the studies were based on different primate species, and methods varied, including the type of anesthetic used during recordings and possibly the criteria for judging responsiveness to

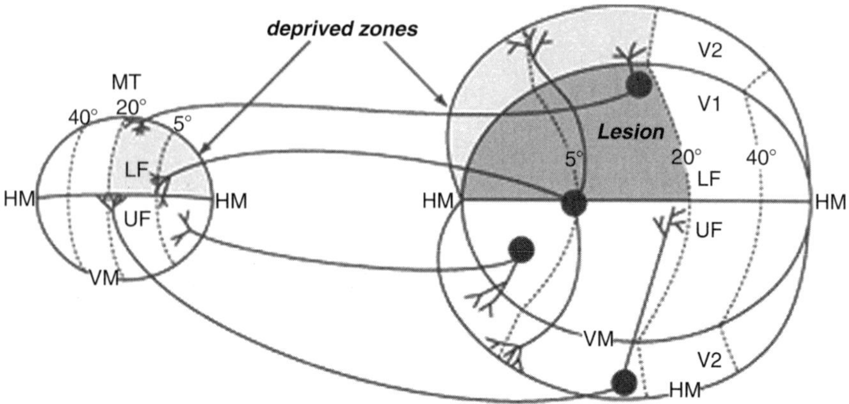

FIGURE 10.3. The projection pattern of V1 to V2 and MT of monkeys, and the effects of a partial lesion of V1. V1 projects topographically to V2 and MT, but in different patterns. The representation of the contralateral visual hemifield in V1 is matched in a representation in MT that is a mirror image reversal of V1. The representation in V2 is "split" along the horizontal meridian (HM) so that V2 forms a long belt around most of V1 along congruent representations of the vertical meridian (VM). V1 projects to retinotopic locations in V2 and MT, with some divergent scatter (Fig. 10.1). A lesion of the central 20° of the representation of the lower visual quadrant (lesion, shading) in V1 deprives corresponding portions of the representations in V2 and MT of a major source of visual activation (the deprived zones). While there is widespread agreement that neurons would be totally deprived of visual activation over most of the deprived zone in V2, studies on MT differ in that some provide evidence for a source of visual activation that is independent of V1. One proposal is that visual inputs from the superior colliculus to the inferior pulvinar are relayed to MT. However, the portion of the inferior pulvinar with dense projections to MT does not receive inputs from the superior colliculus. Other experimental results suggest that some, but only limited filling-in occurs in MT, and this is based on direct or indirect (via V2 & V3) inputs from V1 to MT. The unresponsive zones in V2 and MT appear to be proportionally smaller than the lesioned part of V1, as projections from V1 diverge from precise topography in V2, and especially MT. Thus, neurons along the outer fringe of the deprived zones in V1 and MT are unlikely to be totally deprived of V1 inputs, and, the deprived zones in V2 and MT may partially fill in due to the potentiation of preserved but weak inputs. More extensive filling-in could occur with axon growth, or if horizontal connections in V2 and MT become potentiated or grow. Dots; projection neurons with axons and axon arbors in V2 and MT. LF, lower field; UF, upper field.

visual stimuli. It can be difficult in such experiments to accurately estimate the portion of the representation in V1 that has been removed or deactivated, and deactivations may be variably effective.

The persistence or recovery of responsiveness in MT of humans and macaques to an alternative source of activation after lesions of V1 is a postulated source of "blindsight" (Cowey and Stoerig, 1993; Moore et al., 1998), the ability to make judgments about the attributes of visual stimuli without conscious awareness of those stimuli (Weiskrantz, 1986; Barbur et al., 1993). There is some evidence from functional magnetic imaging in one patient with blindsight that MT is activated by visual stimuli (Barbur et al., 1993). However, some instances of blindsight have been attributed to the preservation of islands of cells in V1 that provide near-threshold vision, and above chance performance without perceptual certainty (Fendrich et al., 1992; Wessinger et al., 1997). One possibility for these mixed results is that humans and macaques differ from New World monkeys (however, see Rosa et al., 2000) and prosimian galagos in having an alternative source for activating MT and contributing to blindsight. Another possibility is that this alternative source exists in all primates, but it is easily disrupted by anesthetics and other conditions. Alternatively, MT may depend on V1 for visual activation in all primates, and there are other explanations for the apparent activation of deprived portions of MT by other sources. In the face of these possibilities, the role of plasticity in reorganization of MT after V1 lesions is difficult to evaluate. Our studies suggest that over a period of months, reactivation of large, deprived portions of MT is incomplete, and based on preserved inputs from V1. While the superior colliculus has been postulated as an alternative source of activation of MT after V1 lesions, such lesions not only directly and indirectly deprive MT neurons of cortical sources of visual activation, but also may greatly depress the evoked visual activity in the superior colliculus (Rushmore and Payne, 2003) and in the targets of the superior colliculus in the pulvinar complex (Bender, 1983).

References

Albright TD, Stoner GR. 2002. Contextual influences on visual processing. *Annu Rev Neurosci* 25:339–379.

Allman JM, Kaas JH. 1971. A representation of a visual field in the caudal third of the middle temporal gyrus of the owl monkey (*Aotus trivirgatus*). *Brain Research* 31:85–105.

Angelucci A, Levitt JB, Walton EJ, Hupe JM, Bullier J, Lund JS. 2002. Circuits fro local and global signal integration in primary visual cortex. *J Neurosci* 22:8633–8646.

Arckens L, Eysel UT, Vanderhaeghen J-J, Orban GA, Vandesande F. 1998. Effect of sensory deafferentation on the GABAergic circuitry of the adult cat visual system. *Neurosci* 83:381–391.

Bair W, Cavanaugh JR, Movshon JA. 2003. Time course and time-distance relationships for surrond suppression in macaque V1 neurons. *J Neurosci* 23:7690–7701.

Baisden RH, Polley EH, Goodman DC, Wolf ED. 1980. Absence of sprouting by retinogeniculate axons after chronic focal lesions in the adult cat retina. *Neurosci Lett* 17:33–38.

Barbur JL, Watson JDG, Frackowiak RSJ, Zeki SM. 1993. Conscious visual perception without V1. *Brain* 116:1293–1302.

Bender DB. 1983. Visual activation of neurons in the primate pulvinar depends on cortex but not colliculus. *Brain Res* 279:258–261.

Bringuier V, Chavane F, Glaeser L, Fregnac Y. 1999. Horizontal propagation of visual activity in the synaptic integration field of area 17 neurons. *Science* 283: 695–699.

Bullier J. 2004. Hierarchies of cortical areas, in: *The Primate Visual System*, J.H. Kaas, C.E. Collins, eds., CRC Press, Boca Raton, pp. 181–204.

Bullier J, Girard P, Salin P-A. 1994, The role of area 17 in the transfer of information to extrastriate visual cortex, in: *Cerebral Cortex*, A. Peters, K. S. Rockland, eds., Plenum Press, New York, pp. 301–330.

Calford M, Schmid LM, Rosa MG. 1999, Monocular focal retinal lesions induce short-term topographic plasticity in adult cat visual cortex. *Proc Royal Soc Lond B Biol Sci* 266:499–507.

Calford M, Wang C, Taglianetti V, Waleszezyk RJ, Burke W, Dreher B. 2000, Plasticity in adult cat visual cortex (area 17) following circumscribed monocular lesions of all retinal layers, *J Physiol* 524:587–605.

Calford MB, Wright LI, Metha AB, Taglianetti V. 2003. Topographic plasticity in primary visual cortex is mediated by local corticocortical connections. *J Neurosci.* 23:6434–6442.

Casagrande VA, Kaas JH. 1994. The afferent, intrinsic, and efferent connections of primary visual cortex in primates, in: A. Peters, K. S. Rockland, eds., *Cerebral Cortex*, Plenum Press, New York, pp. 201–259.

Cavanaugh JR, Bair W, Movshon JA. 2002a, Nature and interaction of signals from the receptive field center and surround in macalue V1 neurons, *J Neurophysiol* 88:2530–2546.

Cavanaugh JR, Bair W, Movshon JA. 2002b. Selectivity and spatial distribution of signals from the receptive field surround in macaque V1 neurons. *J Neurophysiol* 88:2547–2556.

Chapman B, Stone LS. 1996. Turning a blind eye to cortical receptive fields. *Neuron* 16:9–12.

Chino Y, Smith EL, Zhang B, Matsuura K, Mori T, Kaas JH. 2001. Recovery of binocular responses by cortical neurons after early monocular lesions. *Nat Neurosci* 4:689–690.

Chino YM, Kaas JH, Smith EL, III., Langston AL, Cheng H. 1992. Rapid reorganization of cortical maps in adult cats following restricted deafferentation in retina. *Vis Res* 32:789–796.

Chino YM, Smith EL, Kaas JH, Sasaki Y, Cheng H. 1995. Receptive-field properties of deafferentated visual cortical neurons after topographic map reorganization in adult cats. *J Neurosci* 15:2417–2433.

Chowdhury SR, Rasmusson DD. 2003. Corticocortical inhibition of peripheral inputs within primary somatosensory cortex: The role of $GABA_A$ and $GABA_B$ receptors. *J Neurophysiol* 90:851–856.

Collins CE, Lyon DC, Kaas JH. 2003. Responses of neurons in the middle temporal visual area after long-standing lesions of the primary visual cortex in adult new world monkeys. *J Neurosci* 23:2251–2264.

Collins CE, Xu X, Khaytin I, Kaskan PM, Casagrande VA, Kaas JH. 2005. Optical imaging of visually evoked responses in the middle temporal area after deactivation of primal visual cortex in adult primates. *Proc.Natl.Acad. Sci.*USA 102:5594–5599.

Cotman CW, Nieto-Sampedro M. 1982. Brain function, synapse renewal, and plasticity. *Annu Rev Psych* 33:371–401.

Cowey A, Stoerig P. 1993. Insights into blindsight. *Curr Biol* 3:236–238.

Cusick CG, Gould HJI, Kaas JH. 1984. Interhemispheric connections of visual cortex of owl monkeys (*Aotus trivirgatus*), marmosets (*Callithrix jacchus*), and galagos (*Galago crassicaudatus*). *J Comp Neurol* 230:311–336.

Cusick CG, Kaas JH. 1988. Surface view patterns of intrinsic and extrinsic cortical connections of area 17 in a prosimian primate. *Brain Res* 458:383–388.

Darian-Smith C, Gilbert CD. 1994. Axonal sprouting accompanies functional reorganization in adult cat striate cortex. *Nature* 368:737–740.

Darian-Smith C, Gilbert CD. 1995. Topographic reorganization in the striate cortex of the adult cat and monkey is cortically mediated. *J Neurosci* 15:1631–1647.

Das A, Gilbert CD. 1995. Long-range horizontal connections and their role in cortical reorganization revealed by optical recording of cat primary visual cortex. *Nature* 375:780–784.

Davis KD, Kiss ZHT, Luo L, Tasker RR, Lozano AM, Dostrovsky JO. 1998. Phantom sensations generated by thalamic microstimulation. *Nature* 391:385–387.

De Angelis GC, Anzai A, Ohzawa I, Freeman RD. 1995. Receptive field structure in the visual cortex: Does selective stimulation induce plasticity? *Proc Natl Acad Sci U S A* 92:9682–9686.

De Weerd P, Gattass R, Desimone R, Ungerleider LG. 1995. Responses of cells in monkey visual cortex during perceptual filling-in of an artifical scotoma. *Nature* 377:731–734.

Dykes RW. 1997. Mechanisms controlling neuronal plasticity in somatosensory cortex. *Can J Physiol Pharm* 75:535–545.

Ergenzinger ER, Glasier MM, Hahm JO, Pons TP. 1998. Cortically induced thalamic plasticity in the primate somatosensory system. *Nat Neurosci* 1:226–229.

Eyding D, Schweigart G, Eysel UT. 2002. Spatio-temporal plasticity of cortical receptive fields in response to repetitive visual stimulation in the adult cat. *Neurosci* 112: 195–215.

Eysel UT. 1982. Functional reconnections without new axonal growth in a partially denervated visual relay nucleus. *Nature* 299:442–444.

Eysel UT, Schweigart G. 1999. Increased receptive field size in the surround of chronic lesions in the adult cat visual cortex. *Cereb Cortex* 9:101–109.

Eysel UT, Gonzalez-Aguillar F, Mayer U. 1980. A functional sign of reorganization in the visual system of adult cats: lateral geniculate neurons with displaced receptive fields after lesions of the nasal retina. *Brain Res* 191:285–300.

Eysel UT, Eyding D, Schweigart G. 1998. Receptive optical stimulation elicits fast receptive field changes in mature visual cortex. *NeuroReport* 9:949–954.

Felleman DJ, Van Essen DC. 1991. Distributed hierarchical processing in the primate cerebral cortex. *Cereb Cortex* 1:1–47.

Fendrich R, Wessinger CM, Gazzaniga MS. 1992. Residual vision in a scotoma: implications for blindsight. *Science* 258:1489–1491.

Florence SL, Wall JT, Kaas JH. 1989. Somatotopic organization of inputs from the hand to the spinal gray and cuneate nucleus of monkeys with observations on the cuneate nucleus of humans. *J Comp Neurol* 286:48–70.

Florence SL, Taub HB, Kaas JH. 1998. Large-scale sprouting of cortical connections after peripheral injury in adult macaque monkeys. *Science* 282:1117–1121.

Frahm HD, Zilles K, Schleicher A, Stephan H. 1998. The size of the middle temporal visual area in primates. *J Hirnforsch* 39:45–54.

Garraghty PE, LaChica EA, Kaas JH. 1991. Injury-induced reorganization of somatosensory cortex is accompanied by reductions in GABA staining. *Somatosens Mot Res* 8:347–354.

Gattass R, Gross CG. 1981. Visual topography of striate projection zone (MT) in posterior superior temporal sulcus of the macaque. *J Neurophysiol* 3:621–638.

Gerrits HJ, Timmerman GJ. 1969. The filling-in process in patients with retinal scotomata. *Vis Res* 9:439–442.

Gilbert CD, Wiesel TN. 1989. Columnar specificity of intrinsic horizontal and corticocortical connections in cat visual cortex. *J Neurosci* 9:2432–2442.

Gilbert CD, Wiesel TN. 1992. Receptive field dynamics in adult primary visual cortex. *Nature* 356:150–152.

Girard P, Salin P-A, Bullier J. 1991. Visual activity in areas V3a and V3 during reversible inactivation of area V1 in the macaque monkey. *J Neurophysiol* 66:1492–1503.

Girard P, Salin P-A, Bullier J. 1992. Response selectivity of neurons in area MT of the macaque monkey during reversible inactivation of area V1. *J Neurophysiol* 67:1437–1446.

Heinen SJ, Skavenski AA. 1991. Recovery of visual responses in foveal V1 neurons following bilateral foveal lesions in adult monkey. *Exp Brain Res* 83:670–674.

Hess G, Donoghue JP. 1994. Long-term potentiation of horizontal connections provides a mechanism to reorganize cortical motor maps. *J Neurophysiol* 71:2543–2547.

Horton JC, Hocking D. 1998. Monocular core zones and binocular border strips in primate striate cortex revealed by the contrasting effects of enucleation, eyelid suture, and retinal laser lesions on cytochrome oxidase activity. *J Neurosci* 18:5433–5455.

Hupe JM, James AC, Payne BR, Lomber SG, Girard P, Bullier J. 1998. Cortical feedback improves discrimination between figure and background by V1, V2 and V3 neurons. *Nature* 394:784–787.

Hupe JM, James AC, Girard P, Bullier J. 2001a. Response modulations by static texture surround in area V1 of the macaque monkey do not depend on feedback connections from V2. *J Neurophysiol* 85:146–163.

Hupe JM, James AC, Girard P, Lomber SG, Payne BR, Bullier J. 2001b. Feedback connections act on the early part of the responses in monkey visual cortex. *J Neurophysiol* 85:134–145.

Jain N, Florence SL, Qi H-X, Kaas JH. 2000. Growth of new brain stem connections in adult monkeys with massive sensory loss. *Proc Natl Acad Sci USA* 97:5546–5550.

Jenkins WM, Merzenich MM. 1987. Reorganization of neocortical representations after brain injury: a neurophysiological model of the bases of recovery from stroke. *Prog Brain Res* 71:249–266.

Jones EG. 1993. GABAergic neurons and their role in cortical plasticity in primates. *Cereb Cortex* 3:361–372.

Kaas JH, Collins CE. 2003. Anatomic and functional reorganization of somatosensory cortex in mature primates after peripheral nerve and spinal cord injury. *Adv Neurol* 93:87–95.

Kaas JH, Florence SL. 2001. Reorganization of sensory and motor systems in adult mammals after injury. In: Kaas JH, ed.. *The Mutable Brain*. Amsterdam: Harwood Academic Publishers.

Kaas JH, Lyon DC. 2001. Visual cortex organization in primates: theories of V3 and adjoining visual areas. *Prog Brain Res* 134:285–295.

Kaas JH, Krubitzer LA, Chino YM, Langston AL, Polley EH, Blair N. 1990. Reorganization of retinotopic cortical maps in adult mammals after lesions of the retina. *Science* 248:229–231.

Kaas JH, Krubitzer LA. 1992. Area 17 lesions deactivate area MT in owl monkeys. *Vis. Neurosci.* 9:399–407.

Kasten E, Wust S, Behrens-Baumann W, Sabel BA. 1998. Computer-based training for the treatment of partial blindness. *Nat Med* 9:1083–1087.

Komatsu H, Kinoshita M, Murakami I. 2000. Neural responses in the retinotopic representation of the blind spot in the macaque V1 to stimuli for perceptual filling-in. *J Neurosci* 20:9310–9319.

Krubitzer LA, Kaas JH. 1990. Cortical connections of MT in four species of primates: Areal, modular, and retinotopic patterns. *Vis Neurosci* 5:165–204.

Matsuura K, Zhang B, Mori T, Smith EL, 3rd, Kaas JH, Chino Y. 2002. Topographic map reorganization in cat area 17 after early monocular retinal lesions. *Vis Neurosci* 19:85–96.

Maunsell JHR, Nealey TA, DePriest DD. 1990. Magnocellular and parvocellular contributions to responses in the middle temporal visual area (MT) of the macaque monkey. *J Neurosci* 10:3322–3334.

McLean J, Palmer LA. 1998. Plasticity of neuronal response properties in adult cat striate cortex. *Vis Neurosci* 15:177–196.

Merzenich MM, Kaas JH, Wall J, Nelson RJ, Sur M, Felleman D. 1983a. Topographic reorganization of somatosensory cortical areas 3b and 1 in adult monkeys following restricted deafferentation. *Neurosci* 8:33–55.

Merzenich MM, Kaas JH, Wall JT, Sur M, Nelson RJ, Felleman DJ. 1983b. Progression of change following median nerve section in the cortical representation of the hand in areas 3b and 1 in adult owl and squirrel monkeys. *Neurosci* 10:639–665.

Moore T, Rodman HR, Gross CG. 1998. Man, monkey, and blindsight. *The Neuroscientist* 4:227–230.

Obata S, Obata J, Das A, Gilbert CD. 1999. Molecular correlates of topographic reorganization in primary visual cortex following retinal lesions. *Cereb Cortex* 9:238–248.

Pernberg J, Jirmann KU, Eysel UT. 1998. Structure and dynamics of receptive fields in the visual cortex of the cat (area 18) and the influence of GABAergic inhibition. *Eur J Neurosci* 10:3596–3606.

Pettet MW, Gilbert CD. 1992. Dynamic changes in receptive field size in cat primary visual cortex. *Proc Natl Acad Sci USA* 89:8366–8370.

Rockland KS. 2004. Feedback connections: splitting the arrow. In: Kaas JH, Collins CE, eds., The Primate Visual System. Boca Raton: CRC Press. p 387–405.

Rodman HR, Gross CG, Albright TD. 1989. Afferent basis of visual response properties in area MT of the macaque: I. Effects of striate cortex removal. *J. Neurosci* 9:2033–2050.

Rodman HR, Gross CG, Albright TD. 1990. Afferent basis of visual response properties in area MT of the macaque. II: Effects of superior colliculus removal. *J Neurosci* 10:1154–1164.

Rosa MGP, Tweedale R. 2004. Maps of the visual field in the cerebral cortex of primates: functional organization and significance. In: Kaas JH, Collins CE, eds., *The Primate Visual System*. Boca Raton: CRC Press. p 261–288.

Rosa MGP, Tweedale R, Elston GN. 2000. Visual responses of neurons in the middle temporal area of New World monkeys after lesions of striate cortex. *J Neurosci* 20:5552–5563.

Rosier AM, Arckens L, Demeulemeester H, Orban GA, Eysel UT, Wu YJ, Vandesande F. 1995. Effect of sensory deafferentation on immunoreactivity of

GABAergic cells and on GABA receptors in the adult cat visual cortex. *J Comp Neurol* 359:476–489.

Rushmore RJ, Payne BR. 2003. Bilateral impact of unilateral visual cortex lesions on the superior colliculus. *Exp Brain Res* 151:542–547.

Sceniak MP, Hawken MJ, Shapley R. 2001. Visual spatial characterization of macaque V1 neurons. *J Neurophysiol* 85:1873–1887.

Schiller PH, Malpeli JG. 1977. The effect of striate cortex cooling on area 18 cells in the monkey. *Brain Res* 126:366–369.

Schmid LM, Rosa MGP, Calford MB. 1995. Retinal detachment induces massive immediate reorganization in visual cortex. *NeuroReport* 6:1349–1353.

Schmid LM, Rosa MGP, Calford MB, Ambler JS. 1996. Visuotopic reorganization in the primary visual cortex of adult cats following monocular and binocular retinal lesions. *Cereb Cortex* 6:388–405.

Schweigart G, Eysel UT. 2002. Activity-dependent receptive field changes in the surround of adult cat visual cortex lesions. *Eur J Neurosci* 15:1585–1596.

Sergent J. 1988. An investigation into perceptual completion in blind areas of the visual field. *Brain* 111:347–373.

Stelzner DJ, Keating EG. 1977. Lack of intralaminar sprouting of retinal axons in monkey LGN. *Brain Res* 126:201–221.

Stepniewska I, Kaas JH. 1996. Topographic patterns of V2 cortical connections in macaque monkeys. *J Comp Neurol* 371:129–152.

Stepniewska I, Qi H-X, Kaas JH. 1999. Do superior colliculus projection zones in the inferior pulvinar project to MT in primates? *J Eur Neurosci* 11:469–480.

Stepniewska I, Sakai ST, Qi H-X, Kaas JH. 2003. Somatosensory input to the ventrolateral thalamic region (VL) in the macaque monkey. A potential substrate for Parkinsonian tremor. *J Comp Neurol* 455:378–395.

Stettler DD, Das A, Bennett J, Gilbert CD. 2002. Lateral connectivity and contextual interactions in macaque primary visual cortex. *Neuron* 36:739–750.

Toth LJ, Rao SC, Kim DS, Somers D, Sur M. 1996. Subthreshold facilitation and suppression in primary visual cortex revealed by intrinsic signal imaging. *Proc Natl Acad Sci U S A* 93:9869–9874.

Tucker TR, Fitzpatrick D. 2003. Contributions of vertical and horizontal circuits to the response properties of neurons in primary visual cortex. In: *The Visual Neurosciences*. Cambridge, MA: MIT Press. p 733–764.

Tusa RJ, Palmer LA, Rosenquest AC. 1978. The retinotopic organization of area 17 (striate cortex) in the cat. *J Comp Nuerol* 177:213–236.

Van Essen DC, Maunsell JH, Bixby JL. 1981. The middle temporal visual area in the macaque: myeloarchitecture, connections, functional properties and topographic organization. *J Comp Neurol* 199:293–326.

Wall PD. 1977. The presence of ineffective synapses and the circumstances which unmask them. *Phil Trans B Royal Soc, London* 278:361–372.

Weiskrantz L. 1986. *Blindsight: A case study and implications*. Oxford: Oxford Univ Press.

Weiskrantz L, Cowey A. 1967. Comparison of the effects of striate cortex and retinal lesions in visual acuity in the monkey. *Science* 155:104–106.

Weller RE, Kaas JH. 1983. Retinotopic patterns of connections of area 17 with visual areas V-II and MT in macaque monkeys. *J Comp Neurol* 220:253–279.

Wessinger CM, Fendrich R, Gazzaniga MS. 1997. Islands of residual vision in hemianopic patients. *J Cogn Neurosci* 9:203–221.

Wolff JR, Missler M. 1992. Synaptic reorganization in developing and adult nervous systems. *Ann Anat* 174:393–403.

Wu CW, Kaas JH. 2002. The effects of long-standing limb loss on anatomical reorganization of the somatosensory afferents in the brainstem and spinal cord. *Somatosens Mot Res* 19:153–163.

Zepeda A, Saengpiel F, Guagnelli MA, Vaca L, Arias C. 2004. Functional reorganization of visual cortex maps after ischemic lesions is accompanied by changes in expression of cytoskeletal proteins and NMDA and $GABA_A$ receptor subunits. *J Neurosci* 24:1812–1821.

Ziemann U, Corwell B, Cohen LG. 1998. Modulation of plasticity in human motor cortex after forearm ischemic nerve block. *J Neurosci* 18:1115–1123.

Zihl J, von Cramon D. 1985. Visual field recovery from scotoma in patients with postgeniculate damage. *Brain* 108:335–365.

11

Intra-cortical Inhibition in the Regulation of Receptive Field Properties And Neural Plasticity in the Primary Visual Cortex

LIISA A. TREMERE[1] AND RAPHAEL PINAUD[2]

[1]CROET and [2]Neurological Sciences Institute, OHSU, Portland, OR, USA

Introduction

The goal of this chapter is to review critical roles for gamma-amino butyric acid (GABA) in regulating receptive field (RF) properties and some forms of experience-dependent neuronal plasticity via GABA-A and GABA-B receptors in the mammalian primary visual cortex (V1). This chapter will not provide an extensive review of the literature on this topic as several thorough review papers have been written in the past years, including one from us (Tremere et al., 2003), that will be cited throughout this text, where appropriate, for readers interested in pursuing more information on these subjects.

GABA Neurotransmission and GABAergic Receptors

GABA is the principal inhibitory neurotransmitter in the central nervous system (CNS). Neurons that release GABA were once believed to act on the cortex in a broad fashion described as inhibitory tone. In general terms, inhibitory tone can be thought of as a cortex wide mechanism for lowering the gain on all on-going activity without strong temporal or spatial relationships to processing. Subsequently, it was reported that cells in the visual cortex exhibit spatially overlapping excitatory, as well as inhibitory, RFs that compete for functional expression in response to an adequate visual stimulus (Benevento et al., 1972; Blakemore and Tobin, 1972; Creutzfeldt et al., 1974). More contemporary views of GABAergic transmission are that these inhibitory mechanisms are activated in parallel with excitation as part of sensory processing. Furthermore, balances of inhibition and excitation may be critical in determining if a cell moves towards a state of heightened plasticity or becomes more functionally rigid. The roles that GABAergic mechanisms

R. Pinaud, L. A. Tremere, P. De Weerd(Eds.), Plasticity in the Visual System:
From Genes to Circuits, 229-243, ©2006 Springer Science + Business Media, Inc.

may play in normal sensory processing as well as neural plasticity are the subjects of this chapter and are discussed below.

GABAergic Cells and Receptors

GABAergic neurons represent close to 30% of the overall neuronal population in sensory cortices, including V1 (Jones, 1993). Horizontally, the distribution of GABAergic neurons appears to be uniform for same areas of the visual cortex (e.g., V1 and V2), defined on cytoarchitectonic features. There is, however, considerable variation in the proportion of GABAergic neurons when comparisons are made for cortical layers. Highest density of GABAergic neurons appear to be found in layers II/III (supragranular layers), as well as the thalamo-recipient layer IV (Gabbott and Somogyi, 1986). Conversely, significantly smaller numbers of GABAergic neurons are found in the infragranular layers of the mammalian visual cortex (Gabbott and Somogyi, 1986; Hendry et al., 1987; Hendry et al., 1990). This differential distribution of GABAergic neurons in the vertical organization of visual cortex has been suggested to be important in defining the response range for individual cells during visual stimulation, as well as determining the conditions that enable some forms of neuroplastic changes (discussed below).

The effects of GABA are mediated by three classes of GABA receptors: the GABA-A, GABA-B and GABA-C receptors. GABA-A is the dominant and by far most studied class of receptor expressed in the visual cortex. GABA-A receptors are composed of putatively 5 isoforms (heteropentameric) of four different subunit families: alpha, beta, gamma, delta (Macdonald and Olsen, 1994; Luddens et al., 1995). It has been proposed that the most common assembly for GABA-A receptors is 2 alpha subunits, 1 beta and 2 gamma subunits (Barnard, 1995). GABA-A mediated transmission, has been shown to constitute the principal source of fast acting cortical inhibition in the visual cortex (Krnjevic, 1984; Hendry et al., 1990; Jones, 1993). This receptor is very well positioned to heavily influence both the progression and lateral spread of visual signals as it is expressed throughout the horizontal and vertical extent of V1. Binding of GABA at GABA-A receptors immediately enables an inward chloride conductance that hyperpolarizes the post-synaptic cell, removing it farther from the threshold of action potential firing.

In the mammalian V1, GABA-A receptor subunits are found in all cortical layers in both young and adult animals. Receptor distribution varies as a function of cortical layer and in a species-specific manner. For example, in the adult rat V1, highest density of benzodiazepine binding sites were detected in the thalamorecipient layer IV and in cortical layer VI, while lowest distribution was detected for supragranular layers (Rothe and Schliebs, 1989). In the monkey V1, GABA-A receptors were detected throughout all cortical layers, with highest concentration detected in layers II, III, IVa, IVc beta and cortical layer VI; lowest GABA-A immunolabeling was restricted to layers I, IVb and V (Hendry et al., 1990). These patterns were similar to those observed for the human V1, where highest receptor density was located in cortical layers II,

III and IVc and lowest concentrations were detected in layers I, V and VI (Albin et al., 1991; for a detailed description of the ontogeny of GABA-A receptor distribution and the distribution of specific subunits across V1, please refer to Shaw et al., 1984; Schliebs and Rothe, 1988; Hendry et al., 1990; Shaw et al., 1991; Kumar and Schliebs, 1993; Hendrickson et al., 1994; Hendry et al., 1994; Huntsman et al., 1994).

The second type of GABAergic receptor is the metabotropic GABA-B receptor, which is less prevalent in the visual cortex and often located anatomically within the terminals of the thalamic afferents (Munoz et al., 2001). The kinetics of this receptor is slower as compared to GABA-A and GABA-C receptors given that GABA-B receptors are G-protein-coupled (Mody et al., 1994). Presynaptically, the main action of GABA-B receptors is to inhibit voltage-gated calcium channel activity, which in turn decreases synaptic neurotransmitter release. Activation of GABA-B receptors post-synaptically has been shown to increase an inwardly rectifying potassium conductance, via Kir3 channels, which leads to membrane hyperpolarization. In addition, receptor activation post-synaptically has been shown to inhibit adenylate cyclase, thus decreasing cAMP levels (Gahwiler and Brown, 1985; Misgeld et al., 1995; Luscher et al., 1997; Jones et al., 1998; Kaupmann et al., 1998; White et al., 1998; Kuner et al., 1999).

Anatomical characterization of GABA-B receptor distribution revealed that subtypes 1a and 1b are differentially regulated in the primate V1, with highest mRNA expression found in cortical layers IVc and VI, and lowest levels detected in layers I, IVb and V. Modest expression levels were detected in layers IVa and II (Munoz et al., 2001). In the human V1, GABA-B receptors appear to be uniformly distributed throughout cortical layers (Albin et al., 1991). To our knowledge, distribution of GABA-B receptor subunits in the feline and rodent V1 has not yet been characterized.

The third type of receptor, GABA-C, shares many features with the GABA-A receptors. For example, it is a ligand-gated ionotropic receptor that allows for a chloride conductance upon activation (Qian et al., 2005). GABA-C receptors differ from GABA-A receptors by the presence of rho subunits in their structures; the earliest reports on this receptor resulted from experiments conducted in the retina, however, it is still a question of debate whether brain GABA-C receptors are homomeric and whether their molecular composition is identical to those found in the retina (for review see Zhang et al., 2001). Irrespective of the detailed molecular nature of this receptor, it appears that it may provide a mechanism for shunt-inhibition (Palmer MJ, Hull C, Tremere L, von Gersdorff H, unpublished data). Because little evidence exists for the presence of GABA-C receptors in V1 (Wegelius et al., 1998), this chapter will focus on the functional roles of GABA-A and GABA-B receptors.

Sources of GABAergic Input

There are two main sources for GABA in V1. The majority of GABAergic synapses belong to cortical inhibitory interneurons of which there are several morphologically classified types. These cells and their relative positions in V1 have been described in detail in many previous works and will not be discussed further in this chapter (Somogyi et al., 1981; Fairen et al., 1984; Jones, 1993). However, as briefly mentioned above, the greatest numbers of GABAergic neurons per unit area are localized within supragranular layers and layer IV (Gabbott and Somogyi, 1986; Jones, 1993). Interestingly, similar distributions of inhibitory neurons across cortical layers have been described for other primary sensory areas including the somatosensory and auditory areas (Peters and Jones, 1984; Akhtar and Land, 1991; Doetsch et al., 1993; Prieto et al., 1994a, b). The conservation of this general organizational scheme across sensory modalities suggests that inhibitory mechanisms are a central part of sensory processing, particularly in computations associated at the thalamo-recipient and intra- and inter-cortical levels.

A second source of GABAergic inputs to visual cortical neurons are inhibitory projection neurons from the basal forebrain (Freund and Meskenaite, 1992; Gritti et al., 1993; Gritti et al., 2003). The basal forebrain region contains two populations of projection neurons: cholinergic and GABAergic cells, with some neurons expressing both neurotransmitters (Freund and Meskenaite, 1992). The basal forebrain has been repeatedly involved in a wide variety of arousal, attention and learning related paradigms, which have direct implications for the experience-dependent regulation of RF size and sensory processing.

Interestingly, it has been shown that over 25% of the inhibitory projection neuronal population synapses with local inhibitory interneurons in cortical areas, including the visual cortex. These target cells have been extensively characterized by immunocytochemical approaches directed against calcium binding proteins and specific peptides. Somatostatin- and parvalbumin-positive non-pyramidal cells have been shown to constitute the major targets of the GABAergic forebrain with few, if any, contacts with neuropeptide Y, calbindin or cholecystokinin immunoreactive cells (Freund and Meskenaite, 1992; Hicks et al., 1993). These findings indicate that inhibitory projection neurons from the basal forebrain may participate in disinhibitory effects in sensory cortices, which have been proposed to mediate some forms of plasticity in the visual system, for example, neural interpolation (Tremere et al., 2003).

GABAergic Regulation of RFs

The RF of a neuron in the visual cortex can be defined as the area in the visual field from which activity in that cortical neuron can be evoked. As a visual stimulus passes through the region of visual space to which a given cell responds, it will generate a response to signal the presence of the visual stimulus. The classic RF exhibits a concentric organization where stimuli that are placed in the center of the RF will elicit one type of response, usually excitatory, and stimuli placed towards the RF periphery will have an antagonistic effect, inhibitory in this example. While this concentric organization has become the RF prototype, there are many variations of RF organization with complex features in both the spatial and temporal domains (Benevento et al., 1972; Blakemore and Tobin, 1972; Creutzfeldt et al., 1974).

Neurons in V1 exhibit both orientation and direction preference. The role of GABAergic mechanisms in structuring response properties of V1 neurons was proposed following descriptions of direction and orientation selectivity or criteria for optimal stimulation of cortical neurons in V1 (Sillito, 1975b, 1977, 1979). Other stimulus characteristics to which V1 neurons will respond include spatial frequency and dynamic changes in contrast (Foster et al., 1985; Molotchnikoff et al., 1994).

One technique that has been used extensively to identify contributions of inhibitory inputs onto cortical neurons in V1 is microiontophoresis. Given that the dominant receptor sub-type of GABA in visual cortex is the GABA-A, the majority of studies have characterized roles for GABA at V1 cells with bicuculline (BIC), a competitive antagonist for this receptor (Macdonald and Olsen, 1994). The possibility that inhibitory mechanisms could be responsible for features of RFs of visual cortical neurons was first raised by Hubel and Wiesel (1962). This line of questioning was pursued more aggressively by Adam Sillito and colleagues, as well as other groups, starting in the early 1970's. In the following paragraphs we present the main results of these experiments that detail the role of intra-cortical GABAergic transmission in regulating RF properties in V1 neurons.

Local iontophoretic application of BIC prevented the action of GABA at primary visual cortical neurons in the cat. This effect was demonstrated to be qualitatively different from that of increased cell excitability resulting from glutamate application (Sillito, 1975a, b), indicating that the effects of GABA do not merely reflect enhanced neuronal activity. With the utility of BIC established as a selective antagonist to GABA-A mediated neurotransmission in V1, the specific ways in which inhibition shaped the response properties were characterized primarily in two functional classifications of cortical neurons in V1: simple and complex cells.

Simple cells often exhibit small RFs with clear "on" and "off" domains, low baseline activity and sharp orientation preference. Conversely, complex cells are characterized by broader orientation preference curves, high baseline activity and the absence (in most cases) of clearly defined "on" and "off" RF domains (Hubel and Wiesel, 1962, 1968).

Intra-cortical microiontophoretic application of BIC markedly altered firing behavior of V1 simple cells. For example, static flashing stimuli-evoked responses that allowed for the clear delineation of on and off RF domains in this type of cells. Under blockade of GABAergic transmission, however, simple cells in V1 tended to fire robustly for the onset and for the offset of stimulus presentation, but not during the interval between (thus classified as on-off responses), regardless of whether the stimulus was placed on the originally "on" or "off" subfields, or covered the full extend of the RF (Sillito, 1975b).

Normal GABAergic transmission was also shown to significantly contribute to shape RF area. Antagonism of GABA-A receptors by local BIC treatment was shown to significantly expand RF area in V1 neurons, suggesting that GABAergic transmission contributes to the functional pruning of anatomical connections leaving only a subset of inputs expressed (Ramoa et al., 1988; Eysel et al., 1998).

The contributions of GABAergic transmission to orientation and directional selectivity in simple cells were also tested. As indicated above, simple cells exhibit sharp orientation and direction preferences. Blockade of GABA-A receptors by microiontophoretic application of BIC affected both properties. Orientation selectivity tended to become much broader, as compared to the pre-BIC condition (Sillito, 1975b). Similar results were obtained by BIC delivery through micro-osmotic pump (Ramoa et al., 1988). In some infrequent cases, responses were obtained with stimuli oriented 90° from the originally optimal orientation (pre-BIC), although responses would significantly decrease after 22.5° (Sillito, 1975b). Likewise, directional selectivity was completely abolished during BIC application (Sillito, 1975b). Both directional and orientation selectivity were recovered once BIC infusion was stopped; cells tended, however, to exhibit decreased responsiveness for approximately half-hour after pharmacological treatment was interrupted.

BIC application also significantly affected RF properties in V1 complex cells. For example, orientation specificity was significantly decreased upon BIC treatment (Sillito, 1975b). In contrast to what was observed at simple cells, robust responses from complex cells were easily detectable at large angles (including 90°) away from the preferred orientation. Interestingly, BIC treatment in this cell type rarely altered directional preference (Sillito, 1975b). Characterization of the time courses of both application and recovery from BIC treatment revealed that BIC effects on simple and complex cells were usually achieved within 5-15 minutes of initiation of microiontophoretic drug application, while cells tended to return to their "original" (pre-BIC) response state approximately 10 minutes after interruption of BIC treatment. Finally, as was observed for simple cells, recovery from BIC treatment was often accompanied by an overall decrease in excitability in complex cells. In addition, application of glutamate failed to trigger the same modifications in RF properties, suggesting that enhanced excitability does not account for the results observed (Sillito, 1975b).

These findings suggest that orientation selectivity at both simple and complex cells results from intra-cortical GABAergic transmission. Another interesting finding revealed in these studies was that the most powerful inhibitory influence on cortical neurons was not detectable for the center of the RF field but rather displaced to its flanking domains. This finding supports the view that a primary function of GABAergic transmission in V1 is edge definition, either as a control for spatial resolution or to enhance contrast.

Adam Sillito's group also tested the role of GABAergic transmission to shaping RF properties in hypercomplex cells in the supragranular layers of the cat V1. These cells are characterized based on their response to a stimulus of appropriate length and orientation placed within their RFs. Increase in length (or variations in orientation) beyond the optimal level suppresses neuronal responses in this cell type (Hubel and Wiesel, 1965, 1968). In addition, hyper-complex cells often display directional preference and low spontaneous activity.

Microiontophoretic BIC application to hypercomplex cells severely affected RF properties. For example, this treatment significantly decreased orientation preference; cells under BIC effect often responded with the same magnitude to visual stimuli that were shifted 45°-90° from the optimal orientation (pre-BIC) (Sillito and Versiani, 1977). This treatment also significantly decreased length preference in hypercomplex cells. Interestingly, directional preference was not affected by BIC treatment. These results suggest that in hypercomplex cells, GABAergic transmission contributes to the generation of orientation selectivity and length preference, but not directional preference.

Studies conducted by Bolz and Gilbert (1986) aimed at detailing the circuitry that participates on the generation of end-inhibition in the cat V1. It was found that inhibition of layer VI by application of GABA abolished end-inhibition of hypercomplex cells in the granular and supragranular layers (Bolz and Gilbert, 1986). Interestingly, both directional and orientation preferences were not affected by layer VI inhibition. These findings suggested that length preference is regulated by inputs that arise from layer VI and are directed at cells in cortical layer IV. Projections from the latter to supragranular layers would be involved in the generation of end-inhibition in layers II/III. Controversy regarding the conclusions of the previous study resurfaced in subsequent investigations using the same experimental paradigm. Grieve and Sillito found that microiontophoretic application of GABA, and GABA-A receptor agonist muscimol, to layer VI significantly decreased visually-evoked responses in hypercomplex cells of layers II/III and IV, and suggested that inhibition of layer VI cells did not lead to increased responses to sub-optimal stimulus length, but rather reflected a significant decrease in responses to the optimal stimulus (Grieve and Sillito, 1991).

Irrespective of how the cytoarchitecture of V1 enables the formation of complex RF properties, it is clear that control of inhibitory transmission via GABA-A receptors plays a direct role in the dynamic regulation of RF properties in simple, complex and hypercomplex cells in the mammalian V1.

Few experiments have directly investigated the contributions of metabotropic GABA-B receptors to shaping RF properties in the mammalian V1 but some evidence can be found for their participation in visual processing.

Iontophoretic application of saclofen, a specific GABA-B receptor antagonist, in the cat V1 was shown to affect orientation selectivity of a subpopulation of simple and complex cells (Allison et al., 1996). Interestingly, it was reported that the effects of saclofen on orientation selectivity were localized to the sustained, but not transient (approximately the first 120 msec), response that results from grating stimulation. Authors suggested that orientation selectivity triggered by inhibitory mechanisms in V1 could potentially be mediated by two distinct mechanisms involving a faster (GABA-A-mediated) and a slower (GABA-B mediated) component (Allison et al., 1996). Systematic investigations on the contributions of GABA-B receptors to directional (in simple and complex cells) as well as length preference (in hypercomplex cells) remain to be conducted. However, the potential for GABA-B-mediated regulation to these processes appears to be a likely possibility, especially given that highest expression of this receptor appears to be confined to layers IVc and VI, the sites proposed to house many of the mechanisms that underlie the generation of end-inhibition (Bolz and Gilbert, 1986; but see also Grieve and Sillito, 1991).

GABAergic Transmission in Ocular Dominance Maintenance and Plasticity

Primary visual cortical cells often exhibit ocular dominance (OD). This distribution ranges from cells that are driven exclusively by the ipsilateral or contralateral eyes, to cells that are equally driven by both eyes (Hubel and Wiesel, 1962, 1972; Hubel et al., 1977; Blakemore et al., 1978). Efforts that were initiated in the early 1960's provided clear evidence that the normal development of OD relied on experience. Neurons in V1 of animals that had been monocularly deprived underwent "shifts" in ocular preference; with time the large majority of neurons would respond to the non-deprived eye (Wiesel and Hubel, 1963; Blakemore et al., 1978). Two dominant hypotheses were forwarded to account for the OD shifts. The first one proposed that activity-dependent synaptic re-weighting and morphological rewiring programs were likely to account for this phenomenon (Hubel et al., 1977; Shatz and Stryker, 1978), while the second took into consideration a possible GABA-mediated restraint of activity associated with the deprived eye (Duffy et al., 1976).

Anatomical evidence that supported the first hypothesis was obtained by ocular infusion of radioactive markers. Distribution of marked cells suggested a morphological rearrangement of thalamocortical projections (Shatz and Stryker, 1978). The direct test of the second hypothesis was conducted by Sillito and colleagues (1980). By recording from neurons in the intact cat V1, these authors identified cells that responded preferentially and strongly to a single eye. Microiontophoretic application of BIC triggered a reversible shift of ocular preference towards the non-preferred eye in a sub-population of simple and complex cells (Sillito et al., 1980). Similar results were obtained and expanded subsequently by other groups. For example, strabismus triggered by surgical intervention led to the development of a population of primarily monocular cells in the cat V1. Intra-cortical BIC application in this paradigm was shown to reinstate binocularity in a large fraction of cells that originally exhibited monocular responses (Mower et al., 1984). Delivery of BIC to V1 of monocularly-deprived cats for 1 week, by means of micro-osmotic pumps, revealed a marked suppression of OD shifts compared to saline-treated cells (in opposite hemisphere) (Ramoa et al., 1988).

A series of more recent studies further implicate normal GABAergic transmission to OD plasticity in the mammalian visual system. It has been shown that knock-out mice with a targeted disruption aimed at the 65 KDa glutamic acid decarboxylase (GAD65) gene do not exhibit the normal plastic effects triggered by monocular deprivation (Hensch et al., 1998). Reinstatement of OD plasticity in GAD65 knock-out animals was achieved by local infusion of diazepam, a benzodiazepine agonist, providing further evidence for the participation of GABA in the regulation of experience-dependent plasticity in the visual cortex (Hensch et al., 1998). One implication of these experiments is that normal GABAergic transmission is needed for the weakening of inputs that are associated with the deprived eye. Subsequent studies conducted in a visual cortex slice preparation of GAD65 knock-out mice revealed that the induction of long-term depression (LTD) in this experimental preparation is markedly impaired; however, chronic treatment with diazepam rescues normal LTD in these knock-out animals (Choi et al., 2002). The development of the critical period also appears to be correlated with alterations of GABA-mediated transmission. For example, in the rat visual cortex, the strength of GABAergic inhibition onto supragranular pyramidal cells undergoes a significant increase (quantified to approximately 300%) from eye-opening to the end of the formal critical period. Interestingly, normal developmental increase in GABAergic strength is blocked by dark-isolation from birth (Morales et al., 2002). This finding is in agreement with dark-induced alterations in the expression of GABA-A receptor subunits in the rat and cat visual cortices (Benevento et al., 1995; Chen et al., 2001). Together these findings suggest that visual input might allow for the normal development of GABAergic networks within the visual cortex, which in turn participate in visual cortical plasticity possibly by weakening unused connections via LTD-like forms of synaptic plasticity. To our knowledge the contributions of GABA-B receptors to the maintenance and plasticity of ocular dominance columns has not been investigated.

Final Comments

Perhaps a unifying theme in the discussion of the roles of GABA in the visual cortex is the tendency for its use in spatial and temporal filtering of sensory information. Far from acting generally as an activity suppressant, GABAergic synaptic activity appears to be precisely controlled, both in strength and timing. At all three categories of V1 neurons discussed in the preceeding paragraphs-simple, complex and hypercomplex cells-the removal of GABAergic inputs reduced the selectivity of responsiveness for at least one response dimension of that neuron. At simple cells, orientation selectivity became seriously compromised with the removal of inhibition

(Sillito, 1975b). Likewise, both complex and hypercomplex cells were shown to respond less selectively to either orientation or length, respectively, when GABA binding was prevented with the use of BIC (Sillito, 1975b, 1977; Sillito and Versiani, 1977). In the OD plasticity paradigm, this general concept is reaffirmed with the demonstration that adjustments to the effective strength of inhibition predicted the appearance or loss of plasticity within that system (Morales et al., 2002).

As discussed in the previous paragraphs, in the absence of filtering imposed by inhibitory mechanisms, the visual system must process information that would normally be suppressed (e.g., respond to a non-optimal orientation). One of the functions of GABAergic mechanisms therefore appears to be to shunt away extraneous or inappropriate inputs, thereby affording the network great economies in processing. In V1, this filtering capability can be interpreted as enhancing the resolution of the sensory signal and, by extension, improving computation and/or enabling more exact response choices.

Roles of inhibition in plasticity may be more complicated to interpret as there appears to be an inverse relationship between the ease of inducing changes, like synaptic potentiation or depression, and the amount of functional inhibition. In deafferentation models where plasticity is observed as reorganization, the greatest changes appear to occur prior to a reinstatement of inhibitory markers to expected levels within sensory cortices (Akhtar and Land, 1991; Rosier et al., 1995; Eysel et al., 1999). Developmental studies (both molecular and functional) and computer modeling are in agreement with the notion that GABAergic inhibition may, to some extent, counter the expression of plasticity in primary sensory areas (Xing and Gerstein, 1996a, b; Choi et al., 2002; Morales et al., 2002). Data that supports this view was discussed above in light of the observation that the end of the critical period, hence the reduction of the period of high plasticity, coincided with the increased expression and strength of the inhibitory system (Morales et al., 2002). A similar trend for reduced plasticity with the return of inhibitory mechanisms has been described in deafferented and reorganized cortex (Rosier et al., 1995; Tremere et al., 2001a, b; Chowdhury and Rasmusson, 2002).

Interesting future directions in this area are likely to involve the characterization of excitation/inhibition rationing and the tendencies for a system to express plasticity-driven changes. The earliest indications are that stimulus pattern repetition, coincidence detection, and neuromodulatory influences are all likely to drive and sustain plasticity in the normal organism. Understanding this balance and controlling the expressed plasticity in a normal system may offer insights in how to improve internally generated learning states, perhaps leading to interventions for those who have neurochemically-determined learning impairments.

In deafferentation-induced plasticity paradigms, it appears that a transient decrease of inhibitory strength that follows sensory deafferentation may generate conditions where invading activity has greater relative influence to alter the network, resembling organizational schemes observed in development during the critical period. Understanding how to prolong, or maximally exploit, the conditions of the disinhibitory reorganizational window afforded by initial loss of sensory drive may hold great promise for therapeutic interventions directed at pathological states.

References

Akhtar ND, Land PW (1991) Activity-dependent regulation of glutamic acid decarboxylase in the rat barrel cortex: effects of neonatal versus adult sensory deprivation. J Comp Neurol 307:200–213.

Albin RL, Sakurai SY, Makowiec RL, Higgins DS, Young AB, Penney JB (1991) Excitatory amino acid, GABA(A), and GABA(B) binding sites in human striate cortex. Cereb Cortex 1:499–509.

Allison JD, Kabara JF, Snider RK, Casagrande VA, Bonds AB (1996) GABAB-receptor-mediated inhibition reduces the orientation selectivity of the sustained response of striate cortical neurons in cats. Vis Neurosci 13:559–566.

Barnard EA (1995) The molecular biology of GABAA receptors and their structural determinants. Adv Biochem Psychopharmacol 48:1–16.

Benevento LA, Creutzfeldt OD, Kuhnt U (1972) Significance of intracortical inhibition in the visual cortex. Nat New Biol 238:124–126.

Benevento LA, Bakkum BW, Cohen RS (1995) gamma-Aminobutyric acid and somatostatin immunoreactivity in the visual cortex of normal and dark-reared rats. Brain Res 689:172–182.

Blakemore C, Tobin EA (1972) Lateral inhibition between orientation detectors in the cat's visual cortex. Exp Brain Res 15:439–440.

Blakemore C, Garey LJ, Vital-Durand F (1978) The physiological effects of monocular deprivation and their reversal in the monkey's visual cortex. J Physiol 283: 223–262.

Bolz J, Gilbert CD (1986) Generation of end-inhibition in the visual cortex via interlaminar connections. Nature 320:362–365.

Chen L, Yang C, Mower GD (2001) Developmental changes in the expression of GABA(A) receptor subunits (alpha(1), alpha(2), alpha(3)) in the cat visual cortex and the effects of dark rearing. Brain Res Mol Brain Res 88:135–143.

Choi SY, Morales B, Lee HK, Kirkwood A (2002) Absence of long-term depression in the visual cortex of glutamic Acid decarboxylase-65 knock-out mice. J Neurosci 22:5271–5276.

Chowdhury SA, Rasmusson DD (2002) Effect of GABAB receptor blockade on receptive fields of raccoon somatosensory cortical neurons during reorganization. Exp Brain Res 145:150–157.

Creutzfeldt OD, Kuhnt U, Benevento LA (1974) An intracellular analysis of visual cortical neurones to moving stimuli: response in a co-operative neuronal network. Exp Brain Res 21:251–274.

Doetsch GS, Norelle A, Mark EK, Standage GP, Lu SM, Lin RC (1993) Immunoreactivity for GAD and three peptides in somatosensory cortex and thalamus of the raccoon. Brain Res Bull 31:553–563.

Duffy FH, Burchfiel JL, Conway JL (1976) Bicuculline reversal of deprivation amblyopia in the cat. Nature 260:256–257.

Eysel UT, Shevelev IA, Lazareva NA, Sharaev GA (1998) Orientation tuning and receptive field structure in cat striate neurons during local blockade of intracortical inhibition. Neuroscience 84:25–36.

Eysel UT, Schweigart G, Mittmann T, Eyding D, Qu Y, Vandesande F, Orban G, Arckens L (1999) Reorganization in the visual cortex after retinal and cortical damage. Restor Neurol Neurosci 15:153–164.

Fairen A, DeFelipe J, Regidor J (1984) Non pyramidal neurons: general account. In: Cellular components of the cerebral cortex (Peters A, Jones EG, eds), pp 201–253. New York: Plenum Press.

Foster KH, Gaska JP, Nagler M, Pollen DA (1985) Spatial and temporal frequency selectivity of neurones in visual cortical areas V1 and V2 of the macaque monkey. J Physiol 365:331–363.

Freund TF, Meskenaite V (1992) gamma-Aminobutyric acid-containing basal forebrain neurons innervate inhibitory interneurons in the neocortex. Proc Natl Acad Sci U S A 89:738–742.

Gabbott PL, Somogyi P (1986) Quantitative distribution of GABA-immunoreactive neurons in the visual cortex (area 17) of the cat. Exp Brain Res 61:323–331.

Gahwiler BH, Brown DA (1985) GABAB-receptor-activated K+ current in voltage-clamped CA3 pyramidal cells in hippocampal cultures. Proc Natl Acad Sci U S A 82:1558–1562.

Grieve KL, Sillito AM (1991) A re-appraisal of the role of layer VI of the visual cortex in the generation of cortical end inhibition. Exp Brain Res 87:521–529.

Gritti I, Mainville L, Jones BE (1993) Codistribution of GABA-with acetylcholine-synthesizing neurons in the basal forebrain of the rat. J Comp Neurol 329:438–457.

Gritti I, Manns ID, Mainville L, Jones BE (2003) Parvalbumin, calbindin, or calretinin in cortically projecting and GABAergic, cholinergic, or glutamatergic basal forebrain neurons of the rat. J Comp Neurol 458:11–31.

Hendrickson A, March D, Richards G, Erickson A, Shaw C (1994) Coincidental appearance of the alpha 1 subunit of the GABA-A receptor and the type I benzodiazepine receptor near birth in macaque monkey visual cortex. Int J Dev Neurosci 12:299–314.

Hendry SH, Schwark HD, Jones EG, Yan J (1987) Numbers and proportions of GABA-immunoreactive neurons in different areas of monkey cerebral cortex. J Neurosci 7:1503–1519.

Hendry SH, Fuchs J, deBlas AL, Jones EG (1990) Distribution and plasticity of immunocytochemically localized GABAA receptors in adult monkey visual cortex. J Neurosci 10:2438–2450.

Hendry SH, Huntsman MM, Vinuela A, Mohler H, de Blas AL, Jones EG (1994) GABAA receptor subunit immunoreactivity in primate visual cortex: distribution in macaques and humans and regulation by visual input in adulthood. J Neurosci 14:2383–2401.

Hensch TK, Fagiolini M, Mataga N, Stryker MP, Baekkeskov S, Kash SF (1998) Local GABA circuit control of experience-dependent plasticity in developing visual cortex. Science 282:1504–1508.

Hicks TP, Albus K, Kaneko T, Baumfalk U (1993) Examination of the effects of cholecystokinin 26–33 and neuropeptide Y on responses of visual cortical neurons of the cat. Neuroscience 52:263–279.

Hubel DH, Wiesel TN (1962) Receptive fields, binocular interaction and functional architecture in the cat's visual cortex. J Physiol 160:106–154.

Hubel DH, Wiesel TN (1965) Receptive Fields and Functional Architecture in Two Nonstriate Visual Areas (18 and 19) of the Cat. J Neurophysiol 28:229–289.

Hubel DH, Wiesel TN (1968) Receptive fields and functional architecture of monkey striate cortex. J Physiol 195:215–243.

Hubel DH, Wiesel TN (1972) Laminar and columnar distribution of geniculo-cortical fibers in the macaque monkey. J Comp Neurol 146:421–450.

Hubel DH, Wiesel TN, LeVay S (1977) Plasticity of ocular dominance columns in monkey striate cortex. Philos Trans R Soc Lond B Biol Sci 278:377–409.

Huntsman MM, Isackson PJ, Jones EG (1994) Lamina-specific expression and activity-dependent regulation of seven GABAA receptor subunit mRNAs in monkey visual cortex. J Neurosci 14:2236–2259.

Jones EG (1993) GABAergic neurons and their role in cortical plasticity in primates. Cereb Cortex 3:361–372.

Jones KA, Borowsky B, Tamm JA, Craig DA, Durkin MM, Dai M, Yao WJ, Johnson M, Gunwaldsen C, Huang LY, Tang C, Shen Q, Salon JA, Morse K, Laz T, Smith KE, Nagarathnam D, Noble SA, Branchek TA, Gerald C (1998) GABA(B) receptors function as a heteromeric assembly of the subunits GABA(B)R1 and GABA(B)R2. Nature 396:674–679.

Kaupmann K, Malitschek B, Schuler V, Heid J, Froestl W, Beck P, Mosbacher J, Bischoff S, Kulik A, Shigemoto R, Karschin A, Bettler B (1998) GABA(B)-receptor subtypes assemble into functional heteromeric complexes. Nature 396:683–687.

Krnjevic K (1984) Some functional consequences of GABA uptake by brain cells. Neurosci Lett 47:283–287.

Kumar A, Schliebs R (1993) Postnatal ontogeny of GABAA and benzodiazepine receptors in individual layers of rat visual cortex and the effect of visual deprivation. Neurochem Int 23:99–106.

Kuner R, Kohr G, Grunewald S, Eisenhardt G, Bach A, Kornau HC (1999) Role of heteromer formation in GABAB receptor function. Science 283:74–77.

Luddens H, Korpi ER, Seeburg PH (1995) GABAA/benzodiazepine receptor heterogeneity: neurophysiological implications. Neuropharmacology 34:245–254.

Luscher C, Jan LY, Stoffel M, Malenka RC, Nicoll RA (1997) G protein-coupled inwardly rectifying K+ channels (GIRKs) mediate postsynaptic but not presynaptic transmitter actions in hippocampal neurons. Neuron 19:687–695.

Macdonald RL, Olsen RW (1994) GABAA receptor channels. Annu Rev Neurosci 17:569–602.

Misgeld U, Bijak M, Jarolimek W (1995) A physiological role for GABAB receptors and the effects of baclofen in the mammalian central nervous system. Prog Neurobiol 46:423–462.

Mody I, De Koninck Y, Otis TS, Soltesz I (1994) Bridging the cleft at GABA synapses in the brain. Trends Neurosci 17:517–525.

Molotchnikoff S, Durand V, Casanova C (1994) Visual cortical neuron responses to drifting sine-wave gratings in rabbits. Int J Neurosci 77:99–115.

Morales B, Choi SY, Kirkwood A (2002) Dark rearing alters the development of GABAergic transmission in visual cortex. J Neurosci 22:8084–8090.

Mower GD, Christen WG, Burchfiel JL, Duffy FH (1984) Microiontophoretic bicuculline restores binocular responses to visual cortical neurons in strabismic cats. Brain Res 309:168–172.

Munoz A, DeFelipe J, Jones EG (2001) Patterns of GABA(B)R1a,b receptor gene expression in monkey and human visual cortex. Cereb Cortex 11:104–113.

Peters A, Jones EG (1984) Cellular components of the cerebral cortex. New York: Plenum Press.

Prieto JJ, Peterson BA, Winer JA (1994a) Laminar distribution and neuronal targets of GABAergic axon terminals in cat primary auditory cortex (AI). J Comp Neurol 344:383–402.

Prieto JJ, Peterson BA, Winer JA (1994b) Morphology and spatial distribution of GABAergic neurons in cat primary auditory cortex (AI). J Comp Neurol 344:349–382.

Qian H, Pan Y, Zhu Y, Khalili P (2005) Picrotoxin accelerates relaxation of GABAC receptors. Mol Pharmacol 67:470–479.

Ramoa AS, Paradiso MA, Freeman RD (1988) Blockade of intracortical inhibition in kitten striate cortex: effects on receptive field properties and associated loss of ocular dominance plasticity. Exp Brain Res 73:285–296.

Rosier AM, Arckens L, Demeulemeester H, Orban GA, Eysel UT, Wu YJ, Vandesande F (1995) Effect of sensory deafferentation on immunoreactivity of GABAergic cells and on GABA receptors in the adult cat visual cortex. J Comp Neurol 359:476–489.

Rothe T, Schliebs R (1989) Laminar distribution of benzodiazepine receptors in visual cortex of adult rat. Gen Physiol Biophys 8:371–380.

Schliebs R, Rothe T (1988) Development of GABAA receptors in the central visual structures of rat brain. Effect of visual pattern deprivation. Gen Physiol Biophys 7:281–291.

Shatz CJ, Stryker MP (1978) Ocular dominance in layer IV of the cat's visual cortex and the effects of monocular deprivation. J Physiol 281:267–283.

Shaw C, Needler MC, Cynader M (1984) Ontogenesis of muscimol binding sites in cat visual cortex. Brain Res Bull 13:331–334.

Shaw C, Cameron L, March D, Cynader M, Zielinski B, Hendrickson A (1991) Pre- and postnatal development of GABA receptors in Macaca monkey visual cortex. J Neurosci 11:3943–3959.

Sillito AM (1975a) The effectiveness of bicuculline as an antagonist of GABA and visually evoked inhibition in the cat's striate cortex. J Physiol 250:287–304.

Sillito AM (1975b) The contribution of inhibitory mechanisms to the receptive field properties of neurones in the striate cortex of the cat. J Physiol 250:305–329.

Sillito AM (1977) Inhibitory processes underlying the directional specificity of simple, complex and hypercomplex cells in the cat's visual cortex. J Physiol 271:699–720.

Sillito AM (1979) Inhibitory mechanisms influencing complex cell orientation selectivity and their modification at high resting discharge levels. J Physiol 289:33–53.

Sillito AM, Versiani V (1977) The contribution of excitatory and inhibitory inputs to the length preference of hypercomplex cells in layers II and III of the cat's striate cortex. J Physiol 273:775–790.

Sillito AM, Kemp JA, Patel H (1980) Inhibitory interactions contributing to the ocular dominance of monocularly dominated cells in the normal cat striate cortex. Exp Brain Res 41:1–10.

Somogyi P, Cowey A, Halasz N, Freund TF (1981) Vertical organization of neurones accumulating 3H-GABA in visual cortex of rhesus monkey. Nature 294: 761–763.

Tremere L, Hicks TP, Rasmusson DD (2001a) Expansion of receptive fields in raccoon somatosensory cortex in vivo by GABA(A) receptor antagonism: implications for cortical reorganization. Exp Brain Res 136:447–455.

Tremere L, Hicks TP, Rasmusson DD (2001b) Role of inhibition in cortical reorganization of the adult raccoon revealed by microiontophoretic blockade of GABA(A) receptors. J Neurophysiol 86:94–103.

Tremere LA, Pinaud R, De Weerd P (2003) Contributions of inhibitory mechanisms to perceptual completion and cortical reorganization. In: Filling-in: from perceptual completion to cortical reorganization (Pessoa L, De Weerd P, eds), pp 295–322. New York: Oxford University Press.

Wegelius K, Pasternack M, Hiltunen JO, Rivera C, Kaila K, Saarma M, Reeben M (1998) Distribution of GABA receptor rho subunit transcripts in the rat brain. Eur J Neurosci 10:350–357.

White JH, Wise A, Main MJ, Green A, Fraser NJ, Disney GH, Barnes AA, Emson P, Foord SM, Marshall FH (1998) Heterodimerization is required for the formation of a functional GABA(B) receptor. Nature 396:679–682.

Wiesel TN, Hubel DH (1963) Single-Cell Responses in Striate Cortex of Kittens Deprived of Vision in One Eye. J Neurophysiol 26:1003–1017.

Xing J, Gerstein GL (1996a) Networks with lateral connectivity. I. dynamic properties mediated by the balance of intrinsic excitation and inhibition. J Neurophysiol 75:184–199.

Xing J, Gerstein GL (1996b) Networks with lateral connectivity. III. Plasticity and reorganization of somatosensory cortex. J Neurophysiol 75:217–232.

Zhang D, Pan ZH, Awobuluyi M, Lipton SA (2001) Structure and function of GABA(C) receptors: a comparison of native versus recombinant receptors. Trends Pharmacol Sci 22:121–132.

12

Plasticity in V1 Induced by Perceptual Learning

PETER DE WEERD[1], RAPHAEL PINAUD[2]
AND GIUSEPPE BERTINI[3]

[1]Department of Psychology, University of Maastricht, The Netherlands
[2]Neurological Sciences Institute, OHSU, Portland, OR, USA
[3]Department of Morphological & Biomedical Sciences, University of Verona, Italy

1. Introduction

Primary visual cortex (V1) has been shown to be highly malleable during early development (Blakemore et al., 1978; Cynader et al., 1980; Hubel & Wiesel, 1965; Hubel et al., 1977; Olson & Freeman, 1978, 1980; Shatz & Striker, 1978), but V1 remains plastic during adulthood. One of the most spectacular forms of plasticity described in adult V1 takes place after restricted retinal lesions, which locally deafferents V1 neurons. A number of studies (for review, see Kaas et al., 2003) have demonstrated that within minutes after the lesion, neurons just inside the deafferented zone in V1 expanded or shifted their receptive fields (RFs) to include regions in the visual field just outside the retinal scotoma. In the course of several months, this reorganization often extended to neurons localized increasingly towards the middle of the deafferented region in V1, until all neurons acquired RFs outside the retinal scotoma. Thus, the initial shifts of RFs occurring immediately after a retinal lesion represent the beginning of a more permanent remapping of visual space onto V1.

Plasticity in the adult cortex is not limited to the visual system. For example, Pons et al. (1991) demonstrated large topographical reorganization in somatosensory cortex after deafferentation of the arm and hand. Several years after deafferentation, the cortical region in SI that normally responded to stimulation of arm and hand could be activated by facial stimuli. Many studies have reported related findings in somatosensory and other sensory modalities (for review, see Buonomano & Merzenich, 1998; Tremere et al., 2003). In addition, skill learning has been shown to produce topographical reorganization in sensory cortex. For example, violinists show an enlarged representation in somatosensory cortex of the hand used to press the strings (Elbert et al., 1995). Physiological and fMRI studies have shown that skill

R. Pinaud, L. A. Tremere, P. De Weerd(Eds.), Plasticity in the Visual System:
From Genes to Circuits, 245-283, ©2006 Springer Science + Business Media, Inc.

learning can result in an expansion of task-relevant topographic representations in lower-order motor (Karni et al., 1995; Nudo et al., 1996), somatosensory (Jenkins et al., 1990; Recanzone et al., 1992a-c), and auditory cortex (Recanzone et al., 1993; Weinburger et al., 1993; Scheich et al., 1993). Hence, a correlate of use (during skill learning) is an expansion of representations, while the correlate of under-use (for example, after deafferentation) is a shrinking due to invasion by other representations. Furthermore, in motor cortex, training has been demonstrated to induce neuronal morphological changes (Kleim et al., 2004) and changes in horizontal connectivity, which could form the basis for topographical changes induced by deafferentation or skill learning (Donoghue, 1995; Hess & Donoghue 1994, 1996; Rioult-Pedotti et al., 1998, 2000).

As shown in other sensory modalities, training in a visual skill (perceptual learning) can also induce plasticity in the visual system. While evidence for functional plasticity at higher-order, integrative stages of vision has long been available, the possibility that training can induce modifications in the circuitry underlying early visual processing, including V1, traditionally considered hard-wired in adulthood, has emerged only recently. The present chapter will focus on physiological studies in monkeys that have attempted to link changes in the properties of V1 neurons with training-induced improvements in the perception of specific stimuli. In these studies, monkeys are trained in a visual discrimination task in which improved performance is hypothesized to depend upon plasticity in V1 (and possibly other early visual areas). A correlation between increased discrimination performance and plasticity in V1 is seen as evidence that the memory for the trained perceptual skill is laid down at least in part in V1, or requires V1. In these physiological studies, V1 plasticity is assessed by measuring changes in RF properties and topography. Some fMRI studies and psychophysical in human subjects that suggest learning-induced plasticity in V1 or in related low-level visual areas will be reviewed as well.

Several authors have proposed by now classic cognitive principles that permit a prediction of the site of plasticity in the visual system based on the behavioral properties of the learning and the type of stimuli used. The goal of this review is to compare studies that have used different stimuli and tasks in order to help establish which types of perceptual training are most likely to induce plasticity in V1, and assess whether well-accepted principles of perceptual learning fit the empirical evidence. We will show that slight differences in attention conditions, stimuli, behavioral tasks, and in the levels of expertise, will determine whether there will be learning-induced plasticity in V1, and which RF properties might be modified. Furthermore, we will argue that the data do not always easily fit generally accepted principles of perceptual learning. In the discussion, we propose and illustrate that the addition of a molecular analysis of perceptual learning to current neuroscience approaches will greatly enhance our insight into this type of learning.

2. The anatomical and physiological basis for perceptual learning in V1

2.1 Factors contributing to RF definition in V1

Before discussing the effects of perceptual learning on the properties of V1 neurons, a review of the circuitry that produces the standard properties of V1 neurons is in order. The majority of sensory inputs are relayed by thalamic afferents into layer IV of primary sensory cortices (including V1), after which information is sent to superficial layers (II & III), and subsequently to deep layers (V & VI). Along this processing pathway, to which inhibitory interneurons contribute importantly, the size and complexity of RF characteristics increase progressively (Bolz et al., 1989; Douglas et al., 1989; Armstrong-James et al., 1992). Primary visual cortex forwards its outputs to layer IV of the next cortical processing stage, from where it also receives feedback, which predominantly originates in deep cortical layers. There is substantial horizontal connectivity within cortical areas, predominantly in superficial layers (Schwark & Jones, 1989; McGuire et al., 1991; Tanifuji et al., 1994; see also Levitt et al., 1994). RFs are the product of the balance between excitatory and inhibitory inputs provided by all their anatomical connections.

From a functional point of view, visual RFs can be subdivided into a classical RF (CRF), and a RF surround. Single stimuli presented against a uniform background drive visual neurons only if presented within the CRF. The best stimulus for a typical V1 neuron is a small, optimally oriented luminance bar presented in the CRF center. The CRF of V1 neurons is surrounded by a region within which the presentation of a small, single stimulus elicits little or no change in firing rate, although the surround stimulus can modulate responses to stimuli concurrently presented in the CRF. Stimuli in the surround generally suppress responses to stimuli inside the CRF, suggesting GABAergic contributions to center-surround interactions (Maffei & Fiorentini, 1976; Nelson & Frost, 1978, Allman et al., 1985; Desimone & Schein, 1987; Schein & Desimone, 1990; Knierim & Van Essen, 1992; Levitt & Lund, 1997; Bringuier et al., 1999; Hupe et al., 2001).

The overall structure of a CRF and its surround can be approximated by a 'Difference of Gaussians' model (Enroth-Cugell & Robson, 1966). In this model, a broad Gaussian distribution of inhibition is subtracted from a narrower Gaussian distribution of excitation, with a higher maximum, to yield an excitatory classical RF enveloped in a suppressive surround (Laskin & Spencer, 1979; Marcelja, 1980; Stork & Wilson, 1990; Sceniak et al., 2001). The CRF is derived to a significant extent from thalamic input, or inputs from lower-order cortical neurons, while lateral cortico-cortical connections have been proposed as the anatomical substrate for the suppressive surround. In visual cortex, it has been shown that axon collaterals of pyramidal cells can extend laterally for 5mm or more (Gilbert & Wiesel, 1989; McGuire et al., 1991;

Levitt et al., 1994). They make direct excitatory connections on other pyramidal cells, but they can also make connections on local GABAergic neurons (Winfield et al., 1981; McGuire et al., 1991) which may inhibit nearby pyramidal cells. In addition to a contribution of horizontal connections, feedback influences have been reported to contribute to the extent and properties of the surround (Lamme, 1995; Zipser et al., 1996; Levitt & Lund, 2002; Bair et al., 2003).

Through the action of local, inhibitory interneurons, GABA contributes significantly to the definition of both the CRF and the surround of V1 neurons. In a recent study, Ringach et al. (2002) found that V1 neurons most narrowly tuned to orientation or spatial frequency were also the ones that showed the strongest suppression when non-optimal stimuli were shown in their CRFs. Sillito (1975a; 1975b; 1977; 1979) and others (Eysel et al., 1998; Bolz et al., 1989; Pernberg et al., 1998) directly observed the role of inhibition as a significant loss of direction and orientation selectivity and changes in RF profile in V1 neurons during microiontophoretic application of BMI (bicuculline methiodide, a $GABA_A$ antagonist). BMI application primarily affected the properties of neurons in superficial cortical layers (Eysel et al., 1998), suggesting a dominance of GABAergic cells in these layers. Related findings have been reported in other sensory modalities (for review, see Tremere et al., 2003).

Studies of perceptual learning have reported changes in the properties of both the CRF and the surround. Below, we will review a number of these studies to gain insight in the perceptual learning conditions that produce one or the other type of RF change. Before we do so, the concept of perceptual learning must be introduced.

2.2 What is perceptual learning?

Perceptual learning is a form of skill learning, which, in turn, is part of the broader class of implicit or procedural learning. Indeed, learning to ride a bike, to play the piano, to tune a violin, or to perceive the precise orientation of a line segment are all examples of skill learning. Acquiring and perfecting a skill requires extensive practice, but once attained, the performance gains are typically retained over extended periods of time even in the absence of further training. Thus, skill learning is to be distinguished from incidental learning (e.g., the formation of episodic or factual memory), which may require a single trial to establish and, unlike skill learning, is strongly dependent on the hippocampus and adjacent cortices (McClelland et al., 1995; Reber & Squire, 1994; Nadel & Moscovitch, 2001; O'Reilly & Rudy, 2001).

In a typical visual perceptual learning task, subjects are asked to detect a simple visual stimulus or to discriminate one of its features. While conceptually very simple, the task is made perceptually challenging by progressively decreasing the strength of the signal to be detected. Performance thresholds are measured during training, which usually consists of a few days to a few weeks of daily sessions.

Skill learning and perceptual learning are characterized by a specific time course (Karni & Sagi, 1993; Karni, 1996; Karni & Bertini, 1997; Karni et al., 1998). When confronted with a novel task, subjects usually show large increments in performance during one or more initial sessions. Later on, learning rate slows down gradually until a learning-asymptote is reached. fMRI studies suggest that independent of the type of skill being learned, there is a common sequence in the type of brain structures that is recruited during initial-to-late learning phases. During fast, initial learning, many structures are active, which may include primary motor and sensory areas and related cortex, frontal and parietal cortex, the cerebellum, and the basal ganglia (Ahissar & Hochstein, 1993, 2002; Willingham et al. 1998; Doyon et al., 2003), dependent on the task and stimuli. During initial learning, subjects distill the rules and set up strategies to solve the task; processes which usually require attention and working memory. During later, asymptotic learning, when performance becomes automated, the contribution of many, but not all, of those structures decreases. Depending on the type of skill, long-term retention may involve the basal ganglia or the cerebellum (Fernandez-Ruiz et al., 2001; Doyon et al., 2003). It is during this late, asymptotic learning phase that many behavioral studies have reported strong specificities of skilled performance (Karni & Bertini, 1997; Karni et al., 1995; Schwartz et al., 2002; Karni & Sagi, 1991; Fahle, 1997; Schoups et al., 1995), and that physiological and neuroimaging studies have shown expansions of task-relevant topographic representations in lower-order motor (Karni et al., 1995; Nudo et al., 1996), somatosensory (Jenkins et al., 1990; Recanzone et al., 1992a, b, c), auditory (Recanzone et al., 1993; Weinberger, 1993; Scheich et al., 1993), and visual cortex (Vania et al., 1998; Schwartz et al., 2002; Furmanski et al., 2004).

Four principles have been proposed to characterize visual cortical reorganization and memory formation during the training of perceptual (and other) skills. A first principle (*lowest-level memory formation*), proposed by Karni (1996), states that the lowest processing level that is required for the skilled performance of a task will also be a crucial site for memory formation. By extension, the procedural memory resulting from training will show characteristics that reflect the properties of the processing level where the memory is stored. Matching the stimuli to processing capacities of neurons in one (or more) areas makes it most likely, therefore, that those areas will also be a site for memory formation. For example, if a given retinotopic area were crucially involved in memory formation for a specific perceptual skill acquired with stimuli presented in a fixed visual field location, then the training-induced performance increments in the perceptual task should be specific to the trained location, and performance in other locations should remain largely unskilled (position specificity). Likewise, ocular specificity (or monocularity) of skilled performance might suggest that memory formation took place specifically in V1, which is the only cortical region with monocular neurons (in layer IV). An important, general implication of this line of reasoning is that the observation of learning curves obtained while manipulating task and stimulus variables can provide insight into the mechanisms and help localize the neuronal substrates of perceptual learning (Fiorentini & Berardi, 1981; Karni & Sagi, 1991). A second, closely related principle (*the reverse hierarchy hypothesis*) has been formulated by Ahissar and colleagues (Ahissar & Hochstein, 1997, 2002; 2004) who proposed that plasticity during initial learning will occur at higher levels in the visual system, and that plasticity at the lowest levels of the system will only be induced during later, asymptotic learning. In agreement with the previous principle, the lowest level that could process the stimuli to support skilled performance will be a site of memory formation. This implies that the specificities that will appear depend on the stimuli and task, and that they will only appear towards the end of learning, when the system is being pushed to maximize performance. A third principle (*the requirement of attention*) is based on the observation that skill learning requires practice and effort, suggesting that attention is a necessary gateway to learning. In most reported types of perceptual learning, repeatedly presenting stimuli will not lead to improved performance unless attention is directed to them (Ahissar et al., 1992; Ahissar & Hochstein, 1993; Karni & Sagi, 1991; 1993; but see also Watanabe et al., 2001). A fourth principle (*extended memory consolidation*) is that progress in visual discrimination learning (and skill learning in general) predominantly occurs in-between daily training sessions rather than during the sessions. This indicates that memory formation depends on processes that extend beyond the duration of a training session (Karni & sagi, 1993).

These four principal themes will be discussed and evaluated in the review below. In Sections 2.3-5, studies will be presented that lend themselves to a discussion of the idea of storage at the lowest level, the reverse hierarchy hypothesis, and the requirement of attention for learning. In Section 2.6, studies will be presented that reveal mechanisms of plasticity that support the extended consolidation idea.

2.3 Changes of the classical RF induced by perceptual learning

Schoups et al. (1995) performed a straightforward test of the hypothesis that complete training in a sensory task should produce a skill that requires the lowest level of the sensory system both for skilled performance and memory formation. They used an identification task in a visual psychophysical study, in which human subjects had to report whether the orientation of a grating matched the orientation of an internalized reference orientation. In this task, skilled performance, characterized by thresholds in the order of $1°$ or less, was specific to both the position of the stimuli and the reference orientation. After the subjects had reached a stable orientation threshold in a particular location with a $2°$ grating, moving the stimulus to a neighboring location (only 2° away) required complete retraining before a new, low and stable threshold was re-attained. Similarly, after complete training around a particular reference orientation, a rotation by 90° required complete relearning. These findings are in agreement with other studies in humans, cats, and monkeys (Vogels & Orban, 1985; De Weerd et al., 1990; Schoups et al., 2001).

Schoups et al. (2001) then trained monkeys in the same task used with human subjects. Monkeys discriminated the orientation of small grating stimuli always presented in the same visual field location ('trained' location) around a particular reference orientation, while similar stimuli presented in a different visual quadrant were ignored ('untrained' location). The authors reported that neurons with CRFs in the trained location, but not in the untrained location, showed an increase in the steepness of the flanks of their orientation tuning curves. Interestingly, this effect was limited to neurons whose preferred orientations were about 15° away from the reference orientation used for behavioral training. That result fits with the fact that for a given orientation difference, the differential response will be maximized along the flank of the orientation tuning curve. Training thus selectively altered the single aspect of orientation tuning curves that most strongly contributes to fine orientation discrimination. Tuning width of V1 neurons (and steepness of flanks of tuning curves) strongly depends on intra-cortical connectivity within V1 (Section 2.1). The reported modification of tuning curves, therefore, is a neurophysiological correlate of training-induced long-term plasticity in V1.

To evaluate the improvement in estimating orientation in the population of trained neurons, Schoups et al. (2001) constructed neurometric curves in selected sets of trained and untrained neurons. A neurometric curve plots the success of a neuron in discriminating two orientations (% correct transformed into Z-scores) as a function of the orientation difference. It was found that the slope for the neurometric curve was about 25% steeper in the trained compared to the untrained neurons for orientations that fell on the flanks for the orientation tuning curves.

The study from Schoups et al. (2001) is the only demonstration to date of significant changes in tuning properties of V1 neurons induced by training. However, the 25% training-induced increase in slope of neurometric curves is insufficient to explain the entire decrease in orientation thresholds (from roughly 30° to 1°). In monkeys, however, the initial thresholds are likely to largely reflect non-specific factors that do not reflect sensory limitations or even perceptual learning (such as simply not 'understanding' the task). To link the changes in V1 neurons with improvements in sensory performance, these non-specific factors must be removed. Non-specific factors can be isolated by testing the transfer of performance from a trained to an untrained position. After complete training in a first location, moving the stimuli to a second location led to initial thresholds in the 5-10° range. Hence, the training in the very first location required to bring thresholds down from 30° to a 5-10° range reflected aspecific factors, and only threshold decreases below the

5-10° range should be linked with the neuronal changes in V1. Accordingly, it is possible that position and orientation specificity of behavioral performance were induced only when training pushed subjects to discriminate stimuli below the 5-10° range of orientation differences. Nevertheless, even after removing aspecific factors, the 25% training-induced increase in slope of neurometric curves is insufficient to explain training-induced decreases in orientation thresholds. In human subjects, the effects of aspecific factors at the beginning of training are limited, and hence position-specificity of learning is almost complete (Schoups et al., 1995). It is interesting then that in humans, training-induced increases in the slope of Z-transformed psychometric curves are on the order of 50%, which remains higher than the 25% increase in slope of Z-transformed neurometric curves in monkey V1 neurons. Hence, the data suggest that part of the improvements in orientation discrimination was not related to the training-induced modifications of orientation selective neurons in V1.

The findings by Schoups et al. (2001) were complemented by a similar study conducted simultaneously by Ghose et al. (2001), who trained two monkeys for more than half a year in a match-to-sample orientation discrimination task, but failed to find changes in any properties of the CRFs (including size and orientation tuning) in the trained region of the visual field. Furthermore, although the training was specific for reference orientation, they found that there was little position specificity. To investigate the possibility of changes in cortical representation, cortical magnification was compared in trained and un-trained regions of cortex, and again, no difference was observed. The authors concluded that profound changes in orientation discrimination can be induced by training, without appreciable changes in physiological properties of V1 neurons. The authors pointed out that the lack of representational changes, and the lack of changes in CRF size (and in other CRF properties) contrasts with the high degree of plasticity found in other early sensory areas in other modalities. Accordingly, the authors suggested that V1 might be less plastic than other primary sensory cortices. Furthermore, they suggested that the presence of strong orientation and position specificities would be insufficient to conclude that the memory for the trained skill is limited to V1. There are two aspects to this idea that deserve further commentary. First, unless one can show that the learning is strongly monocular, memory-formation limited to V1 is unlikely, given the fact that other lower-order areas, which are heavily interconnected with V1, also show retinotopic organization, and contain orientation-selective neurons. In other words, the idea of thinking in terms of isolated 'levels' in a system may be hard to maintain. Further, even when skilled performance of a perceptual task is strongly specific for the trained eye, it does not imply that the memory trace is limited to V1. It does show that the network that represents the skill includes strongly monocular neurons (in layer IV) of V1, but it does not exclude that other neurons within and outside V1 would be included in a network that underwent plastic changes to represent the memory for the skill (see Section 2.4). Second, if the effect of training were to purely change the orientation-filtering properties of V1 neurons, its effect should be present independent of behavioral task. This could be tested, for example, in an experiment in which subjects are first familiarized with both an identification task and a match-to-sample task, and then fully trained in orientation discrimination with one type of task. After completion of training in one task, training effects should immediately transfer to the other task. Such a result would indicate that the plasticity induced by learning was limited to a 'multi-purpose' modification of filtering properties, and such plasticity is likely to be limited to lower-order visual areas. By contrast, a lack of transfer of performance between different tasks despite the use of identical stimuli would suggest that the network that stores the memory for the perceptual skill extends beyond V1, and includes a representation of the task (see Ahissar & Hochstein, 1993).

Despite these interesting considerations, there are a number of straightforward differences between the Schoups et al. (2001) and Ghose et al. (2001) studies that might explain the lack of position specificity in the latter study. In that study, stimuli were presented in two locations (the fovea, during a preliminary training phase, and subsequently a peripheral location), which could have triggered a different, more general type of orientation discrimination learning. Consider in this respect that after training of a task in two locations, neurons in higher-order areas with RFs that encompass both locations will have been exposed twice as many times to the stimulus than neurons in lower-order areas with small RFs. Thus, even for stimuli for which an early-level solution might be selected by most subjects, training in two locations could constitute a sufficient bias towards memory formation in higher order areas, which of course would result in more generalized learning-effects. In addition, in one of the monkeys, the target grating was surrounded by distracter stimuli, which, again, might have induced a more global type of learning. Furthermore, orientation discrimination was measured by means of a match-to-sample task, in which an unpredictable sample orientation is presented first, followed by a test orientation, to be matched by the monkey to the previously presented sample. This is a complex task, with an important working memory component. Schoups et al. (2001) used an identification task, in which monkeys were required to indicate the clockwise or anticlockwise tilt of a single stimulus relative to an internalized criterion. While match-to-sample tasks are dependent on prefrontal working memory mechanisms, the identification task is dependent on striatal mechanisms. The monkey's task in Ghose et al. (2001) was probably also more difficult because in that study sample and test were Gabor stimuli with different spatial frequency, while Schoups et al. (2000) used circular sinusoidal gratings of fixed spatial frequency. It is striking that the monkeys in the Schoups et al. (2001) study reached orientation thresholds in the order of less than a degree in about half of the time required for the monkeys in the Ghose et al. (2001) study to reach thresholds in the order of 5°. Possibly, difficulties specific to the task and stimuli used in Ghose et al.'s (2001) study prevented the monkeys from reaching a sensory limit, despite lengthy training (but see Section 3). Accordingly, a lack of position specificity was found, which makes the absence of plasticity in V1 unsurprising.

The comparison of results from Schoups et al. (2001) and Ghose et al. (2001) shows that changes of CRF properties in V1 neurons will only be obtained under very specific conditions, and that it is crucial to push the sensory limits of the system before plasticity of the CRF will be induced in neurons especially suited to encode the stimuli. Plasticity after training to about 5° thresholds in a match-to-sample task, however, is sufficient to induce changes in orientation tuning curves in V2 and V4 neurons (Yang & Maunsell, 2004), in agreement with a reverse hierarchy hypothesis. In conclusion, the behavioral properties of perceptual learning with single oriented stimuli and its physiological correlates conform to the broad principles of perceptual learning outlined in Section 2.2 (*lowest-level memory formation, reverse hierarchy hypothesis*).

2.4 Changes in the RF surround induced by perceptual learning

In this section, three types of experiments will be discussed that involve perceptual learning with contextually defined target stimuli. In bisection and Vernier tasks, subjects are required to determine, respectively, the position or the alignment of a target stimulus relative to two reference stimuli (flankers). In contextual contrast discrimination tasks, the contrast of a target stimulus must be estimated in the presence of flanker stimuli. In texture segregation tasks, a target region, defined by one or more texture elements that are different from those in the otherwise homogeneous texture background, must be detected or discriminated.

With training, humans as well as monkeys become very proficient in bisection tasks, in which subjects judge whether the position of the central of three juxtaposed, parallel line segments deviates from the middle. Crist et al. (1997) found in human subjects that skilled performance of this task was highly dependent on the trained location and stimulus orientation. Based on these observations, the authors predicted the involvement of V1 in the learning of this skill, and proceeded to training two monkeys in the bisection task (Crist et al., 2001). They compared properties of V1 neurons with CRFs that overlapped with the trained location in the visual field, to the properties of V1 neurons from the other hemisphere where no training had taken place. They found that the training had not modified any of the properties of the CRF. CRF size and orientation tuning were unchanged, and no changes in visual topography (cortical magnification) could be discerned in the trained region of cortex.

Crist et al. (2001), however, found that training changed the effects of stimuli presented in the RF surround on the neuronal responses to CRF stimulation. To demonstrate this, they presented a probe stimulus within the CRF, in the 'trained' portion of the visual field, and a second probe stimulus in the RF surround, at varying distances from the CRF, while the monkey performed the bisection task in a nearby location. Neuronal responses in this condition were compared to those obtained while the monkey ignored the 'trained' visual field location and performed a neutral fixation task. A significantly larger modulation of the flanking probe on the central probe was found when the monkey performed the bisection task, compared to when it performed the fixation task. In a further control condition, the monkey continued to do the bisection task in the trained region of the visual field, while the effect of a flanking probe on responses to the central probe was measured for neurons with RFs in the opposite, 'untrained' hemifield. In this condition, the flanking probe stimulus showed modulatory influences that could not be distinguished from those obtained during the fixation task.

In order to dissociate the effects of learning from those related to spatial attention directed towards the trained location, the authors showed that if the probe flanker in the surround was collinear, rather than parallel, with the probe stimulus in the CRF, neuronal responses to the probe stimuli in the trained vs. untrained locations were not different. Similar experiments from the same group (Li et al., 2004) demonstrated that the amount of surround modulation exerted by parallel or collinear line segments on the responses to a central segment in the CRF was task-dependent. Indeed, training on the bisection task revealed modulatory effects limited to parallel flanking stimuli. Likewise, training in a Vernier task, in which subjects judge whether the central of three collinear line segments deviates from perfect alignment with the outer two segments, revealed modulatory effects limited to collinear flanking stimuli. In these experiments, it was shown again that the task-dependent effects of flanking stimuli vanished when the task was performed at a location distant from the RFs of the recorded neurons.

Taken together, the bisection and Vernier experiments permit three conclusions. First, training with specific contextual stimuli produces behavioral and physiological effects limited to these contextual stimuli. The physiological effects can only be revealed when the task is performed in a location close enough to probe stimuli in the trained location. By contrast, the modulation of tuning properties of V1 neurons after orientation discrimination in the study by Schoups et al. (2001) was measured while the monkeys' attention was directed away from the RFs of the recorded cells, and engaged in an irrelevant fixation task. Therefore, training in the studies from the Gilbert group produced training effects that were specific not only to stimulus position and orientation, but also to behavioral context.

Second, the studies from Gilbert's group show that the inhomogeneity of surround influences (see Sceniak et al., 2001) can be shaped by training. Thiele (2004) has suggested that the focus of attention adopts the particular elongated shape required to encompass the central element and the flanking elements relevant for a given contextual task. The combination of repeated stimulation during training and selective attention might be the basis for a long-lasting modulation of connection strengths in a local network of neurons engaged in a specific contextual task. However, the connections between these local networks and higher-order neural networks involved in attention might be modified as well during training. As a result, the memory formation for the skill might involve lateral connectivity changes of which the effects are only revealed when the task is being executed.

Third, bisection and Vernier skills can be position and orientation specific in the absence of changes in CRF properties. Indeed, in the above studies there was no evidence of training-induced changes in RF size, in orientation tuning, or in visual topography.

Another set of perceptual learning studies from the Gilbert group has used luminance discrimination as a paradigm. It is well-known that a line segment will be detected more readily if it is flanked by one or two collinear line segments (Dresp 1993, Polat & Sagi, 1994, Wehrhahn & Dresp, 1998; Kapadia et al., 1995), and a corresponding enhancement of neural responses has been reported in V1 when a central line in the CRF is flanked by a collinear line segment in the RF surround (Kapadia et al., 1995; Polat et al., 1998). These enhancements are thought to be instrumental in contour perception (Field et al., 1993; Dresp & Grossberg, 1997; Li & Gilbert, 2002). Similarly, in a human study, Ito et al. (1998) showed under distributed attention conditions that collinear flanking lines produced a significant shift in brightness perception of target line segments. The visual display used in this experiment consisted of a central fixation spot, with next to it a reference line against which the brightness of a peripherally presented target line was to be compared. There were four potential target lines, one in each quadrant, and all four were flanked by a single collinear segment. Subjects had to indicate whether a single odd target (that deviated in luminance from the three other potential targets) was brighter or dimmer than the reference line next to fixation. The task was carried out while subjects were cued before stimulus onset about the position of the odd target (focused attention) or not cued (distributed attention). In the distributed attention condition, the flanking stimuli led to a significant overestimation of target brightness, but in the focused attention condition no such overestimation was present. With training, the effect of the flanking line segments in the distributed attention condition disappeared. Hence, in the absence of focused attention, the perception of a target can be influenced strongly by its context, but with training, the effect of contextual interactions can be reduced. In a recording study from Ito and Gilbert (1999), corresponding changes were observed in the firing rates of V1 cells in two monkeys. After initial training in focused and distributed attention conditions, a collinear line element in the surround facilitated the neuronal response to a line segment in the RF under distributed (or away) attention conditions, but not when attention was focused on the RF element. Interestingly, the effects of several weeks of training and further experience during recording produced a different outcome in the two monkeys. In one monkey, behavioral facilitation remained in the distributed attention condition and none remained in the focal attention condition, whereas the opposite pattern was found in a second monkey. These effects were reflected in the data recorded from V1. In the first monkey, adding the collinear flank in the RF surround led to a much larger response enhancement to a target in the CRF in the distributed attention condition than with attention focused on the RF. In the second monkey, the opposite pattern of physiological results was observed.

These data taken together do suggest a contribution of V1 to the training-induced reductions of contextual influences under distributed attention conditions. However, its role was clearly modulated by individual learning strategies. Furthermore, the contribution of V1 to this type of learning was complex, as illustrated by a remarkable pattern of position specificities found by Ito et al. (1998) in humans (see also Gilbert et al., 2000). After training in the diffuse attention condition, four additional stimulus sets (target elements with neighboring collinear flankers) were placed in-between the four already trained sets (leading to a total of 8 sets in the visual field). As expected, the flankers influenced perception during distributed attention in the new locations, but not in the already trained locations. However, when the pattern of four target lines and flankers was rotated as a whole by 45° (only 4 sets in the visual field), training effects generalized to the new, untrained locations. Thus, the memory for this skill was stored in a reference frame that was not retinotopic. Rather, if the specific configuration of 4 sets of targets and flankers is considered an object, it appears that the effects of the training may possible have been stored in an object-centered reference frame. Therefore, memory formation for this type of skill learning (the reduction of the effects of contextual stimuli during distributed attention) is unlikely to be limited to V1.

The importance of changes in the RF surround to explain training effects in bisection and Vernier tasks, and in contextual contrast discrimination tasks suggests that changes in RF surround may also underlie training effects with other types of target stimuli that are defined by virtue of their embedding in surrounding stimuli. This is illustrated by studies from Bertini et al. (1995, 1996), who trained two monkeys in a texture segregation paradigm similar to the one used by Karni and Sagi (1991) in human subjects. In Karni and Sagi's original study, subjects performed a dual task. Firstly, they had to correctly perform a difficult letter discrimination task at fixation. Secondly, they discriminated a horizontal from a vertical bar-shaped region defined by 3 oblique line elements on a background of either horizontal or vertical line elements. Difficulty was manipulated by changing the stimulus onset asynchrony (SOA) between the texture and a mask. Training led to pronounced decreases in SOA thresholds. Expertise in this task was highly specific to the orientation of the background elements, to the location of the stimulus, and to the eye used during training (monocular learning). The latter finding most strongly implies V1 as a participant in memory formation, because it is the only visual area that contains monocular neurons (in layer IVa) (but see below). The two monkeys in the study from Bertini et al. (1995, 1996) performed a texture segregation task similar to that used in humans, but were not required to perform a simultaneous task at fixation. Learning in the two monkeys was specific for target position and orientation of background elements, but not monocular (one might speculate that this was due to the lack of a dual task requirement). The physiological recordings in the two monkeys were carried out, therefore, using binocular stimuli, while the monkeys performed a color discrimination task presented away from the recorded RFs. In one of the two monkeys, it was found that in the trained location and for the trained background orientation, neurons showed differential responses between texture backgrounds with and without the target, provided that the target was shown in the RF surround. This suggested that the specific training effects at the trained location and background orientation exclusively depended on plastic changes in the RF surround. In the second monkey, however, the differential response was found for targets inside the CRF, thus complicating the interpretation of the data. The complexity of contextual stimuli may have permitted the use of different strategies to solve the task, resulting in individual differences.

The above studies raise a number of issues. First, it appears that subtle differences in conditions of perceptual learning can lead to pronounced differences in the specificity of the learning. A condition in which attention was distributed (due to the dual task) led to monocular learning (in human subjects), and a condition in which attention was focused on the region where the target could be expected (a narrow range of eccentricities 2.5 to 5° in a single visual field quadrant) did not lead to monocular learning (in monkeys). Note also that in experiments from Schoups et al. (1995), the human subjects did not perform a task at fixation, and skilled discrimination of grating orientation was not monocular. This suggests that the amount of attention available at the target location might strongly influence the behavioral properties of the learning. Furthermore, the lack of a task at fixation is likely to produce large individual differences, as subjects are free to allocate attention as they wish (Karni et al., 1996).

A second issue is whether monocularity implies that the learning took place in V1. The finding of monocularity is often used to claim that memory formation occurred exclusively at the lowest level of the visual system. The network carrying the memory might then include neurons in Layer IVa as well as neurons in upper layers that show strong dominance for the trained eye. However, in the above texture segregation studies, and others in which learning modifies the interactions between CRF and surround, it is not straightforward to conclude that learning would be limited to V1. Specifically, monocular input neurons in Layer IV that are stimulated selectively during a learning task using contextual stimuli may increase the strength of their connections with a binocular network where center surround computations take place, thus getting preferential access to that network. Monocular input neurons involved in training would interact differently with that binocular network compared to input neurons dominated by the unexposed eye. Thus, the 'trained' monocular network would become more strongly connected to a binocular network extending beyond V1 where feedback and center-surround computations take place, and plastic changes could take place in that entire network. Learning then could be monocular, while the neural network in which connectivity would have changed might extend beyond primary cortex.

In addition, although strong ocular, position, and orientation specificities can be an indication of the involvement of early visual areas, the absence of such specificities does not exclude a role of early visual areas in memory formation. Sigman & Gilbert (2000) trained human subjects on a search task in which a triangle in a top-down orientation had to be detected in an array of differently oriented distracter triangles. They found that the learning (percent correct) was specific to the target shape and its orientation, but not to the background shapes. In addition, the learning was specific to the area around the fixation spot within which the array was confined. Within that array (4°by 4°, centered on fixation), however, the target was detected correctly at high rates at the end of learning, independent of location. Because the target was presented at random at any location within the array, and because higher-order areas such as V4, TEO and TE have CRFs encompassing (large parts of) the array (Gross et al., 1972; Gattass et al., 1988), it is natural to assume that these areas were a likely site of memory formation of the perceptual skill (see also Lueschow et al., 1994). A recent lesion study indeed supports a role of V4 and TEO in the generalization of learning over space after localized perceptual learning of a shape discrimination task (De Weerd et al., 2003). However, when Sigman and Gilbert (2000) investigated the time course with which the skill increased in each of the 24 locations in the array, it appeared that skill increased first in one particular location, and that the next location to improve was not another random location, but, rather, was most likely to be a location next to the previous location. This suggests that the generalization of the skill across visual space was driven by lateral connectivity within lower-order areas, where neurons have CRFs that are small enough to encompass single elements in the array (Gattass et al., 1981). This, in turn, most likely implies that these lower-order areas are also an integral part of the neuronal network that supported memory formation. The absence of position specificity of a skill, therefore, does not exclude an involvement of early visual areas such as V1 and V2. In fact, one might hypothesize that as the skill transferred to increasingly large parts of the visual field, the skill also became increasingly dependent on higher-order neurons with large RFs. Such a finding would go against a reverse hierarchy hypothesis as it would be an example of a type of learning in which continued training moves the critical substrate for memory formation from lower to higher levels in the system.

Sigman and Gilbert's (2000) study also showed that whether or not a stimulus is 'matched' to the properties of neurons in a given area does not always predict where plasticity will take place. It is likely that the responses to the triangles and their global orientation are more vigorous in extrastriate areas, and nevertheless it appears that learning induces plasticity in V1 or V2. Another study that strengthens this point (Lee et al., 2002) used a search array with elements whose shape was defined by shading (Ramachandran, 1988). Since we have a strong tendency to assume that light comes from above, a circular aperture in which luminance gradually decreases from top to bottom is perceived as convex. For the same reason, a circular aperture in which luminance gradually increases from top to bottom is perceived as concave. Further, a convex element placed against a homogeneous background of concave elements 'pops out'. Lee et al. (2002) recorded from V1 neurons of naïve monkeys, and found that a single shape-from-shading (SS) element in the CRF produced the best response in V1, while embedding it in a background of other SS elements produced suppression of the response. The suppression was slightly less when the SS element in the CRF was different from the surrounding elements. Analogous findings have been reported by Knierim & Van Essen (1991) for line elements. The reduction of suppression when elements in the RF surround are different from the CRF element, rather than the same, has been proposed as the basis for pop-out. Lee et al. (2002) then trained their monkeys for 15 sessions in a target detection task (1200 trials per session), and found a significant increase in the differential suppression between homogeneous arrays and arrays with an oddball in the CRF. The authors used a range of different SS elements, and found that after training, those elements that behaviorally triggered the fastest responses also produced the largest increase in differential suppression between homogeneous and inhomogeneous arrays. A contribution of feedback to these results is evident because the learning-induced physiological effects were present only 100ms after stimulus onset (see also Lamme et al., 1995; Zipser et al., 1996; Super et al., 2001). Nevertheless, learning-induced modifications of contextual modulation in V1 neurons suggest that V1 can be part of a neural network representing the memory for a perceptual skill with complex stimuli.

Combining the data from all experiments in this section, one might conclude that, compared to single CRF stimuli, contextual stimuli lend themselves to more complex, less predictable learning strategies. As a consequence, it is difficult to make inferences about the processing level implicated in learning based on the nature of the stimulus or on the pattern of learning specificities. In particular, the lack of ocular, position, and orientation specificity might be indicative of memory formation at a higher level in the system, but a contribution from lower levels cannot be excluded. The opposite holds as well: the presence of ocular, position, and orientation specificity might be indicative of memory formation at a low level, but a contribution from higher levels cannot be excluded. Despite the fact that most studies reviewed here exclusively reported V1 data (but see Yang & Maunsell, 2004), there are strong indications that training-dependent plastic changes extend beyond area V1. Indeed, any type of perceptual learning that modifies the interactions between the CRF and its surround suggests a role for feedback networks that include higher-level areas, in addition to horizontal connections within V1. Furthermore, the finding that task and attentional requirements determine to a large extent the properties of learning for identical stimuli is in itself an indication that different requirements may trigger different plastic changes within different networks at different processing levels.

Thus, the principle of memory formation at the lowest level, the role of attention as a pure gating mechanism (simply enabling learning), and the reverse hierarchy model do not fully accommodate the data available with contextual stimuli.

2.5 fMRI studies of visual perceptual learning

With the advent of fMRI, it has become possible to investigate changes in early visual cortex resulting from perceptual skill learning in humans. In general, increased skill with a particular set of stimuli leads to increased activity (amplitude or extent) in V1 and other visual areas. This finding was also reported by Furmanski et al. (2004), who determined contrast detection thresholds for sinusoidal grating patches oriented at a left oblique and horizontal orientation. Before training, subjects showed higher thresholds for the oblique than for the horizontal orientation (the oblique effect). Subjects were then trained on a two-alternative forced choice task in which they had to indicate which of two subsequent time intervals contained an oblique stimulus. The training during a period of about 30 days significantly decreased contrast detection thresholds, leading to the disappearance of the oblique effect. The effect of training was specific to the orientation of the stimuli (no improvement for the untrained, horizontal orientation), and to the training location (training at 4° of eccentricity generalized partially to 9°, but not to 20°). fMRI activity was measured with high-contrast stimuli while subjects discriminated a 5-deg difference at the horizontal or the oblique reference orientation. Before training, V1 cortex in the trained location produced a significantly larger fMRI signal in response to the horizontal than to the oblique orientation. After training, the fMRI response to the oblique orientation had increased and had reached the level seen for the horizontal orientation.

A fMRI study by Schwartz et al. (2002) using the already described texture discrimination task (previous section) devised by Karni et al. (1991) has found a similar increase in the signal from V1 after training. Schwartz et al. (2002) reported that fairly limited training produced reductions of 80% correct SOA thresholds that were specific to target position, background orientation, and to the eye used, as had been shown previously by Karni et al. (1991) after much more extensive training. Thus, V1 appears to strongly contribute to the acquisition of this perceptual skill, even after limited training. In a subsequent fMRI experiment, Schwartz et al. (2002) compared activity between trained and untrained conditions in V1 by presenting the same stimuli at a fixed SOA of 200ms through the trained and untrained eyes, while subjects performed the task. A comparison of activity in trained and untrained conditions revealed a single cluster of voxels in V1, in a retinotopic location corresponding to target location. No clusters with increased signal were found in extrastriate visual cortex, or in cortical regions related to attentional processing. The training that preceded this result was limited to a single 1760-trial training session 24h before fMRI scanning. Based on behavioral data reported in Schwartz et al. (2002), 80% SOA thresholds after training would be in the order of 200-300ms, which indeed would correspond to initial learning in Karni' s studies (see Karni & Sagi, 1991, 1993). Karni and colleagues have shown that with extended training (several weeks) SOAs in very comparable stimulus conditions can decrease till about 40-50ms. A baseline fMRI session before training did not reveal any difference in activity elicited by stimuli presented monocularly through the two eyes. Connectivity analysis suggested increased connectivity with attention-related structures during untrained performance. It is unclear whether this fast form of plasticity in V1 can be equated with plastic changes in V1 observed after extensive training. Nevertheless, the finding that initial, limited training can lead to fast plasticity in V1 and strong learning specificities raises some questions about the assumptions underlying the reverse hierarchy model.

A third study has reported an increase in the extent of fMRI activity in area MT, induced by training on a global motion discrimination task within a single fMRI scanning session (Vania et al. 1998). The total amount of training during the fMRI session totaled roughly 5 minutes (400 trials), which resulted in a strong improvement from random performance to nearly perfect performance for the chosen stimulus parameters. The task consisted of detecting left versus right movement of 20% signal dots embedded in 80% noise dots. Improvement on the task was position and orientation specific. A comparison of activity in the task condition with activity in a fixation task control condition (presenting similar but static dot patterns) showed that the extent of activity at the end of the session was about 5-fold larger compared to the beginning. No such effect was reported in V1. Extrastriate areas and attention-related structures such as the cingulate gyrus showed significant reductions in activity by the end of the session.

These three studies confirm changes in extent or amplitude in fMRI signal in early visual cortex as a function of perceptual learning. The increased fMRI signal is likely to reflect complex changes in the network that is the basis of the CRF or the RF surround measured physiologically. Changes in interactions between CRF and surround may have driven learning-related changes in fMRI signal in the Schwartz et al. (2002) study, because of the contextual stimuli used in that study. In the two other studies, single, isolated stimuli have been used within the CRF, and changes in the fMRI signal might reflect CRF changes. A remarkable finding is that training-related fMRI signals can be found in early visual cortex, including V1 and MT, after very limited training. This finding appears not to fit with a reverse hierarchy theory, in which training on simple discrimination tasks would affect lower levels in the visual system only after pronounced training. The fMRI data presented here suggest, instead, that the lowest levels undergo plastic changes from the beginning of training. By contrast, neurophysiological studies using single stimuli do support the idea of reverse hierarchy. Extended, but incomplete training results in changes in V2 and V4, but not V1, (Ghose et al. 2002; Yang & Maunsell, 2004) while completion of training has been shown to induce changes in V1 (Schoups et al., 2001). Data from Gilbert's group also suggest that limited training does not result in measurable changes in contextual interactions measured in spiking behavior of single V1 neurons (Gilbert et al., 1999; 2000). The difference in findings between neurophysiological and fMRI studies might be due to the nature of the fMRI signal. The fMRI signal is an indirect measure of metabolic activity to which a large number of factors contribute. The fMRI signal reflects both inhibitory and excitatory subthreshold events, as well as spiking activity (Logothetis & Pfeuffer, 2004; Logothetis & Wandell 2004; Logothetis, 2003). These subthreshold events reflect inputs from a network of feedforward, feedback, and horizontal connections, and it is possible that learning induces subtle plastic changes in connectivity that initially remain invisible in the spiking behavior of individual neurons. For example, in an initial training session, a network of neurons relevant for a task might be recruited by selective attention mechanisms, and the memory trace after that initial session might involve subtle changes in connectivity between lower order sensory neurons and neurons delivering the attention feedback signal. These initial subthreshold connectivity changes might facilitate learning in later training sessions, and guide plastic changes within lower-order sensory areas, which would eventually become suprathreshold. According to this reasoning, the idea of plasticity trickling down to increasingly lower levels of the system with training might be an over-simplification. Instead, even the lowest levels in a sensory system might undergo plastic changes from the beginning of learning, although the type of plasticity might evolve during learning (see Section 3). Thus, we suggest that a learning-induced increase of fMRI activity early in learning in lower-order sensory or motor areas reflects subthreshold changes reflecting in part changed connectivity with high-order brain regions or areas. Late in learning, an increase in fMRI activity may reflect predominantly changed connectivity within the lower-order sensory or motor area, accompanied by changes in spiking behavior of individual neurons. In general, of course, the results from fMRI are entirely compatible with the general notion that perceptual learning can induce plasticity in early visual cortex.

2.6 Mechanisms underlying long-term plasticity in V1

Attention appears to contribute in a crucial manner to plasticity in the visual areas where memory formation takes place. In neurophysiological studies, the effects of attention have been studied predominantly using stimuli placed within the CRF. When single stimuli are placed within a CRF, attending to that stimulus often increases the response to that stimulus in a way that resembles the effects of an increase in physical stimulus contrast (McAdams & Maunsell, 1999; Reynolds et al., 2000). When two (or more) stimuli are placed within a CRF, directing attention to the target stimulus will cause that stimulus to dominate the response of the neuron, while effects from distracter stimuli on the neuron's response are minimized (Moran & Desimone, 1985, Chelazzi et al., 1993, 1998; Treue & Maunsell, 1996; Luck et al., 1997; Reynolds et al., 1999). Related findings have been reported in human visual cortex with fMRI (Kastner et al., 1998). Both types of findings are compatible with an input gain model of attention (Reynolds et al., 1999), which also has been referred to as the biased competition theory of attention (Desimone & Duncan, 1995). In this model, somewhat simplifying, the response of a neuron is determined by the balance of all the excitatory and inhibitory inputs provided to that neuron. Attention increases the gain of all inputs provided by the attended stimulus, and this gain increase can explain both the attention-induced response increase to stimuli presented alone, and the dominance of target stimuli over distracters shown together in a CRF. Because gain changes in neural networks are the gateway to learning, it is useful to consider mechanisms that could produce these changes. Two types of neuronal coding potentially responsible for perceptual gain modulations will be considered in the following section: 1) absolute neuronal spiking activity following stimulus presentation, quantified by average firing rates; 2) temporal coherency within neuronal ensembles, captured by measurements of synchronous spiking activity.

To understand the mechanisms behind gain increases, it is instructive to consider the similarities between the effects of contrast and the effects of attention. Recent data suggest that when a single stimulus is placed in the CRF, the effects of attentional and contrast manipulations are indistinguishable (Reynolds et al., 2000). Related findings have been reported by psychophysical (Carrasco et al., 2004) as well as fMRI studies (Ress et al., 2000). Likewise, when multiple stimuli are present within the CRF, a contrast increase of a given stimulus mimics the effect of attending to that stimulus (Reynolds et al., 2000; Reynolds & Desimone, 2003). For example, if two unattended stimuli are presented in the RF, responses typically reflect the average of each stimulus presented individually. However, responses to the stimulus pair can be similarly increased by either attending to or enhancing the contrast of the stimulus preferred by the neuron. Likewise, attention to or contrast enhancement of the non-preferred stimulus will decrease the response of the neuron. Hence, by analogy with the effects of stimulus contrast, attentional gain modulations of (selected) inputs could be achieved through modulation of neuronal firing rates.

However, temporal mechanisms can also play a role. Fries et al. (2003) have shown that attention to a stimulus synchronizes activity in the gamma-frequency range in the population of neurons coding that stimulus. At a higher level of processing, where neurons with large RFs would receive inputs from both an attended and an unattended stimulus, the synchronous inputs from the population coding the attended stimuli puts the processing of the attended stimulus at an advantage, compared to an unattended stimulus whose inputs are non-synchronized. This temporal effect could provide a way to manipulate the gain of inputs provided by an attended stimulus to a higher level of processing.

Perceptual learning implies that an enhancement of sensory processing that initially requires attention becomes hard-wired in the system. As a consequence, some of the ways in which attention modulates the efficacy of stimulus processing might be implemented in the implicit memory for the perceptual skill. For example, learning could reflect a change of connectivity among neurons in lower-order areas that were originally 'selected' by attention (by the requirements of the behavioral task). If these connectivity changes were at least partially implemented by a temporal mechanism, one might speculate that synchronization in visual networks initially steered by attention (top-down), and involving a number of levels of the visual system, eventually becomes replaced by synchronization limited to the specific visual network involved in the learning. Hence, after prolonged training, the trained stimulus would induce synchronized activity in the absence of top-down attention. Such learning mechanisms could provide a basis for the increased automaticity that accompanies perceptual and skill learning in general. It could also provide a basis for the 'capture' of attention by stimuli that have been relevant in previous learning tasks (Moores et al., 2003).

In-vivo intracellular recordings of V1 neurons in cat striate cortex have shown that these neurons are intrinsically equipped to generate bursting activity in the gamma range (Gray & McCormick, 1996) in response to strong stimuli (depolarizing current injections as well as visual stimulation). By combining individual cell staining with the recording method, Gray and McCormick (1996) localized these cells in superficial layers of V1. Because the neurons only showed oscillations in membrane potential during visual stimulation, and not during periods of spontaneous activity, these cells were referred to as 'chattering' cells. Chattering cells may make a substantial intra-cortical contribution to the generation of synchronous cortical oscillations and participate in the recruitment of large populations of cells into synchronously firing assemblies (see also Bringuier et al., 1997). It is possible that attention changes the connectivity in local networks of V1 cells, such that the probability that chattering cells would drive the network into synchronized activity would be increased. According to this reasoning, an effect of training in the study of Schoups et al. (2001) might have been that oriented stimuli around the trained reference orientation would produce synchronized responses in the population of neurons in the trained location of the visual field. This hypothesis was not tested by the authors, but has found support in motor studies (Schieber, 2002; Laubach et al., 2000).

The direct comparability of contrast and attention discussed above for the CRF may also hold for interactions between stimuli in the CRF and the surround. For example, it has been shown that the effect of surround stimuli collinear with a line stimulus within the CRF depends on the contrast threshold of the neuron (Polat et al., 1998). In psychophysical tasks that require the detection of a line element, attention might decrease the threshold for neurons whose CRF contain the target and increase the gain for the lateral inputs. In addition, such gain changes could be achieved, in part, by temporal mechanisms. Engel et al. (1991) have shown that neurons with dispersed CRFs stimulated by the same light bar (collinearity of local CRF stimulation) show synchronous firing in the gamma range. Likewise, other subsequent studies in early extrastriate cortex (Kreiter & Singer, 1992; Castelo-Branco et al., 2000) have shown that dispersed neurons with dispersed CRFs are recruited into synchronously firing neural ensembles when they are stimulated by the same object. Furthermore, top-down attention mechanisms have been shown to influence synchrony of (subthreshold) responses, thereby biasing the processing of bottom-up stimuli (for reviews, see Ritz & Sejnowski, 1997; Engel et al., 2001; Niebur et al., 2002). Although the links between attention, stimulus contrast, and temporal mechanisms are less well understood with respect to the interactions between CRF and surround stimuli, it is likely that the principal ideas described for the effects of attention and learning on CRF stimuli remain valid for perceptual learning involving contextual stimuli. Attentional bias toward certain stimuli will become 'crystallized' in a selected local sensory network when the stimuli are presented repeatedly in the context of a perceptual learning task.

The above review indicates that synchrony and potentially other temporal mechanisms play a significant role in attention and learning. A recent study (Rodrigues et al., 2004) suggests that cholinergic inputs to V1 combined with visual stimulation by a drifting grating (4.5s) can lead to enhanced synchrony of spiking responses in the population of neurons stimulated by the grating. The paradigm used by Rodriguez et al. (2004) employed stimulation of the mesencephalic reticular formation (MRF) just preceding visual stimulation to enhance synchrony of the visual responses. The MRF stimulation was applied in 60-100ms trains of five pulses, starting at 150-300ms before the onset of visual responses. The effects of cholinergic antagonists (scopolamine) and of cholinergic agonists were investigated on neural responses to visual stimuli preceded by MRF stimulation. Iontophoretic administration of scopolamine immediately decreased the synchrony in spiking response to the presented moving grating. Administering cholinergic agonists did not have an immediate effect, but led to an increase in the synchrony of the response after 1-2 hours; an effect that remained present for up to 4 hours (maximum duration measured).

The effects of the cholinergic agonists on the response to the visual stimulus were found after a total of 600 trials delivered in a total of 100 mins during which the cholinergic agonists (acetylcholine and carbachol) were paired with visual stimulation. Control experiments showed that the pairing was critical, and that no delayed effect on synchrony was present when a large number of trials presented cholinergic or visual input alone (an immediate effect was present, however, in the LFP response).

These findings are important, because cholinergic inputs to neocortex originating from the basal forebrain have been implicated in attention and arousal. The basal forebrain can be considered a relay between the MRF and the neocortex (Semba & Fibiger, 1989). MRF stimulation is known to enhance synchronous patterns of EEG activity (Munk et al., 1996; Herculano-Houzel, 1999) that are associated with cognitive abilities including attention, learning, and various kinds of cognition. In the domain of vision, synchrony has been shown to play a role in perception (Rodriguez et al., 1999), selective attention (Fries et al., 2001), and rapid perceptual learning (Gruber et al., 2002). Based on these considerations, Rodriguez et al. (2004) suggest that synchrony-inducing effects produced by MRF stimulation are mediated by enhanced basal forebrain cholinergic inputs to the neocortex.

It is an interesting question then why cholinergic agonists together with visual stimulation (preceded by MRF stimulation) did not cause immediate increases in synchrony, and only late effects. It is possible that the indiscriminate manner in which the agonist diffuses through cortex quelled any clear, immediate effect, or that the agonist influenced a cortical region that was too small. However, the finding that there were immediate effects in the LFP responses, argues against both possibilities. It appears that there were immediate, subthreshold changes in connectivity that were visible in the LFP, but not in spiking behavior. This may be related to the dissociation between spiking data and fMRI data discussed in Section 2.4. The delayed effect of pairing cholinergic agonists with prolonged visual stimulation suggests that cholinergic inputs may help trigger changes in intrinsic V1 circuits that facilitate visual stimulation-induced synchronization of activity. It is likely that these types of modifications of circuitry could play an important role in perceptual learning, and other forms of skill learning. The delayed consolidation processes extending for several hours beyond the training session is a characteristic of skill and perceptual learning that fits well with the data reported by Rodriguez et al. (2004).

3. Discussion

The present chapter has focused on changes in V1 induced by perceptual learning. An important conclusion to be drawn from the material discussed is that the specific stimuli and requirements in the learning task have a profound effect on the properties of learning and the characteristics of the procedural memory. Two types of tasks have been used. In one type of task, a perceptual skill was learned with regard to the discrimination of a feature of a single stimulus. In the other type of task, a perceptual skill was learned that required an assessment of relationships between several stimuli.

In the first type of task (Section 2.3), subjects (monkeys or humans) were trained in orientation discrimination of a single stimulus. Performance of the resulting perceptual skill was selective to retinotopic position and reference orientation only if the training had pushed the subjects to the limits of their sensory capabilities. The orientation and position specificity of the perceptual skill is in agreement with the involvement in memory formation of lower-order visual areas, such as V1, which are retinotopically organized and where neurons show strong orientation selectivity. These specificities suggest that memory formation is restricted to the network of neurons that was selectively stimulated during the training (i.e., at the reference orientation in a given retinotopic position). Schoups et al. (2001) have used training conditions that pushed the limits of sensory performance, and they did indeed find strong position and orientation specificity coupled with plasticity in V1. Ghose et al. (2001) used training conditions that did not push the limits of sensory performance, and they did not find position specificity, nor did they find plasticity in V1, although plasticity was present in V4 (Yang and Maunsell, 2004). It is noteworthy that in Schoup et al.'s (2001) study, the changes in orientation tuning in V1 were measured in the absence of attention. Thus, while attention was required to produce learning and associated changes in V1, the changes at the end of training were 'hard-wired', and could be shown without attention to the stimuli.

These data support the general principles of perceptual learning formulated in Section 2.2. The demonstration of plasticity in V1 after reaching a high level of expertise in orientation discrimination of a single line supports the *lowest-level principle of memory formation* (although it remains an open question whether Schoups et al. (2001) would not have found plasticity in extrastriate cortex as well). The data presented also support the *attentional requirement for learning*, but a modification of the *reverse hierarchy model* must be allowed to account for the fMRI data. While the extracellular recording studies in monkeys suggest that memory formation occurred at lower levels as expertise increased, Furmanski et al. (2004) and Schwartz et al. (2002) have demonstrated changes in V1 after very limited learning. This finding suggests the presence of early plasticity at low levels in the system, which could involve changes in connectivity within those low-level areas or changes in connectivity between lower and higher-order areas.

In a second type of task (see Gilbert, 1996; Gilbert et al., 2000; Section 2.4), the effects of contextual stimuli on a target stimulus were manipulated, by selecting which contextual stimuli were relevant, and by manipulating the amount of attention available to perform the task. The physiological effects of contextual stimuli were modulated by training in three types of tasks: Bisection (and Vernier) tasks, contrast discrimination tasks, and texture discrimination tasks. In the bisection and Vernier tasks, training for a particular set of contextual stimuli (parallel or collinear flankers) increased accuracy of bisection or Vernier performance, and enhanced the impact of the flankers on the neural response to a CRF stimulus. Training effects were specific to the specific flankers chosen. In addition to the specificity of learning and physiological effects for the type of contextual stimuli, the training effects were also specific for the global orientation of the stimuli and retinal position. In contrast to perceptual learning with a single stimulus, it was necessary for the monkey to perform the task to reveal the physiological effects of training with contextual stimuli. This suggests that for V1 neurons in the trained region of cortex, training-induced changes in center-surround interactions remained below threshold when non-attended stimuli were presented in their CRFs. Thus, the functional consequences of the changed architecture in trained cortex in V1, which might involve a modification of the strength of selected lateral connections, was only revealed with the help of attention. This means that behavioral context had become part of the memory trace. In contrast discrimination tasks, collinear flankers were used to enhance the perceived contrast of a central target, and the effect of the flankers was modulated by training and attention manipulations. Early in training, collinear flankers increased the perceived contrast of a target when little attention was available to perform the task, but this effect was minimized when attention could be focused on the target. With training, the effect of the flankers was minimized even in absence of focused attention to the target. Single cell recordings in primates revealed physiological correlates of these training effects in V1. Nevertheless, a number of findings reviewed in Section 2.4 suggest that the neural network whose connections were changed to store the effect of training extended beyond V1. Perceptual learning with contextual stimuli may strongly depend upon feedback connections, which may partly become incorporated into the neural network representing the memory for the skill. Likewise, the finding that enhanced performance in a texture discrimination task after training can depend upon changed interactions between CRF and surround in V1 neurons leaves the possibility open that the neural network representing the memory for this perceptual skill extended beyond V1.

Thus, memory formation in perceptual learning tasks that involve contextual stimuli requires more than V1, even after extended training that pushes the limits of the sensory system. Although strong stimulus and position specificities, and related physiological effects in V1 have been reported, these specificities were not necessarily indicative of storage of the skill in a network limited to V1. The fact that attention was required to reveal the specific learning effects underscores this point. In addition, the finding that the plastic changes in V1 associated with training-induced changes in performance occurred most often in the RF surround, suggests that feedback connections contributing to the surround might have undergone plastic changes as a result of learning. The general idea that this complex type of procedural memory, which includes storage of behavioral context, requires plasticity in areas beyond V1 is in agreement with the *lowest-level principle for memory formation*, and with the *reverse hierarchy model*. The level of areas that contribute to the procedural memory, however, cannot be easily predicted by the nature of the elements used in the contextual stimulus. Furthermore, the lack or the presence of learning specificities is not related in any straightforward way with the hierarchical level of the brain regions involved in processing the stimuli and in memory formation.

Finally, the principle of *attention as a simple gating mechanism* must be modified to allow for the different effects of focused and distributed attention on learning. It is worth mentioning that Watanabe et al. (2001) also demonstrated a form of perceptual learning (resulting in lowered motion detection thresholds) that occurred for specific, repeated stimuli presented subliminally, while attention was directed to other (target) stimuli. The authors argued that plasticity in this condition might have been induced by diffuse reinforcing signals triggered by the presentation of the attended targets, which might be mediated by subcortical neurotransmitters such as acetylcholine, noradrenaline and dopamine (Seitz & Watanabe, 2003). Another perspective on these findings can be found in Chapter 7.

There are limits to the precision with which theoretical frameworks allow to predict the site of plasticity in the visual system for different types of perceptual learning. It is useful, therefore, to also consider the fundamental biological factors underlying neuronal plasticity. Three interacting factors can be identified that play a role in orchestrating the re-wiring during perceptual learning: Hebbian learning and associated gene expression, synchrony of neuronal firing, and disinhibition. These factors have been studied extensively in lower animal species, often in sensory systems other than the visual system.

Hebbian learning refers to the modification of functional connectivity between neurons when there is coincidence of pre- and postsynaptic activity. The best known cellular mechanism of functional plasticity of neuronal connections is termed long-term potentiation (LTP), a persistent facilitation of synaptic signaling triggered by calcium influx consequent to activation of NMDA glutamergic receptors on the postsynaptic membrane (for review, see Merzenich & Buonomano, 1998). Sustained LTP often requires multiple periods of simultaneous (pre- and postsynaptic) stimulation, and requires new protein synthesis and gene expression (Kandel, 2000). LTP has been studied extensively in lower animals, and in the hippocampus of rats, where LTP can be induced easily, but only fairly recently have effective protocols been developed to induce LTP in cortex. To maximize LTP in cortex, data from Trepel and Racone (1998) suggest that stimulation over multiple daily sessions can be required, in line with the time-consuming and asymptotic nature of skill learning. The functional potentiation of a neuronal connection can reflect increases in the number of pre-synaptic neurotransmitter release sites and/or in the number of post-synaptic receptors, but also more drastic changes in morphology, including the sprouting of new axon collaterals (Darian-Smith & Gilbert, 1994; Donoghue, 1995). Neuromodulatory systems such as the noradrenergic system or the cholinergic system control modulations in functional connectivity by influencing NMDA receptor-gated processes (Brocher et al., 1992; Kirkwood et al., 1999; Gu, 2002) (see Chapter 7).

Hebbian learning (and LTP) is critically dependent on co-activation of two or more neurons. It is not a surprise, therefore, that the temporal microstructure of spiking of interconnected neurons might influence connectivity changes. Indeed, there is direct evidence suggesting that the degree of synchrony during oscillatory discharges of interconnected neurons determines the magnitude and direction of Hebbian modifications of connection strength (Lisman, 1989; Markram et al., 1997). Thus, synchrony provides a means to recruit selected neuronal populations into the representation of a perceptual skill. In visual experiments in the monkey, attentive processing of a stimulus leads to synchronization of activity in the gamma-band in populations of neurons coding that stimulus (Fries et al., 2001). The repetition of a task may thus lead to changes in connectivity in the specific neuronal population selected by attention, and required to carry out the task. Furthermore, enhanced connectivity in relevant ensembles may in itself lead to further enhancements of correlated activity or synchrony, strengthening selective plasticity. Laubach et al. (2000) demonstrated in monkeys an association between increased skill in a motor task and increases in correlated firing in an ensemble of M1 motor neurons activated by task performance. A link between attention, synchrony, and cortical plasticity (LTP, Hebbian learning) is in agreement with several behavioral findings that show that repeated stimulation does not result in skill learning unless attention is directed to the stimuli.

The requirement of numerous pairings between visual stimulation and cholinergic agonists to produce (delayed) enhanced synchrony in response to visual stimuli (Rodriguez et al., 2004) provides some further insight into why attention is crucial for skill learning. These findings are in agreement with the known role of the cholinergic basal forebrain in attention (Parasumaran, 2000), perceptual learning (Karni et al., 1994, Walker et al., 2002; 2003), and topographic remapping in sensory cortex (Kilgard & Merzenich, 1998; see also Mercado et al., 2001).

Cortical plasticity is enhanced by disinhibition. The contribution of disinhibition to plasticity is evident from deafferentation experiments, which have shown that after an initial period of silence, deafferented neurons become active again. It has been suggested that the recovery of activity occurs thanks to an initial disinhibition due to the lack of afferent input, and a number of deafferentation experiments in the somatosensory (Jones, 1993; Rasmusson et al., 1993; Tremere et al., 2001a,b) and visual domain (Rosier et al., 1995; Arckens et al., 1998, 2000) support that idea. These studies all suggest that soon after deafferentation, disinhibition occurs, which permits the strengthening of connectivity of deafferented neurons with surrounding neurons through Hebbian learning, until they become driven by their new inputs. There is also evidence suggesting that disinhibition might enhance skill learning. Butefisch et al. (2000) showed in humans that drugs that enhance $GABA_A$ receptor function, prevent the electromyographical changes that can normally be triggered by Trans-Cranial Stimulation directed at M1 after the learning of a simple motor skill. They reported similar results using another drug that blocks NMDA receptors (whose normal function is required for LTP). Attention might modulate inhibition during skill learning through GABAergic projection neurons (intermingled with a majority of cholinergic neurons) in the basal forebrain that project to GABAergic interneurons in sensory cortex (Freund & Meskanaite, 1992; Gritti et al., 1993). The resulting disinhibition and thus malleability of cortical connections might be exploited by other brain regions, such as fronto-parietal cortex and subcortical structures involved in learning and selective attention, such that these connections start to reflect the acquired skill.

Lasting changes in the connectivity in a neuronal ensemble require gene expression (Izquierdo & Cammarota, 2004), and protein synthesis (Chew et al., 1995; Luft et al., 2004). Neurons activated during learning-related stimulation have been shown to rapidly express immediate early genes (IEGs). Many proteins encoded by IEGs are transcription factors that, in turn, control the expression of a range of other, late genes (LGs), many of which appear to be involved in the formation and re-arrangement of cortical circuits (Pinaud et al., 2001, 2002). A number of studies indicate that the process of consolidation of procedural memories continues for several hours after the end of each training epoch (Brashers-Krug, 1996; Shadmehr & Holcomb, 1997; Karni & Sagi, 1993; Karni et al., 1994; Walker et al., 2003a, b; Fenn et al., 2003; Korman et al., 2003; Rodriguez et al., 2004). For some learning paradigms (e.g. Karni & Sagi, 1993), this phenomenon can be inferred from the fact that performance gains are mainly observed between training sessions that are at least several hours apart, rather than within a single session or between sessions separated by a short interval. Interestingly, the latency between training and performance gains is similar to the time window of LG expression. Furthermore, different sets of genes might regulate plasticity at different stages of perceptual learning.

The different sets of genes that might become expressed in different learning stages, may be linked to the different types of circuitry to which sensory neurons have access. Any single V1 neuron, for example, can either be connected to other V1 neurons (short-range connections) or to neurons outside V1 (long-range connections), or both (for review, see Tremere et al., 2003). Changes in short-range connections are likely to be important in the formation of long-term memories for visual skills, as they maintain the very tuning properties of single neurons (Sillito, 1975, 1977, 1979) that are modified by learning (e.g., Schoups et al., 2001). Long-range feedback connections, however, are equally crucial for learning, as changes in local circuitry in sensory areas during perceptual learning are enabled and guided by error-related, reward-related, and attention-related signals. These signals involve complex interactions between, on one hand, dopaminergic, cholinergic and noradrenergic neuromodulatory systems as well as the amygdala and orbito-frontal cortex, and, on the other hand, cortical regions such as parietal and prefrontal cortex known to guide attentional and learning processes in sensory cortex (for reviews, see Ridderinkhof et al., 2004; Schultz, 2004; Kastner & Ungerleider, 2001) Furthermore, neuromodulatory systems can also directly influence processing in sensory areas. For example, cholinergic (Karni et al., 1994; Kilgard & Merxenich, 1998; Rodriguez et al., 2004; Freund & Meskanaite, 1992; Gritti et al., 1993) and noradrenergic projection systems (Pinaud et al., 2000) are likely to be involved in experience-related plasticity in sensory cortex. Many studies indicate a reduction of the role of feedback during asymptotic learning. This suggests that part of the genomic contributions to early learning in V1 may be related to a perhaps temporary strengthening of synapses that carry feedback signals and signals from various neuromodulatory systems, while genomic contributions to late, asymptotic phases of learning may be more related to the modification of local circuitry within V1. Therefore, the gene population that regulates plasticity during fast learning may be at least partially different from the collection of those that putatively control plasticity during local remodeling in V1 induced by asymptotic learning (automation of performance). We speculate that the fast learning component might be regulated by effector IEGs and fast kinetic transcription factors, while the asymptotic learning component may reflect LG regulation.

A study from Pinaud and colleagues (2000) provided insight into how higher order processes such as attention may directly influence experience-dependent genomic mechanisms in the rat visual cortex. It has been known for a number of years that noradrenergic input to the cerebral cortex is critical to attentional processing. It was shown that disruption of noradrenergic input (by lesions in the noradrenergic locus coeruleus) dramatically repressed the light-induced expression of the IEG NGFI-A in the rat visual cortex (Pinaud et al., 2000) (see Chapter 7). NGFI-A is a gene that is very well positioned to coordinate the activation of a large set of LGs involved in experience-dependent network reorganization, via a specific cis-element embedded in the promoter of hundreds of genes expressed in central neurons (see chapter 8). These findings from Pinaud and colleagues suggest that dark-deprivation followed by light-stimulation, a protocol that reliably induces plastic changes in the visual cortex and triggers NGFI-A induction, is dependent on noradrenergic input, a key attentional regulator. Given that this gene encodes a transcriptional regulator, it is likely that the expression of LGs regulated by NGFI-A, and putatively involved in long-term plasticity, would also be impaired during noradrenergic deafferentation. Thus, it appears that the induction of plastic changes may required a combination of neuromodulatory influences associated with the engaging of attentional processing (e.g., noradrenergic input), and repeated sensory stimulation to produce learning. These processes may converge on genomic activation through regulation of IEG-mediated plasticity changes (including LG regulation) (Pinaud, 2004) (see Chapters 7 and 8). Such data indicate that adding a molecular level of inquiry to more traditional neuroscience approaches will add an important dimension to our understanding of perceptual learning, and cognition in general.

References

Ahissar E, Vaadia E, Ahissar M, Bergman H, Arieli A, Abeles M (1992) Dependence of cortical plasticity on correlated activity of single neurons and on behavioral context. *Science* 257:1412–1415.

Ahissar M, Hochstein S (1993) Attentional control of early perceptual learning. *Proc Natl Acad Sci USA* 90:5718–5722.

Ahissar M, Hochstein S (1997) Task difficulty and the specificity of perceptual learning. *Nature* 387:401–406.

Ahissar M, Hochstein S (2002) View from the top: hierarchies and reverse hierarchies in the visual system. *Neuron* 36:791–804.

Ahissar M, Hochstein S (2004) The reverse hierarchy theory of visual perceptual learning. *Trends Neurosci* 8:457–464.

Allman J, Miezin F, McGuinness E (1985) Direction- and velocity-specific responses from beyond the classical receptive field in the middle temporal visual area (MT). *Perception* 14:105–126.

Arckens L, Eysel UT, Vanderhaeghen JJ, Orban GA, Vandesande F (1998) Effect of sensory deprivation on the GABAergic circuitry of the adult cat visual system. *Neurosci* 83:381–391.

Arckens L, Schweigart G, Qu Ying, Wouters G, Pow DV, Vandesande F, Eysel UT, Orban GA (2000) Cooperative changes in GABA, glutamate and activity levels: the missing link in cortical plasticity. *Eur J Neurosci* 12:4222–4232.

Armstrong-James M, Diamond ME, Ebner FF (1994) An innocuous bias in whisker use in adult rats modifies receptive fields of barrel cortex neurons. *J Neurosci* 14: 6978–6991.

Bair W, Cavanaugh JR, Movshon JA (2003) Time course and time-distance relationships for surround suppression in macaque V1 neurons. *J Neurosci* 23:7690–7701.

Bertini G, Karni A, De Weerd P, Desimone R, Ungerleider LG (1995) A behavioral and electrophysiological study of monkey visual cortex plasticity. *Soc. Neurosci. Abstr* 117.13:276.

Bertini G, Karni A, De Weerd P, Desimone R, Ungerleider LG (1996) Plasticity in macaque visual cortex: comparison of learning-related effects in V1 and V2. *Soc. Neurosci. Abstr* 444.2:1122.

Blakemore C, Garey L, Vital-Durand, F (1978). The physiological effects of monocular deprivation and their reversal in the monkey's viosual cortex. *J Physiol (London)* 283:223–262.

Bolz J, Gilbert CD, Wiesel TN (1989) Pharmacological analysis of cortical circuitry. *Trends Neurosci* 12:292–296.

Brashers-Krug T, Shadmehr R, Bizzi E (1996). Consolidation in human motor memory. *Nature* 382:252–255.

Bringuier V, Chavane F, Glaeser L, Fregnac Y (1999) Horizontal propagation of visual activity in the synaptic integration field of area 17 neurons. *Science* 283:695–699.

Bringuier V, Fregnac Y, Baranyi A, Debanne D, Shulz DE (1997) Synaptic origin and stimulus dependency of neuronal oscillatory activity in the primary visual cortex of the cat. *J Physiol* 500 (Pt 3):751–774.

Brocher S, Artola A, Singer W (1992) Agonists of cholinergic and noradrenergic receptors facilitate synergistically the induction of long-term potentiation in slices of rat visual cortex. *Brain Res.* 573:27–36.

Buonomano DV, Merzenich MM (1998) Cortical plasticity: From synapses to maps. *Ann Rev Neurosci* 21:149–186.

Bütefisch CM, Davis BC, Wise SP, Sawaki L, Kopylev L, Classen J, Cohen LG (2000) Mechanisms of use-dependent plasticity in the human motor cortex. *Proc Natl Acad Sci USA* 97:3661–3665.

Carrasco M, Ling S, Read S (2004) Attention alters appearance. *Nat Neurosci.* 7:308–313.

Castelo-Branco M, Goebel R, Neuenschwander S, Singer W (2000) Neural synchrony correlates with surface segregation rules. *Nature* 405: 685–689. Erratum in: Nature 2000; 407:926.

Chew SJ, Mello C, Nottebohm F, Jarvis E, Vicario DS (1995) Decrements in auditory responses to a repeated conspecific song are long-lasting and require two periods of protein synthesis in the songbird forebrain. *Proc Natl Acad Sci USA* 92:3406–3410.

Chelazzi L, Miller EK, Duncan J, Desimone R (1993) A neural basis for visual search in the inferior temporal cortex. *Nature* 363:345–347.

Chelazzi L, Miller EK, Duncan J, Desimone R (1998) Responses of neurons in inferior temporal cortex during memory-guided visual search. *J Neurophysiol* 80: 2918–2940.

Crist RE, Li W, Gilbert CD (2001). Learning to see: experience and attention in primary visual cortex. *Nat Neurosci* 4: 519–525.

Crist RE, Kapadia MK, Westheimer G, Gilbert CD (1997) Perceptual learning of spatial localization: specificity for orientation, position, and context. *J Neurophysiol.* 78(6):2889–2894.

Cynader M, Timney B, Mitchell D (1980) Period of susceptibility of kitten visual cortex to the effects of monocular deprivation extends beyond six months of age.

Darian-Smith C, Gilbert CD (1994) Axonal sprouting accompanies functional reorganization in adult cat striate cortex. *Nature* 368:737–740.

Desimone R, Duncan J (1995) Neural mechanisms of selective attention. *Ann Rev Neurosci* 18:193–222.

Desimone R, Schein SJ (1987) Visual properties of neurons in area V4 of the macaque: sensitivity to stimulus form. *J Neurophysiol* 57:835–868.

De Weerd P, Desimone R, Ungerleider LG (2003) Impairments in spatial generalization of visual skills after V4 and TEO lesions in macaques. *Behav Neurosci* 117:1441–1474.

De Weerd P, Vandenbussche E, Orban GA (1990) Speeding up visual discrimination learning in cats by differential exposure of positive and negative stimuli. *Behav Brain Res* 36:1–12.

Donoghue JP (1995) Plasticity of adult sensorimotor representations. *Curr Opin Neurobiol* 5:749–754.

Donoghue JP, Sanes JN (1987) Peripheral nerve injury in developing rats reorganizes representation pattern in motor cortex. *Proc Natl Acad Sci USA* 84:1123–1126.

Douglas RJ, Martin KAC, Whitteridge D (1989) A canonical microcircuit for neocortex. *Neural Comput.* 1:480–488.

Doyon J, Penhune V, Ungerleider LG. (2003). Distinct contribution of the cortico-striatal and cortico-cerebellar systems to motor skill learning. *Neuropsychologia* 41: 252–262.

Dresp B (1993). Bright lines and edges facilitate the detection of small light targets. *Spat Vis* 7:213–25.

Dresp B, Grossberg S (1997) Contour integration across polarities and spatial gaps: from local contrast filtering to global grouping. *Vision Res* 37:913–924.

Elbert T, Pantev C, Wienbruch C, Rockstroh B, Taub E (1995) Increased cortical representation of the fingers of the left hand in string players. *Science* 270: 305–307.

Engel AK, Fries P, Singer W (2001) Dynamic predictions: oscillations and synchrony in top-down processing. *Nat Rev Neurosci* 2:704–716.

Engel AK, Konig P, Singer W (1991) Direct physiological evidence for scene segmentation by temporal coding. *Proc Natl Acad Sci USA* 1991 88:9136–9140.

Enroth-Cugell C, Robson JG (1966) The contrast sensitivity of retinal ganglion cells of the cat. *J Physiol* 187:517–552.

Eysel UT, Shevelev IA, Lazareva NA, Sharaev GA (1998) Orientation tuning and receptive field structure in cat striate neurons during local blockade of intracortical inhibition. *Neurosci* 84:25–36.

Fahle M (1997) Specificity of learning curvature, orientation and vernier discriminations. *Vision Res* 37:1885–1895.

Fenn KM, Nusbaum HC, Margoliash D (2003) Consolidation during sleep of perceptual learning of spoken language. *Nature* 425:614–616.

Fernandez-Ruiz J, Wang J, Aigner TG, Mishkin M (2001) Visual habit formation in monkeys with neurotoxic lesions of the ventrocaudal neostriatum. *Proc Natl Acad Sci USA* 98:4196–4201.

Field DJ, Hayes A, Hess RF (1993) Contour integration by the human visual system: evidence for local association field. *Vision Res* 37:173–193.

Fiorentini A, Berardi N (1981) Learning in grating waveform discrimination: specificity for orientation and spatial frequency. *Vision Res* 21:1149–1158.

Freund TF, Meskenaite V (1992) gamma-Aminobutyric acid-containing basal forebrain neurons innervate inhibitory interneurons in the neocortex. *Proc Natl Acad Sci USA* 89:738–742.

Fries P, Reynolds JH, Rorie AE, Desimone R (2001) Modulation of oscillatory neuronal synchronization by selective visual attention. *Science* 291:1560–1563.

Furmanski SF, Schluppeck D, Engel SA (2004) Learning strengthens the response of primary visual cortex to simple patterns. *Current Biology* 14:573–578.

Gattass R, Gross CG, Sandell JH (1981) Visual topography of V2 in the monkey. *J Comp Neurol* 201:519–539.

Ghose GM, Yang T, Maunsell JH (2002). Physiological correlates of perceptual learning in V1 and V2. *J Neurophysiol* 87:1867–1888.

Gilbert CD (1996) Plasticity in visual perception and physiology. *Curr Op Neurobiol* 6:269–274.

Gilbert CD, Minami Ito M, Kapadia M, Westheimer G (2000) Interactions between attention, context and learning in primary visual cortex. *Vision Res* 40:1217–1226.

Gilbert CD, Wiesel TN (1989) Columnar specificity of intrinsic cortico-cortical connections in cat visual cortex. *J Neurosci* 9:2432–2442.

Gray CM, McCormick DA (1996) Chattering cells: superficial pyramidal neurons contributing to the generation of synchronous oscillations in the visual cortex. *Science* 274:109–13.

Gritti I, Mainville L, Jones BE (1993) Codistribution of GABA-with acetylcholine-synthesizing neurons in the basal forebrain of the rat. *J Comp Neurol* 329:438–457.

Gross CG, Rocha-Miranda CE, Bender DB (1972) Visual properties of neurons in inferotemporal cortex of the macaque. *J Neurophysiol* 35:96–111.

Gruber T, Muller MM, Keil A (2002) Modulation of induced gamma band responses in a perceptual learning task in the human EEG. *J Cogn Neurosci* 14:732–744.

Gu Q (2002) Neuromodulatory transmitter systems in the cortex and their role in cortical plasticity. Neurosci 111:815–835.

Herculano-Houzel S, Munk MH, Neuenschwander S, Singer W (1999) Precisely synchronized oscillatory firing patterns require electroencephalographic activation. *J Neurosci* 19:3992–4010.

Hess G, Donoghue JP (1994) Long-term potentiation of horizontal connections provides a mechanism to reorganize cortical motor maps. *J Neurophysiol* 71: 2543–2547.

Hess G, Donoghue JP (1996) Long-term depression of horizontal connections in rat motor cortex. *Eur J Neurosci* 8:658–665.

Hubel D, Wiesel T (1965) Binocular interaction in striate cortex of kittens reared with artificial squint. *J Neurophysiol* 28:1041–1059.

Hubel D, Wiesel T, Levay S (1977) Plascitity of ocular dominance columns in monkey striate cortex. *Philos Trans R Soc Lond B Biol Sci* 278:377–409.

Hupe JN, James AC, Girard P, Bullier J (2001) Response modulations by static texture surround in area V1 of the macaque monkey do not depend on feedback connections from V2. *J Neurophysiol* 85:146–163.

Jones EG (1993) GABAergic neurons and their role in cortical plasticity in primates. *Cereb Cortex* 3:361–372.

Ito M, Westheimer G, Gilbert CD (1998) Attention and perceptual learning modulate contextual influences on visual perception. *Neuron* 20: 1191–1197.

Ito M, Gilbert CD (1999) Attention modulates contextual influences in the primary visual cortex of alert monkeys. *Neuron* 22:593–604.

Izquierdo I, Cammarota M. (2004) Zif and the survival of memory. *Science* 304: 829–830.

Jenkins WM, Merzenich MM, Ochs MT, Allard T, Guic-Robles E (1990) Functional reorganization of primary somatosensory cortex in adult owl monkeys after behaviorally controlled tactile stimulation. *J Neurophysiol* 63:82–104, 1990.

Kaas JH, Collins CE, Chino YM (2003) The reactivation and reorganization of retinoptopic maps in visual cortex of adult mammals after retinal and cortical lesions. In: *Filling-in: From Perceptual Completion to Skill Learning*. Pessoa L, De Weerd P (Eds) Oxford University Press.

Kandell ER (2000) Cellular mechanisms of learning and the biological basis of individuality. In Kandell ER, Schwartz JH, Jessell TM (Eds). *Principles of neural science*. NY, McGraw-Hill, pp 1247–1277.

Kapadia MK, Ito M, Gilbert CD, Westheimer G (1995) Improvement in visual sensitivity by changes in local context: parallel studies in human observers and in V1 of alert monkeys. *Neuron* 15:843–856.

Karni A (1996) The acquisition of perceptual and motor skills: a memory system in the adult human cortex. *Cog Brain Res* 5:39–48.

Karni A, Bertini G (1997) Learning perceptual skills: behavioral probes into adult cortical plasticity. *Cur Op Neurobiol* 7:530–535.

Karni A, Bertini G, De Weerd P, Desimone R, Ungerleider LG (1996) Perceptual learning of texture discrimination: subject and task dependent variables. *Soc Neurosci Abs* 22:1122.

Karni A, Meyer G, Rey-Hipolito C, Jezzard P, Adams MM, Turner R, Ungerleider L (1998). The acquisition of skilled motor performance: fast and slow experience-driven changes in primary motor cortex. *Proc Natl Acad Sci USA* 95: 861–868.

Karni A, Meyer G, Jezzard P, Adams M, Turner R, Ungerleider LG (1995) Functional MRI evidence for adult motor cortex plasticity during motor skill learning. *Nature* 377:155–158.

Karni A, Sagi D (1991). Where practice makes perfect in texture discrimination: Evidence for primary visual cortex plasticity. *Proc Natl Acad Sci USA* 88, 4966–4970.

Karni A, Sagi D (1993) The time course of learning a visual skill. *Nature* 365: 250–252.

Karni A, Tanne D, Rubenstein BS, Askenasy JJ, Sagi D (1994) Dependence on REM sleep of overnight improvement of a perceptual skill. *Science* 265: 679–682.

Kastner S, De Weerd P, Desimone R, Ungerleider LG (1998) Mechanisms of directed attention in the human extrastriate cortex as revealed by functional MRI. *Science* 282:108–111.

Kastner S, Ungerleider LG (2001) The neural basis of biased competition in human visual cortex. Neuropsychologia 39(12), 1263–1276.

Kilgard MP, Merzenich MM (1998) Cortical map reorganization enabled by nucleus basalis activity. *Science* 279:1714–1717.

Kirkwood A, Rozas C, Kirkwood J, Perez F, Bear MF (1999) Modulation of long-term synaptic depression in visual cortex by acetylcholine and norepinephrine. *J Neurosci* 19:1599–1609.

Kleim JA, Hogg TM, VandenBerg PM, Cooper NR, Bruneau R, Remple M (2004) Cortical synaptogenesis and motor map reorganization occur during late, but not early, phase of motor skill learning. *J Neurosci* 24:628–633.

Knierim JJ, Van Essen DC (1992) Neuronal responses to static texture patterns in area V1 of the alert macaque monkey. *J Neurophysiol* 67:961–980.

Korman M, Raz N, Flash T, Karni A (2003) Multiple shifts in the representation of a motor sequence during the acquisition of skilled performance. *Proc Natl Acad Sci USA* 100:12492–12497.

Kreiter AK, Singer W (1992) Oscillatory Neuronal Responses in the Visual Cortex of the Awake Macaque Monkey. *Eur J Neurosci* 4:369–375

Lamme VAF (1995) The neurophysiology of figure-ground segregation in primary visual cortex. *J Neurosci* 10:649–669.

Laskin SE, Spencer WA (1979) Cutaneous masking. II. Geometry of excitatory and inhibitory receptive fields of single units in somatosensory cortex of the cat. *J Neurophysiol* 42:1061–1082.

Laubach M, Wessberg J, and Nicolelis MA (2000) Cortical ensemble activity increasingly predicts behavioral outcomes during learning of a motor task. *Nature* 405: 567–571.

Lee TS, Yang CF, Romero RD & Mumford D (2002). Neural activity in early visual cortex reflects behavioral experience and higher-order perceptual saliency. *Nat Neurosci* 5:589–597.

Levitt JB, Yoshioka T, Lund JS (1994) Intrinsic cortical connections in macaque visual area V2: Evidence for interaction between different visual streams. *J Comp Neurol* 342:551–570.

Levitt JB, Lund JS (1997) Contrast dependence of contextual effects in primate visual cortex. *Nature* 387:73–76.

Levitt JB, Lund JS (2002) The spatial extent over which neurons in macaque striate cortex pool visual signals. Vis Neurosci. 19:439–452.

Li W, Gilbert CD (2002) Global contour saliency and local collinear interactions. *J Neurophysiol* 88, 2846–2856.

Li W, Piech V, Gilbert CD (2004). Perceptual learning and top-down influences in primary visual cortex. *Nat Neurosci* 7:651–657.

Lisman J (1989) A mechanism for the Hebb and the anti-Hebb process underlying learning and memory. *Proc Natl Acad Sci USA* 86:9574–9578.

Logothetis NK, Wandell BA (2004) Interpreting the BOLD signal. *Annu Rev Physiol* 66:735–769.

Logothetis NK, Pfeuffer J (2004) On the nature of the BOLD fMRI contrast mechanism. *Magn Reson Imaging* 22:1517–1531.

Logothetis NK (2003). The underpinnings of the BOLD functional magnetic resonance imaging signal. *J Neurosci* 23:3963–3971.

Luck SJ, Chelazzi L, Hillyard S, Desimone R (1997) Neural mechanisms of spatial selective attention in areas V1, V2, and V4 of macaque visual cortex. *J Neurophysiol* 77:24–42.

Lueschow A, Miller EK, Desimone R. Inferior temporal mechanisms for invariant object recognition. *Cereb Cortex* 4:523–531, 1994.

Luft AR, Buitrago MM, Ringer T, Dichgans J, Schulz JB (2004) Motor skill learning depends on protein synthesis in motor cortex after training. *J Neurosci* 24:6515–6520.

Maffei L, Fiorentini A (1976) The unresponsive regions of visual cortical receptive fields. *Vision Res* 16:1131–1139.

Marcelja S (1980) Mathematical description of the responses of simple cortical cells. *J Opt Soc Am* 70:1297–1300.

Markram H, Luebke J, Frotscher M, Sakmann B (1997) Regulation of synaptic efficacy by coincidence of postsynaptic AP's and EPSP's. *Science* 275:213–215.

McAdams CJ, Maunsell JH (1999) Effects of attention on orientation-tuning functions of single neurons in macaque cortical area V4. *J Neurosci* 19:431–441.

McClelland JL, McNaughton BL, O'Reilly RC (1995) Why there are complementary learning systems in the hippocampus and neocortex: insights from the successes and failures of connectionist models of learning and memory. *Psychol Rev* 102:419–457.

McGuire BA, Gilbert CD, Rivlin PK, Wiesel TN (1991) Targets of horizontal connections in macaque primary visual cortex. *J Comp Neurol* 305:370–392.

Mercado E, Bao S, Orduna I, Gluck MA, and Merzenich MM (2001) Basal forebrain stimulation changes cortical sensitivities in complex sound. *Neuroreport* 12:2283–2287.

Merzenich MM, Buonomano DV (1998). Cortical plasticity: From synapses to maps. *Ann Rev Neurosci* 21:149–186.

Moores E, Laiti L, Chelazzi L (2003) Associative knowledge controls deployment of visual selective attention. *Nat Neurosci* 6:182–189.

Moran J, Desimone R (1985). Selective attention gates visual processing in the extrastriate cortex. *Science* 229:782–784.

Munk MH, Roelfsema PR, Konig P, Engel AK, Singer W (1996). Role of reticular activation in the modulation of intracortical synchronization. *Science* 272:271–274.

Nadel L, Moscovitch M (2001) The hippocampal complex and long-term memory revisited. TINS 5:228–230.

Nelson JI, Frost B (1978) Orientation selective inhibition from beyond the classical receptive field. *Brain Res* 139:359–365.

Niebur E, Hsiao SS, Johnson KO (2002) Synchrony: a neuronal mechanism for attentional selection? *Curr Opin Neurobiol* 12:190–194.

Nudo RJ, Milliken GW, Jenkins WM, Merzenich MM (1996) Use-dependent alterations of movement representation in primary motor cortex of adult Squirrel monkeys. *J Neurosci* 15:785–807.

Olson C, Freeman R (1978) Monocular deprivation and recovery during sensitive period in kittens. *J Neurophysiol* 41:65–74.

Olson C, Freeman R (1980) Profile of sensitive period for monocular deprivation in kittens. *Exp Brain Res* 39:17–21.

O'Reilly RC, Rudy JW. Conjunctive representations in learning and memory: principles of cortical and hippocampal function. *Psychol Rev* 108:311–345.

Parasumaran R (Ed) (2000) *The attentive brain*. Cambridge, MIT Press.

Pernberg J, Jirmann KU, Eysel UT (1998) Structure and dynamics of receptive fields in the visual cortex of the cat (area 18) and the influence of GABAergic inhibition. *Eur J Neurosci* 10:3596–3606.

Polat U, Sagi D (1994) The architecture of perceptual spatial interactions. *Vision Res* 34:73–78.

Ramachandran VS (1988) Perception of shape from shading. *Nature* 331:163–166.

Reynolds JH, Chelazzi L & Desimone R (1999) Competitive mechanisms subserve attention in macaque areas V2 and V4. *J Neurosci* 19:1736–1753.

Reynolds JH, Desimone R (2003) Interacting roles of attention and visual salience in V4. *Neuron* 37:853–863.

Reynolds JH, Pasternak T, Desimone R (2000). Attention increases sensitivity of V4 neurons. *Neuron* 26:703–714.

Rodriguez R, Kallenbach U, Singer W, Munk MH (2004) Short- and long-term effects of cholinergic modulation on gamma oscillations and response synchronization in the visual cortex. *J Neurosci* 24:10369–10378.

Ringach DL, Bredfeldt CE, Shapley RM, Hawken MJ (2002) Suppression of neural responses to nonoptimal stimuli correlates with tuning selectivity in macaque V1. *J Neurophysiol* 87:1018–1027.

Pinaud R (2004) Experience-Dependent Immediate Early Gene Expression in the Adult Central Nervous System: Evidence from Enriched-Environment Studies. *Int J Neurosci* 114:321–333.

Pinaud R, Penner MR, Robertson HA, and Currie RW (2001) Upregulation of the immediate early gene arc in the brains of rats exposed to environmental enrichment: implications for molecular plasticity. *Mol Brain Res* 91:50–56.

Pinaud R, Tremere LA, Penner MR, Hess FF, Robertson HA, and Currie RW (2002) Complexity of sensory environment drives the expression of candidate-plasticity gene, nerve growth factor induced-A. *Neurosci* 112:573–582.

Pinaud R, Tremere LA, Penner MR (2000) Light-Induced Zif268 Expression is Dependent on Noradrenergic Input in Rat Visual Cortex. *Brain Res* 882:251–255.

Polat U, Mizobe K, Pettet MW, Kasamatsu T, Antony MN (1998) Collinear stimuli regulate visual responses depending on the cells contrast threshold. *Nature* 391:580–584.

Pons TP, Garraghty PE, Ommaya AK, Kaas JH, Taub E, Mishkin, M (1991) Massive cortical reorganization after sensory deafferentation in adult Macaques. *Science* 252:1857–1860.

Rasmusson DD, Louw DF, Northgrave SA (1993) The immediate effects of peripheral denervation on inhibitory mechanisms in the somatosensory thalamus. *Somatosens Mot Res* 10:69–80.

Reber PJ, Squire LR (1994) Parallel brain systems for learning with and without awareness. *Learn Mem* 1:217–229.

Recanzone GH, Jenkins WM, Hradek GT, Merzenich MM (1992) Progressive improvement in discriminative abilities in adult owl monkeys performing a tactile frequency discrimination task. *J. Neurophysiol* 67:1015–1030.

Recanzone GH, Merzenich MM, Jenkins WM (1992). Frequency discrimination training engaging a restricted skin surface results in an emergence of a cutaneous response zone in cortical area 3a. *J Neurophysiol* 67:1057–1070.

Recanzone GH, Merzenich MM, Jenkins WM, Grajski KA, Dinse HR (1992). Topographical reorganization of the hand representation in cortical area 3b of owl monkeys trained in a frequency discrimination task. *J Neurophysiol* 67:1031–1056.

Recanzone GH, Schreiner GE, Merzenich MM (1993) Plasticity in the frequency representation of primary auditory cortex following discrimination training in adult owl monkeys. *J Neurosci* 13:87–103.

Ress D, Backus BT, Heeger DJ (2000) Activity in primary visual cortex predicts performance in a visual detection task. *Nat Neurosci* 3:940–945.

Ridderinkhof KR, van den Wildenberg WP, Segalowitz SJ, Carter CS (2004) Neurocognitive mechanisms of cognitive control: the role of prefrontal cortex in action selection, response inhibition, performance monitoring, and reward-based learning. *Brain Cogn.* 56(2):129–140.

Rioult-Pedotti MS, Friedman D, Hess G, Donoghue JP (1998) Strengthening of horizontal connections following skill learning. *Nat Neurosci* 1:230–234.

Rioult-Pedotti MS, Friedman D, Donoghue JP (2000) Learning-induced LTP in neocortex. *Science* 290:533–536.

Ritz R, Sejnowski TJ (1997) Synchronous oscillatory activity in sensory systems: New vistas on mechanisms. *Curr Op Neurobiol* 7:536–546.

Rosier AM, Arckens L, Demeulemeester H, Orban GA, Eysel UT, Wu YL, Vandesande F (1995) Effect of sensory deafferentation on immunoreactivity of GABAergic cells and on GABA receptors in the adult cat visual cortex. *J Comp Neurol* 359:476–489.

Sceniak MP, Hawken MJ, Shapley R (2001) Visual spatial characterization of macaque V1 neurons. *J Neurophysiol* 85:1873–1887.

Schein JJ, Desimone R (1990) Spectral properties of V4 neurons in the macaque. *J Neurosci* 10:3369–3389.

Schoups A, Vogels R, Orban GA (1995) Human perceptual learning in identifying the oblique reference orientation: retinotopy, orientation specificity and monocularity. *J Physiol*: 483:797–810.

Schoups A, Vogels R, Qian N, Orban GA (2001) Practising orientation identification improves orientation coding in V1 neurons. *Nature* 412:549–553.

Scheich H, Simonis C, Ohl F, Tillein J, Thomas H (1993) Functional organization and learning-related plasticity in auditory cortex of the Mongolian gerbil. *Prog Brain Res* 97:135–143.

Schultz W (2004) Neural coding of basic reward terms of animal learning theory, game theory, microeconomics and behavioural ecology. *Curr Opin Neurobiol.* 14(2):139–147.

Schwark HD, and Jones EG (1989) The distribution of intrinsic corical axons in area 3b of cat primary somatosensory cortex. *Exp Brain Res* 78:501–513.

Schwartz S, Maquet P, Frith C (2002) Neural correlates of perceptual learning: a functional MRI study of visual texture discrimination. *Proc Natl Acad Sci USA* 99: 17137–17142.

Semba K, Fibiger HC (1989). Organization of central cholinergic systems. *Prog Brain Res* 79:37–63.

Seitz AR, Watanabe T (2003) Psychophysics: Is subliminal learning really passive? *Nature* 422(6927):36.

Shadmehr R, Holcomb HH (1997) Neural correlates of motor memory consolidation. *Science* 277:821–5.

Shatz C, Stryker M (1978). Ocular dominace of layer IV of the cat's visual cortex and the effects of monocular deprivation. *J Physiol (Lond)* 281:267–283.

Sigman M, Gilbert CD (2000) Learning to find a shape. *Nat Neurosci* 3:264–269.

Sillito AM (1975a) The effectiveness of bicuculline as an antagonist of GABA and visually evoked inhibition in the cat's striate cortex. *J Physiol* 250:287–304.

Sillito AM (1975b) The contribution of inhibitory mechanisms to the receptive field properties of neurones in the striate cortex of the cat. *J Physiol* 250:305–329.

Sillito AM (1977) Inhibitory processes underlying the directional specificity of simple, complex and hypercomplex cells in the cat's visual cortex. *J Physiol* 271:699–720.

Sillito AM (1979) Inhibitory mechanisms influencing complex cell orientation selectivity and their modification at high resting discharge levels. *J Physiol* 289:33–53.

Schieber MH (2002) Training and synchrony in the motor system. *J Neurosci* 22:5277–5281.

Super H, Spekrreijse H, & Lamme VAF (2001) Two distinct modes of sensory processing observed in monkey primary visual cortex (V1). *Nat Neurosci* 4:304–310.

Stork DG, Wilson HR (1990) Do Gabor functions provide appropriate descriptions of visual cortical receptive fields? *J Opt Soc Am (A)* 7:1362–1373.

Tanifuji M, Sugiyama T, Murase K (1994) Horizontal propagation of excitation in rat visual cortical slices revealed by optical imaging. *Science* 266:1057–1059.

Thiele A (2004). Percpetual learning: Is V1 up to the task? *Current Biolology* 14:R671–R673.

Tremere L, Hicks TP, Rasmusson DD (2001a) Role of inhibition in cortical reorganization of the adult raccoon revealed by microiontophoretic blockade of GABA(A) receptors. *J Neurophysiol* 86:94–103.

Tremere L, Hicks TP, Rasmusson DD (2001b) Expansion of receptive fields in raccoon somatosensory cortex in vivo by GABA(A) receptor antagonism: implications for cortical reorganization. *Exp Brain Res* 136:447–455.

Tremere L.A, Pinaud R. & De Weerd P (2003) Contributions of Inhibitory Mechanisms to Perceptual Completion. In: *Filling-in: From Perceptual Completion to Skill Learning.* Pessoa L, De Weerd P (Eds) Oxford University Press.

Trepel C, Racine RJ (1998) Long-term potentiation in the neocortex of the adult freely moving rat. *Cereb Cortex* 8:719–729.

Treue S, Maunsell JH (1996) Attentional modulation of visual motion processing in cortical areas MT and MST. *Nature* 382:539–541.

Vania LM, Belliveau JW, des Roziers EB, Zeffiro TA (1998) Neural systems underlying the learning and representation of global motion. *Proc Natl Acad Sci USA* 95: 12657–12662.

Vogels R, Orban GA (1985) The effect of practice on the oblique effect in line orientation judgments. *Vision Res* 25:1679–1687.

Walker MP, Brakefield T, Morgan A, Hobson JA, Stickgold R (2002) Practice with sleep makes perfect: sleep-dependent motor skill learning. *Neuron* 35:205–211.

Walker MP, Brakefield T, Hobson JA, Stickgold R (2003a) Dissociable stages of human memory consolidation and reconsolidation. *Nature* 425:616–620.

Walker MP, Brakefield T, Seidman J, Morgan A, Hobson JA, Stickgold R (2003b) Sleep and the time course of motor skill learning. *Learn Mem* 10:275–84.

Watanabe T, Nanez JE, Sasaki Y (2001) Perceptual learning without perception. *Nature* 413(6858):844–848.

Wehrhahn C, Dresp B (1998) Detection facilitation by collinear stimuli in humans: dependence on strength and sign of contrast. *Vision Res* 38:423–428.

Weinberger NM (1993) Learning-induced changes of auditory receptive fields. *Curr Opin Neurobiol* 3:570–577.

Willingham DB (1998) A neuropsychological theory of motor skill learning. *Psych Rev* 105:558–584.

Winfield DA, Brooke RNL, Sloper JJ, Powell TPS (1981) A combined Golgi electron microscope study of the synapses made by the proximal axon and recurrent collaterals of a pyramidal cell in the somatic sensory cortex of the monkey. *Neurosci* 6:1217–1230.

Yang T, Maunsell JH (2004) The effect of perceptual learning on neuronal responses in monkey visual area V4. *J Neurosci* 24:1617–1626.

Zipser K, Lamme VAF, Schiller PH (1996). Contextual modulation in the primary visual cortex. *J Neurosci* 16:7376–7389.

13

Investigating Higher Order Cognitive Functions in the Dorsal (magnocellular) Stream of Visual Processing

ANTONIO F. FORTES[1] AND HUGO MERCHANT[2]

[1]Department of Neuroscience, University of Minnesota, MN, USA and[2] Instituto de Neurobiología, UNAM campus Juriquilla Queretaro Qro., Mexico

Introduction

Since the pioneering studies of Mountcastle, Goldberg and others in the mid 70s much information has been gathered on both the anatomical and functional properties of the posterior parietal cortex (PPC). This area is a conglomeration of subregions that deal with integration of complex phenomena such as coordinate transformation, attention and sensory input in order to develop a sensorimotor plan. We shall see that although these regions share this unifying factor that many of the regions are somewhat specialized for certain functions. We will first cover the processing pathway of the visual signals originated in V1 and then briefly review some of the anatomical connections of some of the subregions of PPC. We will then take a general outlook at the neurophysiological studies that have taken place in these subregions focusing on the better studied PPC structures including areas 7a and those in the intra-parietal-sulcus.

The dorsal and ventral streams

There are two main visual pathways that originate from the primary visual cortices, and their constituents include different areas and are thought to take part in different types of visual processing (Reid, 1999). These pathways have also been termed the "where and what" pathways due to the idea that they carry out different functions in dealing with objects. The *dorsal* stream or "where" pathway is thought to be involved mainly in spatial representation, and the *ventral* (also called temporal) stream or "what" pathway seems to be

R. Pinaud, L. A. Tremere, P. De Weerd(Eds.), Plasticity in the Visual System:
From Genes to Circuits, 285-306, ©2006 Springer Science + Business Media, Inc.

involved in object recognition (Mishkin and Ungerleider, 1982). Some authors have argued that the difference between the two streams is in the use that the individuals make of the visual information processed in the two streams and not in the resulting percept of object vs. space (Goodale and Milner, 1992; Goodale and Westwood, 2004). In this sense, the ventral stream or 'vision-for-perception' system provides visual information for perception, while the dorsal stream or 'vision-for-action' system provides information necessary for the control of action and is not involved in perception (Rizzolatti and Matelli, 2003).

While the division of the visual signals that arise in the primary visual areas into the dorsal and ventral streams may be advantageous, there is a dispute whether the functional divisions are really as clear cut and meaningful as some authors have proposed. While there may exist shared features of the two pathways, there is clearly some functional difference between the two. In fact, some authors have argued that there is a continuous differentiation of function within the dorsal stream itself and that it be further separated into two distinct functional systems, one encompassing areas V6, V6A and MIP, and the other encompassing the area MT and the visual areas of the inferior parietal lobule. The first would be involved in the "on line" control of actions (and damage to this area leads to optic ataxia) (Rizzolatti and Matelli, 2003), while the other would be involved in the perception of space, action understanding and action organization.

The ventral stream receives information from the primary visual areas and projects to multiple visual areas including the temporal cortex (Ungerleider and Mishkin, 1982). The dorsal stream carries information from the primary visual areas to the parietal cortex, a complex region that carries out many associative functions including visual, auditory and somatosensory integration (LaMotte and Mountcastle, 1979; Mazzoni et al., 1996). The areas of the anterior parietal cortex seem to be involved mainly in dealing with somatosensory input, while those in the more posterior region, the **posterior parietal cortex** (PPC), are involved mainly with the processing of visual signals. This region sits at the top of the hierarchy of the dorsal stream and is the subject matter of this chapter.

The visual processing that occurs in parietal cortex is very complex. Here there are cells that deal with one or more functions such as visual motion, coordinate transformations, attention and other cognitive aspects. The neuronal firing rate of these cells can be modulated by a plethora of different conditions such as head position, visual attention, and motor intention.

Anatomy of the Posterior Parietal Cortex

The various visual regions of the posterior parietal cortex connect to different regions in both visual parietal and prefrontal areas. Topographically, the visual areas of the Rhesus PPC (see Fig. 13.1) are divided into areas 7a in the inferior parietal gyrus, and several areas which lie within the intraparietal

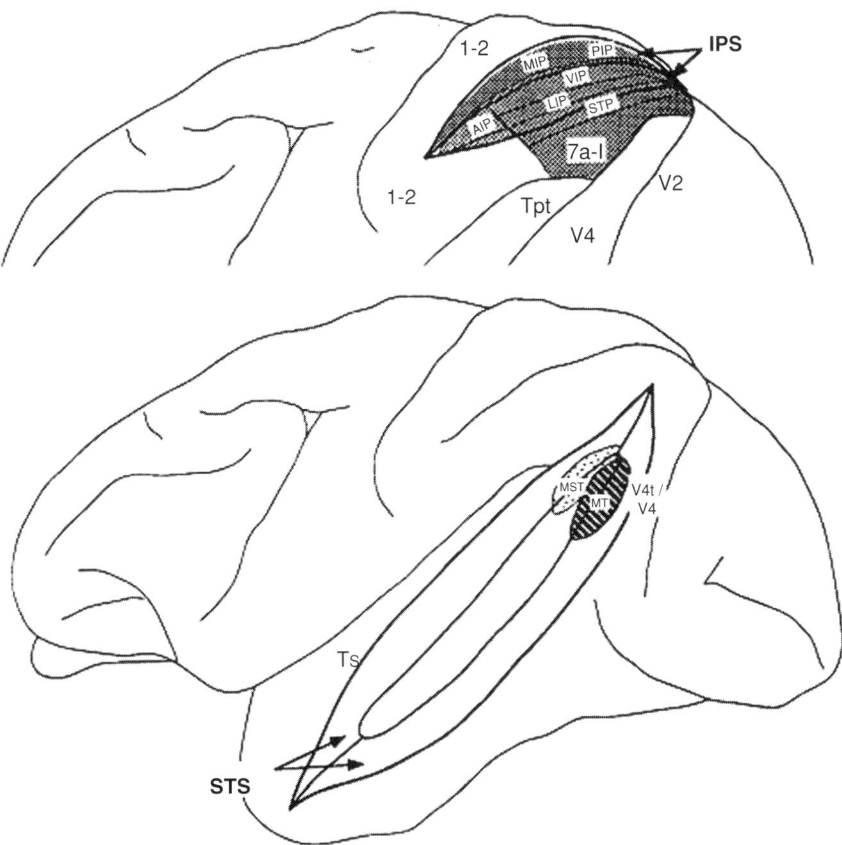

FIGURE 13.1. The organization of the partitions of the posterior parietal cortex in the monkey. **A**. The Intraparietal sulcus is shown as opened. **B**. Areas located within the superior temporal sulcus. Modified with permission from Preuss and Goldman-Rakic (Preuss and Goldman-Rakic, 1991).

sulcus, including, VIP, AIP and LIP (ventral, anterior, and lateral intraparietal areas respectively). The intraparietal sulcus also contains other areas such as MIP and PIP (medial and posterior intraparietal areas). The classification of these areas has been based mainly on their connectivity and cellular responses. Initially the PPC was subdivided architectonically, but recently the parcelation has been redone based largely on the differential connectivity of the different subregions composing the PPC with other visual areas and the prefrontal cortex (Andersen et al., 1990; Baizer et al., 1991; Cavada and Goldman-Rakic, 1993). Interestingly, the architectonic subdivisions of PPC seem to correlate well with those of the connectivity based classification. Of the seventeen architectonic subdivisions identified in parieto-occipital cortex, eight are in regions known from physiological studies to be largely or entirely visual (V3, V3A, PIP, LOP, PO) or visuomotor (7a, LIPd, LIPv) (Lewis and Van Essen, 2000b). By far the best characterized subregions of the PPC are areas 7a, LIP and VIP, which have multiple connections with other visual areas in posterior parietal and prefrontal regions (Cavada and Goldman-Rakic, 1989a, b; Baizer et al., 1991; Felleman and Van Essen, 1991)).

Area 7

Area 7 is situated just posterior to the intra parietal sulcus (IPS) and anterior to the superior temporal sulcus. It is functionally and cytoarchitectonically subdivided into three subregions: 7a, 7b and 7m. Area 7a comprises the posterior-medial extent of the region, and 7b is located more anteriorly and laterally. Area 7m, as the name indicates, is the most medial of the three. Bordering area 7a on the lateral bank of the inferior parietal sulcus is the lateral intraparietal area (LIP). In the fundus of the IPS is the ventral intraparietal area (VIP), and on the medial bank is the medial intraparietal area (MIP). Medial and anterior to the IPS is area 5. Posterior to area 7a is the superior temporal sulcus, containing the middle temporal area (MT) and the medial superior temporal area (MST). Figure 13.2 depicts some of the surface areas of the primary and parietal visual areas.

Area 7m is heavily connected with other sub regions of the parietal cortex such as area MIP, area 5, area 7a, area 7ip, visual area MST, PO, and is interconnected with the prefrontal cortex (Leichnetz, 2001). Recent neurophysiological experiments in this area have shown that neural activity is related to

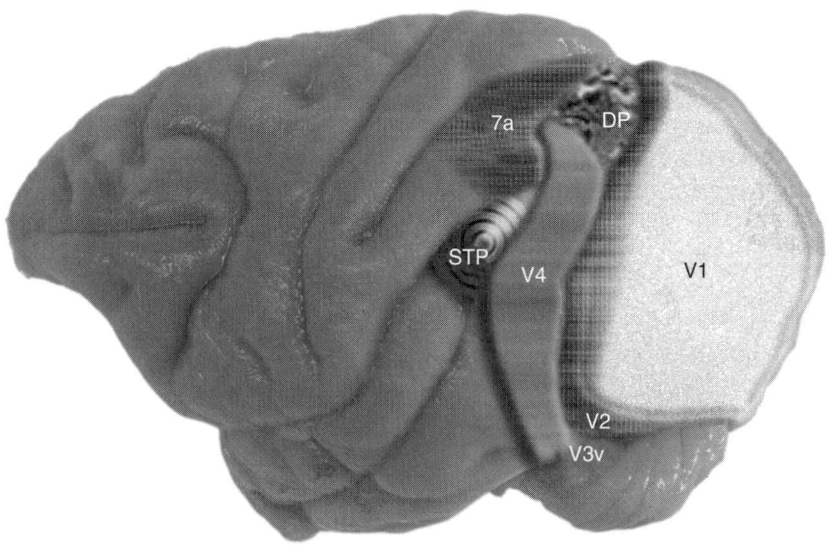

FIGURE 13.2. A schematic coding of the surface areas including the primary visual and visual posterior parietal cortices in the rhesus macaque brain.

the visual control of hand posture and movement during reaching (Ferraina et al., 1997).

Area 7a is heavily interconnected with areas LIP, MST, STP, and PO, as well as with frontal areas such as the frontal eye fields and lateral prefrontal cortex (Cavada and Goldman-Rakic, 1989a). Like area 7m, area 7a is also reciprocally connected with limbic areas, such as the cingulate gyrus and parahippocampal gyrus.

AIP

The anterior intraparietal area (AIP) is situated in the lateral part of the anterior bank of the IPS (see Fig. 13.1). Studies by Sakata and colleagues have implicated this area in configuring the shape of the hands for the grasping of objects (Sakata et al., 1995; Sakata et al., 1998).

Neurons in AIP respond to 3D features of objects that are important for configuring the hand for grasping such as shape, orientation, and size (Taira et al., 1990; Sakata et al., 1995; Murata et al., 1996; Sakata et al., 1998). This area also has a large number of neurons that are active during visually guided manipulation of objects (Murata et al., 2000).

Neurons in area AIP have connections with many sensorimotor-related cortical areas that are similar to those of neighboring area VIP. In contrast to VIP, AIP receives little visual-related input except from higher visual areas and possibly bimodal or polysensory areas (Lewis and Van Essen, 2000a, b).

AIP neurons project to ventral premotor area F5, with which they share functional properties (VIP projects to F4). Each of these circuits is possibly dedicated to specific aspects of sensorimotor transformations. In particular, the AIP-F5 circuit should play a crucial role in visuomotor transformation for grasping, and the VIP-F4 circuit is possibly involved in peripersonal space coding for movement (Luppino et al., 1999). AIP neurons that project to premotor areas such as those mentioned above also receive connections from area LIP[1]. Area LIP receives inputs from area V3A which in turn receives inputs from primary visual areas. These cortical connections between V3A, LIP and AIP are thought to play an important role in the visuomotor transformation for grasping hand movements (Nakamura et al., 2001).

MIP, PIP and VIP

Area VIP was originally identified by Maunsell and Van Essen (1983) (Maunsell and Van Essen, 1983) as the target of projections from areas MT and V2. As in area VIP, connectivity was the original ground for identifying MIP and PIP as specific areas. These were designated by Colby and colleagues (Colby et al., 1988) based on their connections with visual area PO,

---

[1] Neurons in the lateral intraparietal area respond visually to three-dimensional objects.

or V6. Even though Colby, Duhamel, and their colleagues have also studied neuronal activity in VIP and MIP, their boundaries, as well as those of PIP, are still not well defined. Both visual and somatosensory responses can be recorded in VIP and MIP; cells in area VIP are thought to represent perioral space, while those in MIP are thought to represent immediate extrapersonal space (Colby and Duhamel, 1991; Duhamel et al., 1997) (for a review, see Colby and Goldberg, 1999 and Battaglia-Mayer et al., 2001 and 2002).

Parietal Cortex and Sensorimotor Control

Spatial Coordinate Transformations

Over the past few decades, the PPC has become known as an "association" area that integrates information from many sensory modalities to develop cognitive representations of space. As reviewed above, the PPC receives sensory signals from visual, auditory, vestibular and somatosensory modalities. These signals are integrated using a specific gain mechanism[2] that enables the transformation from the input coordinates of receptor surfaces (e.g. the retina) into the coordinates of effectors (e.g. eye, head, or hand).

The first type of gain mechanism has been described in single cells in areas 7a and LIP, which receive a convergence of eye position and visual signals (Andersen and Mountcastle, 1983; Andersen et al., 1985). As a result of this convergence, these are the cells whose retinal receptive fields are modulated in a monotonic fashion by the orbital position of the eyes. This modulation was described as an eye "gain field" because the eye position appeared to modulate the gain of the visual response. The activity of a single neuron with an eye gain field position cannot provide enough information with respect to a head-centered stimulus location. However, the activity across a population of cells with different eye position and retinal position gain fields will have a unique pattern of firing for each head-centered location. Thus the code of head-centered location in the posterior parietal cortex appears to be carried as a distributed population code (Zipser and Andersen, 1988). Other spatially distributed codes are also present in the PPC. There are neurons with head gain fields that are modulated by proprioceptive signals that provide information regarding the location of the head with respect to the body, and that can be used to code the location of a visual stimulus in a body-centered coordinate frame (Snyder et al., 1997). Other PPC neurons such as those with head gain fields are modulated by vestibular signals and can code the location of the head in space using a world-centered coordinate frame (Andersen et al., 1993). Another interesting feature of some PPC neurons is the overlap in their

---

[2] There is a modulation of the firing rate of an individual neuron because of the interaction with other parts of the system.

visual and auditory receptive fields. These types of neurons can represent auditory locations in two coordinate systems: in eye-centered coordinates (the auditory response moves with the eye position), in head-centered coordinates (the auditory response moves with the head position) or in a combination of the previous two (Stricanne et al., 1996). The auditory neurons coding these three coordinate systems have also shown eye gain fields (Stricanne et al., 1996). Thus, these properties indicate that groups of PPC neurons share a common distributed representation of both auditory and visual signals that can be used to represent stimuli in either modality in the same spatial framework.

Using neural network modeling, the laboratory of Andersen and colleagues has trained networks to input retinotopic-visual, head-centered auditory, eye position, and head position signals and to output three separate representations: eye-centered, head-centered, and body-centered (Xing and Andersen, 2000). Interestingly, the hidden layer of the neural network (where most of the processing occurs) develops eye and head gain fields, and single cells develop bimodal auditory-visual fields. This simulation may be analogous to what is occurring *in situ* in the posterior parietal cortex. By using the gain field mechanism, a variety of modalities in different coordinate frames can be integrated into a distributed representation of space. In this way, information is not collapsed or lost, and many coordinate frames can be represented in the same population of neurons. In fact, the coexistence of multiple coordinate frames in the posterior parietal cortex is likely to explain why spatial deficits appear in multiple coordinate frames after lesions to this area in humans (Husain et al., 2000).

Movement Planning

The PPC has also been implicated in higher-level cognitive functions related to action. Among these higher cognitive functions is the formation of intentions, or early plans for movement. These early plans are segregated within the PPC, within regions specialized for the preparation of saccades, reaches, and grasps. In addition, the deficits observed following lesions of the PPC are consistent with its cognitive role in sensory-motor integration.

Patients with PPC lesions may not have primary sensory or motor deficits. However, when an attempt to carry out a behavior involving both of these functions is made, the deficits become apparent. For example, patients with PPC lesions often suffer from optic ataxia; a syndrome in which there is difficulty in estimating the location of stimuli in 3D space and there are pronounced errors in reaching movements (Rondot et al., 1977). Likewise, patients with parietal lobe damage also have difficulty in correctly shaping their hands as they prepare to grasp objects (Perenin and Vighetto, 1988; Goodale and Milner, 1992).

PPC neurons respond when the animal is preparing to execute a movement in delayed memory tasks. It has been reported that during eye movement planning area LIP is more active, and during limb movement planning the parietal reach region (PRR) was more active. PRR includes MIP, 7a, and the dorsal aspect of the parieto-occipital (PO) area. MIP shows the highest concentration of reach-related neurons (Snyder et al., 1997). These neurons show movement fields (regions of the movement space where they respond best). Moreover, the codification of intentions is usually in a eye-centered coordinate frame (Stricanne et al., 1996; Cohen and Andersen, 2000).

An important property of these early plans is their high level of abstraction. In PPC the neural correlates of intentions do not contain information about the details of a movement, such as joint angles, torques and muscle activations required for executing a movement. Neurons are activated every time that the animal is planning to reach a behavioral goal. Moreover, these intentions are evident in the discharge of single PPC neurons even when a specific intention is not carried out (Snyder et al., 1997).

Decision Making

It has been reported that PPC is part of the neural mechanism underlying decision making. In a behavioral task, the prior probability and amount of reward influence the effectiveness of visual stimuli to drive the neural response in PPC, which is consistent with a role for this area in decision making (Platt and Glimcher, 1999). In addition, as monkeys accumulate sensory information to make a movement plan, activity increases for neurons in PPC (Shadlen et al., 1996; Kim and Shadlen, 1999; Leon and Shadlen, 1999). These results are consistent with these areas weighing decision variables for the purpose of planning eye movements (Gold and Shadlen, 2001). Neural activity that follows the probability to make a movement decision has been found also in the supplementary motor cortex, the prefrontal cortex and the basal ganglia (Merchant et al., 1997; Romo et al., 1997; Romo et al., 1998; Hernandez et al., 2000). Overall, these results suggest that decision making is a distributed function that includes the PPC.

Parietal Cortex and Spatial Attention

The PPC not only represents multiple coordinate frames using multimodal information but is also part of an integrated circuitry involved in spatial attention mechanisms, decision making and movement preparation. It is clear, then, that this area is an important interface between the sensory apparatus and the motor system in the frontal lobe.

PPC is intimately related to spatial attention mechanisms. Dramatic spatial attention deficits are produced after lesions in the PPC in humans and monkeys, including neglect. Neglect is characterized by the lack of awareness within the personal and extrapersonal space contralateral to the cerebral hemisphere with the lesion, with the most profound deficits seen with right hemispheric lesions (Rosselli et al., 1985; Hommet et al., 2004). In neurophysiological studies in the monkey, it has been shown that an important group of PPC neurons is activated in relation to the appearance of a stimulus that is pertinent to the animal for completion of a behavioral task. However, these neurons do not respond to stimuli that are not used to drive the behavior of the monkey. Hence, these neurons are probably involved in a mechanism of spatial attention, since the responses to a relevant stimulus are independent of the animal's intended action. The responses occur whether or not the monkey is going to make a saccade to the attended stimulus (Colby et al., 1996), during a delayed saccade task, and during a peripheral attention task where there is a bar release. In this respect, it has been proposed that the activity evoked by a suddenly appearing stimulus is primarily an attentional response rather than a purely visual one. In fact, marginal responses are obtained when a monkey makes a saccade that brings a stable, non-novel stimulus into the receptive field of these PPC neurons (Gottlieb et al., 1998). In contrast, in the fixation task, the same stimulus evokes a strong response when it suddenly appears in the receptive field and is pertinent to the behavior (Gottlieb et al., 1998).

Various studies have specifically investigated the functional properties of PPC neurons is spatial attention tasks. First, it has been shown that the visual responsiveness of PPC neurons, in some attentional contexts, instead of being enhanced is actually reduced at the focus of attention (Steinmetz et al., 1994; Robinson et al., 1995). However, at locations away from the focus of attention the responsiveness is enhanced. The sensitivity for unattended stimuli in this group of neurons suggests that they may play a role in shifting attention, possibly by providing signals of the spatial locations of novel stimuli (Steinmetz and Constantinidis, 1995; Constantinidis and Steinmetz, 2001a, b). In another set of experiments, the spatial and temporal aspects of attention were tracked in monkeys, and a strict correlation was found between the neural activity in PPC and the monkey's attentional performance (Bisley and Goldberg, 2003). The population activity profile described the spatial and temporal dynamics of a monkey's attention. This is to say that there was an increase in the ensemble response that followed the attentional advantage that produces higher contrast sensitivity at the attended spatial locus. This occurred in two attentional situations: before an upcoming saccade (endogenous attention) or after the presentation of a distracter (exogenous attention).

A Long Standing Debate: Attention vs. Intention

The issue of intention versus attention in PPC started long time ago, which is not surprising considering that this area is at the interface between sensory and motor systems. The pioneering study by Mountcastle and colleagues (Mountcastle, 1975) first noted neural activity in the PPC related to the behaviors of monkeys. They suggested that the PPC was involved in the generation of a motor commands used to direct the eyes and arms during exploration. A few years later, Robinson and colleagues (Robinson and Goldberg,

1978) argued that these effects could be due to visual stimulation and attention during movement. Since that time, several groups have supported the attentional or the intentional schools, by adopting the use of ever more sophisticated tasks, quantitative techniques and model testing. It is clear now that these phenomena may not be mutually exclusive and that neural signals associated to these high order behaviors may co-exist in PPC, and is that it is difficult to "tag" one particular function to this associative area. This is especially true because in the PPC the experimental correlation between behavioral events and neural activity is always less precise than in primary sensory or motor areas. The neural responses in PPC are conditional in nature and driven by central state functions of the brain. Thus, interpretations regarding the neural activity in PPC can be as different as visual signals, motor signals, reafferent signals, attentional signals, intentional signals, etc. Experiments designed to dissociate all these components have demonstrated that neurons in PPC can carry multiple signals (Andersen et al., 1987; Gnadt and Andersen, 1988; Barash et al., 1991; Colby et al., 1996). Therefore, PPC is an important node in a series of distributed functions across the parieto-frontal and parieto-temporal systems that pertain to the sensory-motor realm.

High Level Visual Motion Processing

PPC is not only part of the visual dorsal stream, which plays an important role in spatial awareness and coordinate transformations (see above), but is also part of the visual motion stream. The visual motion stream is considered part of the dorsal stream, although it also shares features of the ventral stream. For example, the motion system analyzes visual motion to form percepts of motion patterns and shapes derived from motion, which are features of the ventral stream. In contrast, this pathway also plays an important role in visuomotor transformations, such as spatial awareness based on motion cues (Merchant et al., 2001) and the use of visual motion information to drive behavior, such as during the interception of a moving target (Merchant et al., 2004a). The visual motion information is processed in this stream by a large number of cortical areas, including V1, V2, V3 and V3a, the middle temporal area (MT), the medial superior temporal area (MST), the superior temporal polysensory area (STP) and several areas within the PPC, including LIP, VIP, area 7a and area 7m.

As in other modalities, there is a hierarchical organization of visual motion processing through this pathway, starting in V1 where the cells already possess direction selectivity with respect to stimulus motion. The cells of V1 project to MT where most of the cells are direction selective and show higher levels of motion processing (Andersen, 1997). In the next node, the cells in MST respond to even more complex features of visual motion, such as optic flow stimuli. In this area some cells respond selectively to expansion or contraction of the visual field, and to rotation and spiraling motions (Sakata et al., 1985; Duffy and Wurtz, 1991a, b; Graziano et al., 1994; Lagae et al., 1994). When we reach the PPC, cells respond selectively to the direction of motion of bars and random dots (Motter and Mountcastle, 1981; Siegel and Read, 1997), to optic flow stimuli (Siegel and Read, 1997; Merchant et al., 2001, 2003), to 2D and 3D rotatory stimuli (Sakata et al., 1986; Sakata et al., 1994) and, interestingly, can combine visual motion signals with the angle of gaze (Read and Siegel, 1997). In this hierarchical pathway there is a gradual increase in the receptive field (RF) size from V1 through MT (for a review see Zeki(Zeki, 1993)). At the highest levels of the dorsal stream hierarchy, such as in the PPC, the RFs are quite large and often bilateral (Robinson and Goldberg, 1978; Motter and Mountcastle, 1981; Andersen et al., 1990). Nevertheless, the RFs cannot be considered a "passive" or immutable property of the cells in PPC. The size and structure of RFs in PPC are clearly determined by the attentional state of the monkey (Mountcastle et al., 1981) and the angle of gaze (Andersen and Mountcastle, 1983). Furthermore, the structure and location of the RFs depend on the parameters and the temporal succession of the stimuli used to map them (Motter and Mountcastle, 1981; Motter et al., 1987). In fact, it was found that the RFs could change in size and location depending on the optic flow stimuli used (Merchant et al., 2001 and Merchant et al., 2004b). The characteristic properties of the PPC, the behavioral context, the quality of the stimuli and the internal state are important modulators of visual motion responses in these association areas.

Parietal visual neurons are driven prominently by moving stimuli, and often their responses are dependent on context. The visual receptive fields of some of these neurons are organized so that the neuron responds to radial motion moving in towards or away from the fovea. These neurons were classified as opponent-vector neurons, and it was suggested that they participated in optic flow processing associated with locomotion, or the movement of the hand through peripheral vision toward a foveated target (Motter and Mountcastle, 1981; Motter et al., 1987). More recently, it has been suggested that the latter hypothesis is more likely (Siegel and Read, 1997). Subsequent work has confirmed that 7a neurons are involved in optic flow processing (Siegel and Read, 1997; Merchant et al., 2001, 2003). The responses of 7a neurons to optic flow stimuli appear to be more complex than those in MST. While most MST neurons respond selectively to elementary optic flow components (e.g. expansion, contraction, CW or CCW rotation), some 7a neurons respond similarly to CW and CCW rotations (Siegel and Read, 1997). The dissimilarity matrix of 7a neuronal responses to eight different kinds of motion (right-, left-, up-, down-ward, clockwise, counterclockwise, expansion, contraction) was analyzed using tree clustering and multidimensional scaling (MDS). These analytical techniques were used to reveal possible associations in the activity of neuronal populations driven by elementary optic flow components. Tree clustering analyses showed that [left, right], [upward, downward], and [CW, CCW] motions were clustered in three separate branches (i.e. horizontal, vertical, and rotatory motion, respectively) (Vallar et al., 1999). In contrast, expansion was in a lone branch whereas contraction was also separate but within a larger cluster. The distances among these clusters were then subjected to an MDS analysis to identify the dimensions underlying the tree clustering observed. This analysis revealed 2 major factors in operation. The first factor separated expansion from all other stimulus motions, which seems to reflect the prominence of expansion during the common activity of locomotion. In contrast, the second factor separated planar motions from motion in depth, which suggests that the latter may hold a special place in visual motion processing. Therefore, PPC can process optic flow information in a very complex, but behaviorally meaningful fashion.

PPC is also tightly related to apparent motion perception. For example, using functional MRI it was found that not only the occipitotemporal junction (hMT+/V5), but also PPC was activated during perception of circular motion in the "spinning wheel illusion" (Sterzer et al., 2002). In addition, it was found that the activity of directionally selective cells in three visual motion areas could predict the perceived direction of complex perceptually bistable apparent motion stimuli (Williams et al., 2003). However, these cells were more common in PPC, less common in MST, and practically non-existent in MT. Thus, this pattern suggests that the neural activity is better correlated with the perception of complex apparent motion in higher parietal areas, which fits in with the anatomical hierarchy of these areas in the parietal visual stream (Andersen et al., 1990). Recent findings clearly support this hypothesis. We analyzed the activity of populations of PPC cells that was modulated by the stimulus position in real or apparent circularly moving stimuli (Merchant et al., 2004b) in order to recover the stimulus position over time (Fig 13.3). For this purpose, a multivariate linear was used. Real stimulus motion was decoded successfully from the neural activity of the ensemble at all speeds. In contrast, the decoding of apparent motion was poor at low stimulus speeds but improved markedly above the stimulus speeds which was the threshold for human subjects to perceive continuous stimulus motion in this condition (Merchant et al., 2004c). Hence, there was a tight correlation between the quality of the decoded stimulus position by neural ensembles and the perception of apparent motion. Therefore, ensembles of PPC neurons can create an explicit visual representation of a moving stimulus changing dynamically in the perceived apparent motion path.

All of these experiments support a late instead of an early cortical locus for the perceptual filling-in during apparent motion (Liu et al., 2004). In fact, lesions in the right parietal cortex of human patients produce a bilateral deficit for apparent motion perception, as well as a hemilateral neglect with marked problems in visuospatial and attentional processes (Battelli et al., 2001). Since these lesions do not produce impairments in the discrimination of direction of visual motion, it has been suggested that the perception of apparent moving stimuli depends on a high-level system that includes the parietal cortex and depends on attention mechanisms (Battelli et al., 2001). Indeed, psychophysical experiments have suggested that motion perception depends on two motion systems. A low level system that is effortless and passive, and which probably depends on the activity of neurons in V1 and MT, and a high level system that requires attention and has been linked to posterior parietal areas (Cavanagh, 1992; Lu and Sperling, 2001). In conclusion, the properties of PPC neurons indicate that a high degree of convergence and integration of visual motion information is taking place in this area, which make the PPC a structure well suited to represent a high-level perception such as apparent motion.

In summary, PPC processes complex visual motion inputs from MT, and MST, represents visuospatial information in different coordinate frames and combines this information with attentional and motor planning signals. For example, the visual response to optic flow is modulated in PPC by the retinotopic stimulus position and by the orbital eye position (Read and Segal, 1997). This arrangement provides a neural substrate for representing the direction of heading under different behavioral circumstances, including when the eyes are not looking at the focus of the expanding flow field. Another case of PPC engagement in visual motion processing used to drive actions is our recent work on interceptive behavior. We found that the PPC is part of a parieto-frontal system involved in the interception of moving targets. In a NOGO task we observed groups of PPC neurons that responded in relation to spatio-temporal properties of real and apparent moving targets (Fig. 13.3A) (Merchant et al., 2004b). More interestingly, during an interception task, we found that PPC neurons responded not only to the real and apparent moving targets, but also during the preparation of the interception movement. These findings indicate that the PPC visuomotor transformations occurring during the interception were dependent on the visual motion properties of the target. For real moving targets the encoded variable was mainly the position of the stimulus, while for apparent motion targets the encoded parameter was the temporal profile of the stimulus (Merchant et al., 2004a).

Above the Sensory-Motor Apparatus: Cognition in PPC

Besides its role in space perception, coordinate transformations, early movement planning and complex visual motion processing, PPC is a critical neuronal node for spatial cognition. The first evidence in this regard came from lesion studies. Patients with PPC lesions can suffer from apraxias, a class of deficits characterized by the inability to plan movements (Geshwind and Damasio, 1985). These can range from a complete inability to imitate gestures, to difficulty in performing sequences of movements and the incapacity to solve mazes, copy figures or construct shapes (Balint, 1909; Kleist, 1934; Strub and Black, 1977).

At the neurophysiological level it has been difficult to study cognition in PPC for the following reasons. First, PPC is at the interface between sensory and motor systems; second, conditionality is fundamental; third, the relationship between behavior and neural activity is not crystal clear; and forth there is an important influence of the attentive state and the intentionality to execute a

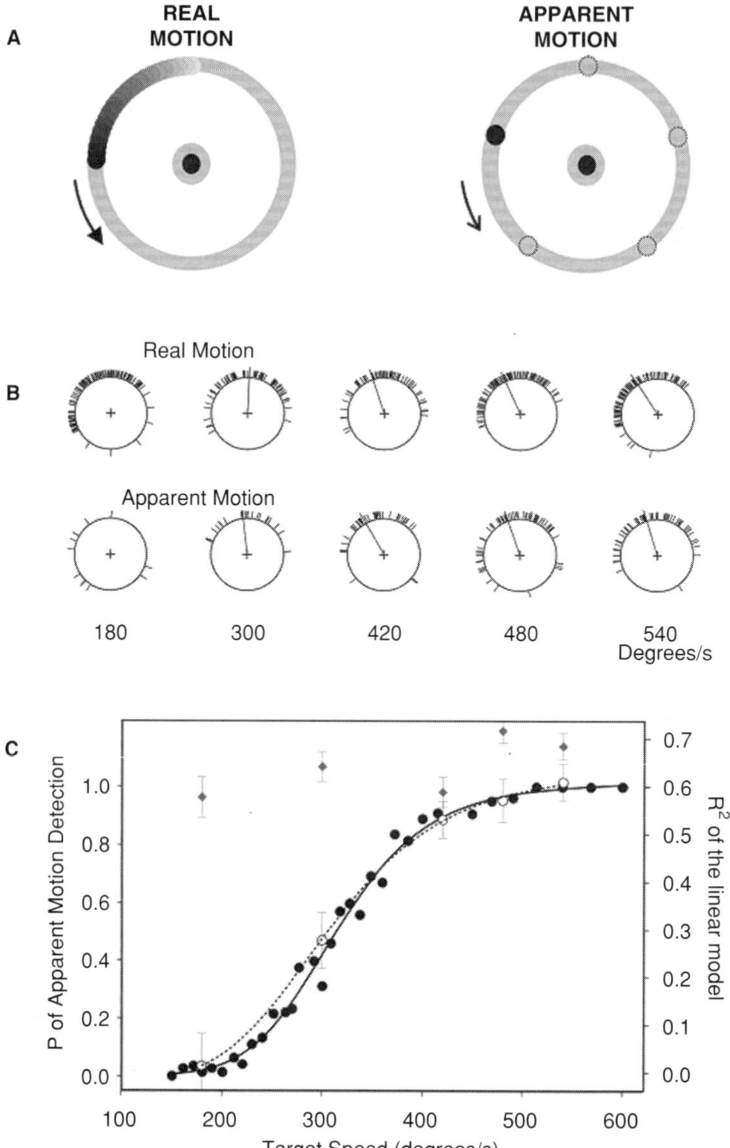

FIGURE 13.3. **A**, Stimuli in the real and apparent motion conditions. A smooth real moving stimulus was produced in the real motion condition. In the apparent motion condition five stimuli were flashed at the vertices of a regular pentagon. All stimuli traveled counterclockwise (CCW), through a circular low contrast path. **B**, Circular rasters of a cell in area 7a that was significantly tuned to the angular position of the stimuli in real and apparent motion for most of the stimulus speeds. **C**, Comparison between the psychometric curve of apparent motion detection as a function of the stimulus speed (9 human subjects, black filled circles) with the sigmoidal function obtained from the R^2 (mean ± SD) from the 20 neurons with the highest fits (black open circles and dashed line). The best R^2 (mean ± SD) of the lineal model during the real motion condition are depicted as solid diamonds.

particular behavior on the neural activity. Thus, the study of higher functions in PPC needs to take into account sensory processing, attentional modulation and motor intentionality before trying to understand the neural mechanisms of cognition in PPC, making these studies extremely difficult to implement.

Recently, Georgopoulos and collaborators investigated the neural underpinnings of maze solving and object construction in the PPC. In the first set of experiments, the PPC neural activity was recorded while monkeys were solving mazes. The animals were required to determine whether the main path of a maze reached an exit of not by executing a motor response. The most important finding was that approximately 25% of the neurons were tuned to direction of the maze path. This directional path tuning was not due to the lower level visual features of the maze stimulus or the early planning to perform a movement in a particular direction. Instead the maze path tuning in PPC was probably associated with maze solution, and as such, reflects a cognitive process applied to a complex visual stimulus (Crowe et al., 2004a). In fact, path tuning did not reflect a covert saccade plan since the majority of active neurons during maze task were not active on a delayed-saccade control task. Moreover, path tuning during maze solution was not due to the locations of visual RFs. For example, it was found that the centers of the mapped RFs were not aligned to the preferred path directions. Finally, no path tuning was observed when a naïve animal viewed the same visual maze stimuli but did not solve them (Crowe et al., 2004a). Interestingly, the dynamic representation of maze solving was determined using population vector analysis. The population representation of path direction traced the main path and changed direction when the path turned. This dynamic neural representation occurred during a period of unchanging visual input, which suggests that the evolution of PPC neural activity is associated with the progression of cognitive operations (Crowe et al., 2004b).

In another study, monkeys were trained to perform a mental object construction task where a model object (consisting of an arrangement of square elements) was presented and memorized for a short delay. Then, a working copy object was shown. This copy was identical to the preceding model except that a single square had been removed (the 'missing square'). Next the monkeys were required to choose between two squares to add at different locations to the working copy object. The addition of one of these squares, replaced the missing square and reproduced the model configuration.

The main finding in this study was that neurons in PPC were activated by the appearance of the working copy object in a manner that depended on the location of the square that had been removed from that object (Chafee et al., 2003). The neurons represented the location of the missing square. This spatial representation was independent from the retinal location of the stimuli, since the working copy was presented in different retinal locations. This was a neural signal linked to a cognitive rule (inferring the location of the missing square) and not strictly to the spatial location. Furthermore, it was clear that this neural signal was linked to the construction behavior because it predicted errors in construction (Chafee et al., 2003).

This important work emphasizes a very interesting feature of PPC, namely that, depending on the cognitive requirements of a task, the neural activity in this area can show sensorimotor independence. This is to say that the neural activity in PPC is related to the spatial aspects of cognition, and under these conditions can be independent of the sensory input and motor outcome.

In future studies of the PPC we expect that, as more cognitively demanding tasks arise, the recording of neural signals from this area may help break the code for the representation of higher order functions.

Conclusion

Anatomically, many of the subregions of the PPC receive input from auditory, tactile visual areas and share many connections between each other and other integrative regions in the prefrontal cortex. Interestingly, there is a clear association between connectivity and functional organization making this area the ideal candidate for multimodal integration.

At the neurophysiological level, there have been two schools of thought on the function of parietal cortical neurons: the intentional and attentional schools; with most published results attributing only one of these functions to the cells in PPC. While it was clearly demonstrated that at least one of these function seemed to exist, it may be that coding for attention or intention are not mutually exclusive properties of these cells. In fact, attention may be a necessary element of a sensorimotor interface. For example, the sensorimotor transformation is parallel at input and serial at output, that is to say that there are many potential targets for a behavior such as reaching while there is usually only one intended behavior (one target to be reached to). Consequently target selection, which is one function of attention, is integral to motor control and the formation of intentions.

Despite the difficulty of studying cognitive aspects of the PPC, it is now clear that this area is intimately involved in spatial cognition. It seems that cells in this area can code for much more abstract concepts than those that have to deal purely with the transformation of sensory input to motor outcome. In fact, such abstract concepts as missing objects and maze path tuning have been found to exist in this region(Chafee et al., 2003; Crowe et al., 2004a).

A variety of factors such as eye position, attention and intention can modulate neurons in PPC. In fact, neuronal responses in this area can be conditional to any one of these or other factors; this is to say that responses in this region may occur in one context and not in another. For example, in trials that are visually identical some neurons may be active in the task only when the monkey is actively attending an object and less active when the animal is not. The modulation and conditionality observed in this area provides clear evidence that this region plays an important role in the plastic modulation of visual signals for the generation of cognitive processes and motor behavior.

In summary, we have seen that the PPC is a complex region receiving inputs from diverse modalities such as somatosensory, auditory and visual signals from many different regions of the cortex. The PPC comprises sub regions that integrate these signals to code location of objects in diverse reference frames, allows behaviors such as shaping the hands for grasping of objects and even abstract cognitive concepts such as mental object construction and maze solving.

Acknowledgements: We would like to thank Mathew Chafee and Alexandra Basford for their help in reviewing this manuscript.

References

Andersen RA (1997) Multimodal integration for the representation of space in the posterior parietal cortex. Philos Trans R Soc Lond B Biol Sci 352:1421–1428.

Andersen RA, Mountcastle VB (1983) The influence of the angle of gaze upon the excitability of the light-sensitive neurons of the posterior parietal cortex. J Neurosci 3:532–548.

Andersen RA, Asanuma C, Cowan WM (1985) Callosal and prefrontal associational projecting cell populations in area 7A of the macaque monkey: a study using retrogradely transported fluorescent dyes. J Comp Neurol 232:443–455.

Andersen RA, Essick GK, Siegel RM (1987) Neurons of area 7 activated by both visual stimuli and oculomotor behavior. Exp Brain Res 67:316–322.

Andersen RA, Asanuma C, Essick G, Siegel RM (1990) Corticocortical connections of anatomically and physiologically defined subdivisions within the inferior parietal lobule. J Comp Neurol 296:65–113.

Andersen RA, Snyder LH, Li CS, Stricanne B (1993) Coordinate transformations in the representation of spatial information. Curr Opin Neurobiol 3:171–176.

Baizer JS, Ungerleider LG, Desimone R (1991) Organization of visual inputs to the inferior temporal and posterior parietal cortex in macaques. J Neurosci 11:168–190.

Balint R (1909) Seelenlahmung des 'Schauens', optische Ataxie, raumliche Storung der Aufmerksamkeit. Monatschrift für Psychiatrie und Neurologie 25:5–81.

Barash S, Bracewell RM, Fogassi L, Gnadt JW, Andersen RA (1991) Saccade-related activity in the lateral intraparietal area. I. Temporal properties; comparison with area 7a. J Neurophysiol 66:1095–1108.

Battaglia-Mayer A, Caminiti R (2002) Optic ataxia as a result of the breakdown of the global tuning fields of parietal neurones. Brain 125:225–237.

Battaglia-Mayer A, Ferraina S, Genovesio A, Marconi B, Squatrito S, Molinari M, Lacquaniti F, Caminiti R (2001) Eye-hand coordination during reaching. II. An analysis of the relationships between visuomanual signals in parietal cortex and parieto-frontal association projections. Cereb Cortex 11:528–544.

Battelli L, Cavanagh P, Intriligator J, Tramo MJ, Henaff MA, Michel F, Barton JJ (2001) Unilateral right parietal damage leads to bilateral deficit for high-level motion. Neuron 32:985–995.

Bisley JW, Goldberg ME (2003) Neuronal activity in the lateral intraparietal area and spatial attention. Science 299:81–86.

Cavada C, Goldman-Rakic PS (1989a) Posterior parietal cortex in rhesus monkey: II. Evidence for segregated corticocortical networks linking sensory and limbic areas with the frontal lobe. J Comp Neurol 287:422–445.

Cavada C, Goldman-Rakic PS (1989b) Posterior parietal cortex in rhesus monkey: I. Parcellation of areas based on distinctive limbic and sensory corticocortical connections. J Comp Neurol 287:393–421.

Cavada C, Goldman-Rakic PS (1993) Multiple visual areas in the posterior parietal cortex of primates. Prog Brain Res 95:123–137.

Cavanagh P (1992) Attention-based motion perception. Science 257:1563–1565.

Chafee MV, Crowe DA, Averbeck BB (2003) Parietal neurons encode object-centered space in a constrauction task. In: SFN 2003. New Orleans: SFN.

Cohen YE, Andersen RA (2000) Reaches to sounds encoded in an eye-centered reference frame. Neuron 27:647–652.

Colby CL, Duhamel JR (1991) Heterogeneity of extrastriate visual areas and multiple parietal areas in the macaque monkey. Neuropsychologia 29:517–537.

Colby CL, Goldberg ME (1999) Space and attention in parietal cortex. Annu Rev Neurosci 22:319–349.

Colby CL, Duhamel JR, Goldberg ME (1996) Visual, presaccadic, and cognitive activation of single neurons in monkey lateral intraparietal area. J Neurophysiol 76: 2841–2852.

Colby CL, Gattass R, Olson CR, Gross CG (1988) Topographical organization of cortical afferents to extrastriate visual area PO in the macaque: a dual tracer study. J Comp Neurol 269:392–413.

Constantinidis C, Steinmetz MA (2001a) Neuronal responses in area 7a to multiple-stimulus displays: I. neurons encode the location of the salient stimulus. Cereb Cortex 11:581–591.

Constantinidis C, Steinmetz MA (2001b) Neuronal responses in area 7a to multiple stimulus displays: II. responses are suppressed at the cued location. Cereb Cortex 11:592–597.

Crowe DA, Chafee MV, Averbeck BB, Georgopoulos AP (2004a) Neural activity in primate parietal area 7a related to spatial analysis of visual mazes. Cereb Cortex 14:23–34.

Crowe DA, Chafee MV, Averbeck BB, Georgopoulos AP (2004b) Participation of primary motor cortical neurons in a distributed network during maze solution: representation of spatial parameters and time-course comparison with parietal area 7a. Neuron 158:28–34.

Duffy CJ, Wurtz RH (1991a) Sensitivity of MST neurons to optic flow stimuli. II. Mechanisms of response selectivity revealed by small-field stimuli. J Neurophysiol 65:1346–1359.

Duffy CJ, Wurtz RH (1991b) Sensitivity of MST neurons to optic flow stimuli. I. A continuum of response selectivity to large-field stimuli. J Neurophysiol 65: 1329–1345.

Duhamel JR, Bremmer F, BenHamed S, Graf W (1997) Spatial invariance of visual receptive fields in parietal cortex neurons. Nature 389:845–848.

Felleman DJ, Van Essen DC (1991) Distributed hierarchical processing in the primate cerebral cortex. Cereb Cortex 1:1–47.

Ferraina S, Garasto MR, Battaglia-Mayer A, Ferraresi P, Johnson PB, Lacquaniti F, Caminiti R (1997) Visual control of hand-reaching movement: activity in parietal area 7m. Eur J Neurosci 9:1090–1095.

Geshwind N, Damasio AR (1985) Apraxia. In: In Handbook of Clinical Neurology (Vinken PJ, Bruyn GW, Klawans KL, eds), pp 423–432. Amsterdam: Elsevier.

Gnadt JW, Andersen RA (1988) Memory related motor planning activity in posterior parietal cortex of macaque. Exp Brain Res 70:216–220.

Gold JI, Shadlen MN (2001) Neural computations that underlie decisions about sensory stimuli. Trends Cogn Sci 5:10–16.

Goodale MA, Milner AD (1992) Separate visual pathways for perception and action. Trends Neurosci 15:20–25.

Goodale MA, Westwood DA (2004) An evolving view of duplex vision: separate but interacting cortical pathways for perception and action. Curr Opin Neurobiol 14: 203–211.

Gottlieb JP, Kusunoki M, Goldberg ME (1998) The representation of visual salience in monkey parietal cortex. Nature 391:481–484.

Graziano MS, Andersen RA, Snowden RJ (1994) Tuning of MST neurons to spiral motions. J Neurosci 14:54–67.

Hernandez A, Zainos A, Romo R (2000) Neuronal correlates of sensory discrimination in the somatosensory cortex. Proc Natl Acad Sci U S A 97:6191–6196.

Hommet C, Bardet F, de Toffol B, Perrier D, Biraben A, Vignal JP, Scarabin JM, Corbineau M, Chauvel P (2004) Unilateral spatial neglect following right inferior parietal cortectomy. Epilepsy Behav 5:416–419.

Husain M, Mattingley JB, Rorden C, Kennard C, Driver J (2000) Distinguishing sensory and motor biases in parietal and frontal neglect. Brain 123 (Pt 8): 1643–1659.

Kim JN, Shadlen MN (1999) Neural correlates of a decision in the dorsolateral prefrontal cortex of the macaque. Nat Neurosci 2:176–185.

Kleist K (1934) Gehirnpathologie. Leipzig: Barth.

Lagae L, Maes H, Raiguel S, Xiao DK, Orban GA (1994) Responses of macaque STS neurons to optic flow components: a comparison of areas MT and MST. J Neurophysiol 71:1597–1626.

LaMotte RH, Mountcastle VB (1979) Disorders in somesthesis following lesions of parietal lobe. J Neurophysiol 42:400–419.

Leichnetz GR (2001) Connections of the medial posterior parietal cortex (area 7m) in the monkey. Anat Rec 263:215–236.

Leon MI, Shadlen MN (1999) Effect of expected reward magnitude on the response of neurons in the dorsolateral prefrontal cortex of the macaque. Neuron 24:415–425.

Lewis JW, Van Essen DC (2000a) Mapping of architectonic subdivisions in the macaque monkey, with emphasis on parieto-occipital cortex. J Comp Neurol 428:79–111.

Lewis JW, Van Essen DC (2000b) Corticocortical connections of visual, sensorimotor, and multimodal processing areas in the parietal lobe of the macaque monkey. J Comp Neurol 428:112–137.

Liu T, Slotnick SD, Yantis S (2004) Human MT+ mediates perceptual filling-in during apparent motion. Neuroimage 21:1772–1780.

Lu ZL, Sperling G (2001) Three-systems theory of human visual motion perception: review and update. J Opt Soc Am A Opt Image Sci Vis 18:2331–2370.

Luppino G, Murata A, Govoni P, Matelli M (1999) Largely segregated parietofrontal connections linking rostral intraparietal cortex (areas AIP and VIP) and the ventral premotor cortex (areas F5 and F4). Exp Brain Res 128:181–187.

Maunsell JH, Van Essen DC (1983) Functional properties of neurons in middle temporal visual area of the macaque monkey. I. Selectivity for stimulus direction, speed, and orientation. J Neurophysiol 49:1127–1147.

Mazzoni P, Bracewell RM, Barash S, Andersen RA (1996) Spatially tuned auditory responses in area LIP of macaques performing delayed memory saccades to acoustic targets. J Neurophysiol 75:1233–1241.

Merchant H, Battaglia-Mayer A, Georgopoulos AP (2001) Effects of optic flow in motor cortex and area 7a. J Neurophysiol 86:1937–1954.

Merchant H, Battaglia-Mayer A, Georgopoulos AP (2003) Functional organization of parietal neuronal responses to optic-flow stimuli. J Neurophysiol 90:675–682.

Merchant H, Battaglia-Mayer A, Georgopoulos AP (2004a) Neural responses during interception of real and apparent circularly moving stimuli in motor cortex and area 7a. Cereb Cortex 14:314–331.

Merchant H, Battaglia-Mayer A, Georgopoulos AP (2004b) Neural responses in motor cortex and area 7a to real and apparent motion. Exp Brain Res 154:291–307.

Merchant H, Battaglia-Mayer A, Georgopoulos AP (2004c) Decoding of path-guided apparent motion from neural ensembles in posterior parietal cortex. Exp Brain Res.

Merchant H, Zainos A, Hernandez A, Salinas E, Romo R (1997) Functional properties of primate putamen neurons during the categorization of tactile stimuli. J Neurophysiol 77:1132–1154.

Mishkin M, Ungerleider LG (1982) Contribution of striate inputs to the visuospatial functions of parieto-preoccipital cortex in monkeys. Behav Brain Res 6:57–77.

Motter BC, Mountcastle VB (1981) The functional properties of the light-sensitive neurons of the posterior parietal cortex studied in waking monkeys: foveal sparing and opponent vector organization. J Neurosci 1:3–26.

Motter BC, Steinmetz MA, Duffy CJ, Mountcastle VB (1987) Functional properties of parietal visual neurons: mechanisms of directionality along a single axis. J Neurosci 7:154–176.

Mountcastle VB (1975) The view from within: pathways to the study of perception. Johns Hopkins Med J 136:109–131.

Mountcastle VB, Andersen RA, Motter BC (1981) The influence of attentive fixation upon the excitability of the light-sensitive neurons of the posterior parietal cortex. J Neurosci 1:1218–1225.

Murata A, Gallese V, Kaseda M, Sakata H (1996) Parietal neurons related to memory-guided hand manipulation. J Neurophysiol 75:2180–2186.

Murata A, Gallese V, Luppino G, Kaseda M, Sakata H (2000) Selectivity for the shape, size, and orientation of objects for grasping in neurons of monkey parietal area AIP. J Neurophysiol 83:2580–2601.

Nakamura H, Kuroda T, Wakita M, Kusunoki M, Kato A, Mikami A, Sakata H, Itoh K (2001) From three-dimensional space vision to prehensile hand movements: the lateral intraparietal area links the area V3A and the anterior intraparietal area in macaques. J Neurosci 21:8174–8187.

Perenin MT, Vighetto A (1988) Optic ataxia: a specific disruption in visuomotor mechanisms. I. Different aspects of the deficit in reaching for objects. Brain 111 (Pt 3):643–674.

Platt ML, Glimcher PW (1999) Neural correlates of decision variables in parietal cortex. Nature 400:233–238.

Preuss TM, Goldman-Rakic PS (1991) Architectonics of the parietal and temporal association cortex in the strepsirhine primate Galago compared to the anthropoid primate Macaca. J Comp Neurol 310:475–506.

Read HL, Siegel RM (1997) Modulation of responses to optic flow in area 7a by retinotopic and oculomotor cues in monkey. Cereb Cortex 7:647–661.

Reid RC (1999) Vision: The Retinogeniculate Pathways. In: Fundamental Neuroscience (Zigmond MJ, Bloom FE, Landis SC, Roberts JL, Squire LR, eds), pp 835–851. San Diego: Academic Press.

Rizzolatti G, Matelli M (2003) Two different streams form the dorsal visual system: anatomy and functions. Exp Brain Res 153:146–157.

Robinson DL, Goldberg ME (1978) Sensory and behavioral properties of neurons in posterior parietal cortex of the awake, trained monkey. Fed Proc 37:2258–2261.

Robinson DL, Bowman EM, Kertzman C (1995) Covert orienting of attention in macaques. II. Contributions of parietal cortex. J Neurophysiol 74:698–712.

Romo R, Merchant H, Zainos A, Hernandez A (1997) Categorical perception of somesthetic stimuli: psychophysical measurements correlated with neuronal events in primate medial premotor cortex. Cereb Cortex 7:317–326.

Romo R, Hernandez A, Zainos A, Salinas E (1998) Somatosensory discrimination based on cortical microstimulation. Nature 392:387–390.

Rondot P, de Recondo J, Dumas JL (1977) Visuomotor ataxia. Brain 100:355–376.

Rosselli M, Rosselli A, Vergara I, Ardila A (1985) Topography of the hemi-inattention syndrome. Int J Neurosci 27:165–172.

Sakata H, Shibutani H, Kawano K, Harrington TL (1985) Neural mechanisms of space vision in the parietal association cortex of the monkey. Vision Res 25:453–463.

Sakata H, Shibutani H, Ito Y, Tsurugai K (1986) Parietal cortical neurons responding to rotary movement of visual stimulus in space. Exp Brain Res 61:658–663.

Sakata H, Taira M, Murata A, Mine S (1995) Neural mechanisms of visual guidance of hand action in the parietal cortex of the monkey. Cereb Cortex 5:429–438.

Sakata H, Shibutani H, Ito Y, Tsurugai K, Mine S, Kusunoki M (1994) Functional properties of rotation-sensitive neurons in the posterior parietal association cortex of the monkey. Exp Brain Res 101:183–202.

Sakata H, Taira M, Kusunoki M, Murata A, Tanaka Y, Tsutsui K (1998) Neural coding of 3D features of objects for hand action in the parietal cortex of the monkey. Philos Trans R Soc Lond B Biol Sci 353:1363–1373.

Shadlen MN, Britten KH, Newsome WT, Movshon JA (1996) A computational analysis of the relationship between neuronal and behavioral responses to visual motion. J Neurosci 16:1486–1510.

Siegel RM, Read HL (1997) Analysis of optic flow in the monkey parietal area 7a. Cereb Cortex 7:327–346.

Snyder LH, Batista AP, Andersen RA (1997) Coding of intention in the posterior parietal cortex. Nature 386:167–170.

Steinmetz MA, Constantinidis C (1995) Neurophysiological evidence for a role of posterior parietal cortex in redirecting visual attention. Cereb Cortex 5:448–456.

Steinmetz MA, Connor CE, Constantinidis C, McLaughlin JR (1994) Covert attention suppresses neuronal responses in area 7a of the posterior parietal cortex. J Neurophysiol 72:1020–1023.

Sterzer P, Russ MO, Preibisch C, Kleinschmidt A (2002) Neural correlates of spontaneous direction reversals in ambiguous apparent visual motion. Neuroimage 15:908–916.

Stricanne B, Andersen RA, Mazzoni P (1996) Eye-centered, head-centered, and intermediate coding of remembered sound locations in area LIP. J Neurophysiol 76: 2071–2076.

Strub RL, Black WF (1977) The Mental Status Examination in Neurology. Philadelphia: F.A.Davis Co.

Taira M, Mine S, Georgopoulos AP, Murata A, Sakata H (1990) Parietal cortex neurons of the monkey related to the visual guidance of hand movement. Exp Brain Res 83:29–36.

Ungerleider LG and Mishkin M (1982) Two cortical visual systems. In: DJ Ingle, MA Goodate and RJW Mansfield (Eds.), Analysis of Visual Behavior, pp. 549–585.

Vallar G, Lobel E, Galati G, Berthoz A, Pizzamiglio L, Le Bihan D (1999) A fronto-parietal system for computing the egocentric spatial frame of reference in humans. Exp Brain Res 124:281–286.
Williams ZM, Elfar JC, Eskandar EN, Toth LJ, Assad JA (2003) Parietal activity and the perceived direction of ambiguous apparent motion. Nat Neurosci 6:616–623.
Xing J, Andersen RA (2000) Models of the posterior parietal cortex which perform multimodal integration and represent space in several coordinate frames. J Cogn Neurosci 12:601–614.
Zeki S (1993) The visual association cortex. Curr Opin Neurobiol 3:155–159.
Zipser D, Andersen RA (1988) A back-propagation programmed network that simulates response properties of a subset of posterior parietal neurons. Nature 331:679–684.

14

Dopamine-Dependent Associative Learning of Workload-Predicting Cues in the Temporal Lobe of the Monkey

BARRY J. RICHMOND

Laboratory of Neuropsychology, National Institute of Mental Health, NIH,
Bethesda, MD, USA

Introduction

Learning to react appropriately to external stimuli within one or a few exposures to them is an amazing feature of normal behavior. We would like to identify both the sites and underlying plastic neural mechanisms that are critical to this ability. To understand how visual stimuli in particular are interpreted, it is important to learn where and how the transformation from information about stimulus identity to information about what an object or stimulus means takes place.

To interpret visual stimuli, the stimuli must become associated with the predicted outcome for subsequent events, be they actions the animals control or not. Studies of vision have largely focused on the first part of this two-step process: how visual scenes are segmented into objects and how those objects are identified. In that domain, questions related to plasticity or learning have been formulated so as to learn how the segmentation and identification of objects might be enhanced. For example, neurons in the rostral parts of the ventral processing stream show differences in response when a stimulus is unattended vs. when it is attended (Gross et al., 1979; Richmond et al., 1983; Moran and Desimone, 1985). Such responses are related to the identity of the stimulus in that they show strong stimulus-selectivity, but the tuning curves shift according to the cognitive state in which the stimulus is presented. This is valuable information in that it must be understood what the role of such modulations might be, for example, increasing the salience of the neural signal in the population, so that the object can be speedily and correctly identified.

Once the object is identified, however, its significance must be interpreted so that appropriate behavior can be generated. Because especially in primates

R. Pinaud, L. A. Tremere, P. De Weerd(Eds.), Plasticity in the Visual System:
From Genes to Circuits, 307-320, ©2006 Springer Science + Business Media, Inc.

and humans behavior is often organized around *interpretations* of visual stimuli (other cognitive variables, though important also, are not considered here), the work from this group has studied the role of stimulus-elicited signals that code for predictions about when a forthcoming reward will be delivered. Our goal is to learn how and where these predictions are learned, and what role these signals play in controlling behavior. Because of the importance of the role of dopamine in reward-contingent behavior, we have focused on dopamine-rich brain regions. In visual processing this led us to the perirhinal cortex.

Background: Single Neuron Responses in the Ventral Striatum Associated with Anticipated Reward

Several single neuron recording experiments from monkey ventral striatum (Bowman et al., 1996; Shidara et al., 1998) have set the stage for our investigations of visual processing in the anterior temporal lobe. In these experiments, the monkeys carry out a behavioral task in which all of the motor acts are identical, and visual stimuli, used as cues, are given different meanings. The visual stimuli are associated with the number of trials remaining to be completed before a reward will be delivered (Fig. 14.1a). The monkeys learn to release a manipulandum within about one second (a short time, but a period not requiring the fastest response possible) after a red target light turns green. After the monkeys learn to perform at greater than 80% correct (generally 2-4 weeks of practice are needed), the task is changed so that the monkeys must perform schedules of 1, 2, or 3 correct trials to receive a reward.

During these visually-cued reward schedules a stimulus used as a visual cue (a patch of light with different intensities in many experiments) appears on the screen being viewed by the monkeys. This cue remains on throughout the trial. The same cue appears in every rewarded trial. In other trials (non-last trials) the cues have unique brightnesses. The brightness is set by the ratio of the current trial number in the current schedule to the current schedule length, thus giving brightnesses of 1/3, 2/3, 1/2, and 1. This cue offers the opportunity for the monkeys to predict exactly how much work (where work is measured in number of trials) remains in the current schedule before a reward is delivered. There is no explicit punishment for releasing the manipulandum too early (before the green go signal) or too late (after the green go signal disappears). When a trial is not completed correctly, the same cue appears in the next trial and the schedule is not advanced, nor is the schedule restarted, i.e., the schedule state remains the same.

In this visually cued reward schedule task, monkeys quickly (within a session) show that they are sensitive to the cues. After 1-4 sessions, the monkeys make almost no 'errors' in trials that, if completed, will elicit an immediate reward. However, as the number of trials remaining before the rewarded trial increases the monkey's reaction times and error rates increase (Fig. 14.1b) (Bowman et al., 1996). Since all the red-green discrimination

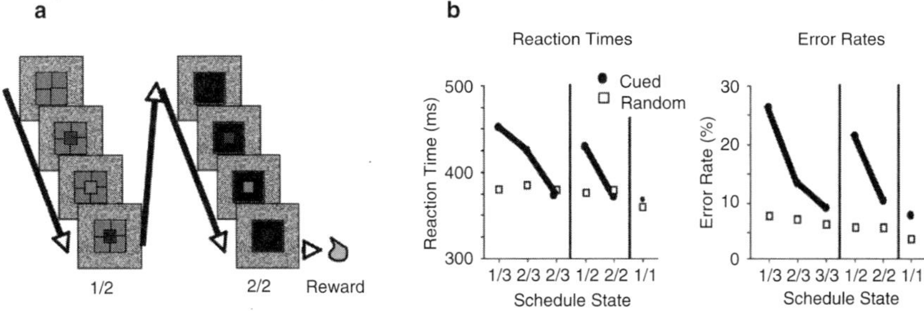

FIGURE 14.1. Panel A. Outline of a two trial reward schedule. The monkey must complete two sequential red-green discrimination trials, here illustrated as two different gray spots in the center of the cues, to obtain a reward. If an error is made the monkey is presented with the same trial, with the same cue. Thus, if the monkey responds correctly to the first red-green transition, and the second trial is missed, the second trial is presented again. The cue underneath the discrimination target is related to the trial in the schedule (see text), allowing the monkeys to know how many trials in the schedule, or whether the current trial will be rewarded if completed correctly. Panel B. Behavioral performance in the reward schedule task. The monkeys perform with short reaction times and few errors in rewarded trials. In other trials, the reaction times and the number of errors increase as the number of trials remaining before the rewarded trial increase (closed squares). Thus, the monkeys are sensitive to the cue. In the case when the cues are chosen randomly, the reaction times are short and the error rates are low in all trials (open squares). In the random cue condition the monkeys presumably have some expectation of receiving a reward in each trial, whereas in the valid cue condition the monkeys know in every trial whether or not they will receive a reward, and if no reward is forthcoming in the current trial, they know how many more trials must yet be completed before receiving a reward. Thus, at a superficial level at least they use the cognitively acquired information to discount the reward's value as a function of the number of trials remaining.

trials are identical, we know that the monkeys can perform the trials without errors. Nonetheless, they make 'errors' in the non-rewarded trials. When the cues are chosen randomly, the monkeys make very few errors. Thus, in a seeming paradox, the differential error rates across the validly cued trials show that the monkeys are attentive to the cues. The monkeys act as if they were discounting their motivation as a function of the number of trials remaining, an effect similar in some ways to temporal difference models for motivation (Sutton and Barto, 1998). Furthermore, because the animals perform equally well, and, often just about perfectly, in every trial of random cue condition we infer that the monkeys effectively ignore the cues in the random condition.

Because of ventral striatum's importance in reward-related behavior, our early recordings were made there (this includes the region of the nucleus accumbens) while monkeys performed this task. Neurons in the ventral striatum are sensitive to the cue, to the green go-signal, and to the reward. Go-signal responses occur in rewarded trials. Cue-related responses occur in some schedule states, and not others. Some neurons respond in first trials (Fig. 14.2), others in non-first trials (including last), and others still in last trials (Shidara et al., 1998). When the cues are picked at random these responses disappear, showing that the responses, although stimulus-elicited, are plastic, and that their presence depends on the monkey interpreting the cue.

Processing of reward-predicting cues in rhinal cortex: evidence from ablation studies, neurophysiology, and molecular biology

The cue-related responses in the ventral striatum led us to look for the source of the visual signal triggering them. Several line of evidence attracted our attention to several cortical areas in the anterior temporal lobe. First, the perirhinal cortex attracted our attention because of its anatomical connectivity. The perirhinal cortex is strongly connected to the ventral tegmental

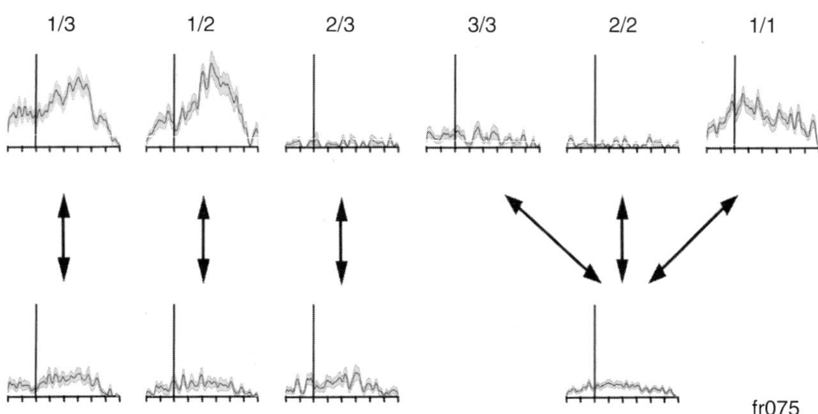

FIGURE 14.2. Cue-elicited responses of a ventral striatal neuron recorded in the valid cue condition (upper panels), and in the random cue condition (lower panels). The fractions above the upper panels show the place in schedule, i.e., the schedule state, with the left number indicating the trial number in the current schedule, and the right number indicating the schedule length. This neuron showed a set of responses to the cues in the first trials in the valid cue condition (top row). In the random cue condition the responses disappeared (bottom row). The neuron shows an immediate and profound change in response when the cue is made invalid, and the monkey's behavior reflected that change. Thus, these neurons show a change in selectivity related to the monkey's interpretation of the cue. These neurons, if tested in the valid cue condition, might have been regarded as 'visual', whereas it is clear from this combination of conditions that the neurons have visually triggered responses, but these responses cannot be regarded as classical stimulus-selective visual responses. Rather, their selectivity is related to the valid association with the outcome. Whether there is some aspect of the responses that is related to the specific stimuli chosen is a question that has yet to be answered.

area (Insausti et al., 1987; Akil and Lewis, 1993). Perirhinal cortex in turn receives a large, direct projection from area TE, an area regarded as the end of the ventral visual processing stream (Suzuki and Amaral, 1994; Saleem and Tanaka, 1996).

Second, rhinal cortex as a whole (entorhinal plus perirhinal cortex) attracted our attention because it contains a relatively large concentration of dopamine containing fibers and axonal terminations, suggesting that it might be sensitive to reward-contingent signals, such as those seen in the ventral striatum (Akil and Lewis, 1993, 1994) (Berger et al., 1988; Richfield et al., 1989).

Finally, rhinal cortex, although not a visual areas in a strict sense, contributes importantly to normal visual perception. It is well-known that the anterior temporal lobe as a whole plays an important role in object identification and memory. An early ablation study showed that monkeys with extensive damage to the anterior temporal cortex were impaired in object identification and/or memory for objects (Iwai and Mishkin, 1968). Subsequent behavioral and single neuronal recording studies suggested not only that the anterior temporal cortex is important for object identification and/or memory (Mishkin, 1982), but also that different regions exist in the anterior temporal lobe that play different roles in the analysis of visual objects (Murray and Bussey, 1999; Bussey et al., 2002, 2003). In particular, after ablations of perirhinal cortex monkeys are impaired in performing a sequential delayed nonmatch sample task, with the severity of the deficit accentuated as the delay between the sample and test stimuli is increased, or especially when the list length between samples is increased from one to ten (Hadfield et al., 2003). It seems as if the more lateral anterior temporal area TE is important for object identification, and the more medial perirhinal cortex plays a role in episodic or semantic-like memory.

As perirhinal cortex gets strong input from area TE, it is not surprising that there are many similarities in the properties of single neurons in both areas. For example, single unit recording studies neurons in area TE on the lateral inferior aspect of the anterior temporal lobe are selective for complex objects, such as hands, faces, and brushes, a result that strongly supports a hierarchical view of visual processing, in which neurons at each successive feed-forward stage of visual processing become selective for increasingly complex stimulus features (Gross et al., 1972). Single neurons in perirhinal cortex show similar stimulus selectivity (Desimone et al., 1984; Baylis and Rolls, 1987; Richmond and Sato, 1987; Nakamura et al., 1994). Because this tissue is thought important for object recognition, awake monkey recordings are often carried out using a delayed matching-to-sample task, where monkeys are shown a visual stimulus and then must indicate when the same stimulus reappears during sequential presentation of one or more test stimuli (Eskandar et al., 1992; Tanaka, 1992; Li et al., 1993; Tanaka, 1993; Nakamura and Kubota, 1995). In the context of this match-to-sample task, neurons in both areas again show striking stimulus selectivity, similar to what had been reported from the very first single neuronal recordings in inferior temporal cortex. Early studies also showed that the responses of neurons in area TE are modulated by attentional factors, even to the degree that the sizes of the receptive fields themselves could apparently change (Richmond and Sato, 1987). In our studies, we saw firing changes depending on whether the stimulus is presented in the sample, nonmatch, or match condition, although the relative tuning of the neurons to different stimuli did not change (Eskandar et al., 1992; Li et al., 1993).

However, as the differences in the connections between area TE and perirhinal cortex have become better appreciated, it seemed possible that perirhinal cortex might play a specific role in the establishment of visual associative memory, which would set it apart from area TE Indeed, both behavioral studies using selective ablations and single neuronal studies began to suggest that the perirhinal cortex plays a role in associative memory (Murray et al., 1993; Higuchi and Miyashita, 1996). Several single neuron recording studies indicate that perirhinal neurons change their firing such that their responses become more similar across associated stimuli, whether or not the animals are required to learn the pairings for any task requirement (Miyashita et al., 1996; Erickson and Desimone, 1999; Holscher et al., 2003).

The presence of associative mechanisms in perirhinal cortex suggested that this tissue might play a general role in the association of stimuli with their meaning. Specifically, we hypothesized that the transformation from stimulus identity to stimulus meaning might take place as information passes from area TE on the lateral, anterior temporal cortex of the monkey, to the perirhinal cortex, located more medially on the anterior temporal cortex. Cue-related responses in the ventral striatum discussed in the previous section might reflect at least in part associations already carried out in perirhinal cortex, or might depend on interactions with perirhinal cortex.

In one set of experiments, we used the same task that we had used when recording from the ventral striatum, except that the red-green sequential discrimination trials were replaced with sequential matching-to-sample trials so that we could assess the stimulus selectivity to the stimuli that had to be actively remembered in the trials (Liu et al., 2000). When we recorded single neurons in area TE and in perirhinal cortex, the similarity of selectivity to the stimuli in the match-to-sample trials (the black and white Walsh patterns) reported previously was reproduced in this study. However, we also found striking differences in the neuronal responses of the two areas. First, for both the stimuli used in both the match-to-sample part of the task and for the stimuli used as cues, the median latencies were about 65 ms longer in perirhinal cortex than those seen in area TE. This seems a surprisingly large time difference given that there are strong projections from about 2/3 of area TE to perirhinal cortex (Suzuki and Amaral, 1994; Saleem and Tanaka, 1996). It is usually expected that the latencies of directly connected areas differ by 5-10 milliseconds (Rousselet et al., 2004). The long delay between the activation of TE and the activation of perirhinal cortex strongly suggests that the direct feed-forward connections are not sufficient by themselves to activate perirhinal cells. Exactly what sources of input are required to activate perirhinal neurons is not yet known.

The responses to the cues were strikingly different in the two areas as well. The cues in the experiment were patches of light, so all of the stimuli were the same except for their brightness. In area TE, a substantial number of neurons responded to the cues. Given that the cues were simple patches of light, this seems somewhat surprising because we have come to expect area TE neurons to be activated by complex stimuli. When a TE neuron responded to the cues, it responded to all of them. However, perirhinal neurons that responded to the cues, responded to a subset of the cues only; never to all of them. It can be shown by example that these neurons are responding to something other than the cue's configuration or identity. In one such example (Fig. 14.3), the neuron responded to the cue in all trials except the ones ending the two and three trial schedules. Because the cue for the rewarded trial in the one-trial schedule was the same as for the other two rewarded trials, it is unlikely that the responses were related to the cue's configuration. The monkeys performed equally well in the rewarded trials, making it unlikely that there was an attentional difference across these conditions. Because there was a response in one rewarded trial type, but not in the other two, it seems unlikely that the neural response occurred just because the monkey was anticipating the reward. Finally, if the cues were picked at random so that they did not

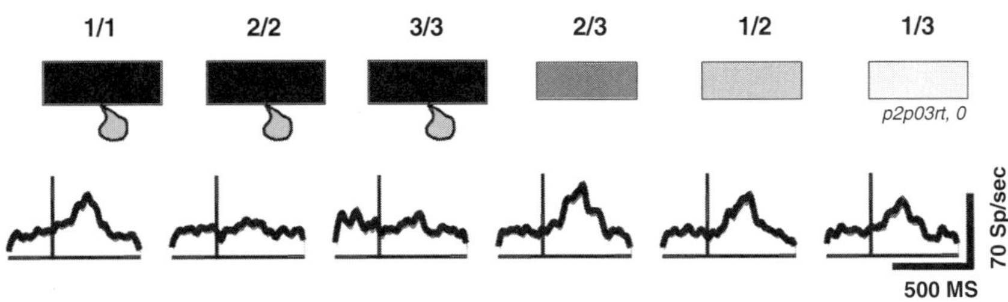

FIGURE 14.3. Responses of a perirhinal neuron to the cues in the reward schedule task. The cue triggered a responses in 4 of the six schedule states; there was no response in the rewarded trials of the two and three trial schedules. Because the neuron responded in the one trial schedule, but not in the 2/2 and 3/3 conditions, we can conclude that the response of the neuron did not depend on the identity of the cue since the same cue appeared in each of these conditions, that the response did not depend on the attentional or arousal state of the animal because the behavioral performances were indistinguishable across rewarded trials (almost no errors made), and that it did not depend on the expectation of a reward in the current trial, since the 1/1, 2/2 and 3/3 states all yield a reward when correctly completed. Thus, the neuron's responses are related to the dynamics of the cognitive state of the monkey.

contain any information about the schedule, the responses either disappeared (not shown), or they lost their selectivity. Thus, the transformation of signals about stimulus identity to signals about the motivational significance of a stimulus appears to occur across the TE-perirhinal interface.

Ablations have suggested that perirhinal cortex is important for various aspects of visual memory, such as recognition memory and stimulus-stimulus associations (Murray et al., 1993; Higuchi and Miyashita, 1996). When normal monkeys that have already been exposed to one set of cues (the darkening patches) in the visually cued reward schedule task are exposed to a second set of cues in the same task, their behavior quickly adapts to the cues in the same manner as had been shown for the first set. If, however, monkeys are first given a bilateral rhinal cortex ablation, they do not learn the reward-related predictions made by the new cues (Liu et al., 2000). They perform as if the cues were not meaningful. They seem unable to learn either the association of the cues to the predicted workload, or perhaps the emotional meaning of the cue, since the cue can be carry an aversive message (more work will be required to obtain a reward). Thus, the selective signal about the schedule conveyed by the cue and carried by perirhinal neurons, and the information predicting the remaining workload seem related. The perirhinal cortex is either the site of the plasticity needed for this learning, or its signal is required for the plasticity to occur at one of its targets.

Since one of the major reasons we chose the perirhinal cortex for study was that it has a relatively high concentration of dopamine containing fibers, we hypothesized that dopamine might play an important role in learning the meanings of the cues. To test this hypothesis, we adapted a technique developed in rodent research wherein a DNA expression vector is constructed that would target the messenger RNA coding for the D2 receptor (Weiss et al., 1997; Davidkova et al., 1998; Weiss et al., 1999), leading to decreased presence of that receptor. In this experiment, the monkeys were first given a unilateral rhinal cortex lesion, after which they were trained and tested in the reward schedule task (using the red-green sequential discrimination trials). All of the monkeys learned the red-green trials at the normal rate, and learned the significance of the cues at the same rate as unoperated monkeys. Thus, one intact rhinal cortex is sufficient to learn to interpret the cues, such that the workload remaining can be predicted by the cues in the visually cued reward schedules (Liu et al., 2004).

After learning the initial cue set, the intact perirhinal cortex was treated with the DNA expression vector through a series of injections (two with D2 material alone, and two with a mixture of D2 and NMDA targeting material). After two weeks of recovery from surgery, the monkeys were tested with a new cue set. Monkeys injected with control material or a DNA expression vector targeting NMDA receptor protein alone learned to react to the new cue set at the same rate as normal monkeys. However, after 10 or more weeks of exposure, the monkeys injected with the D2 targeting DNA expression vector learned to interpret the new cues as normal monkeys would

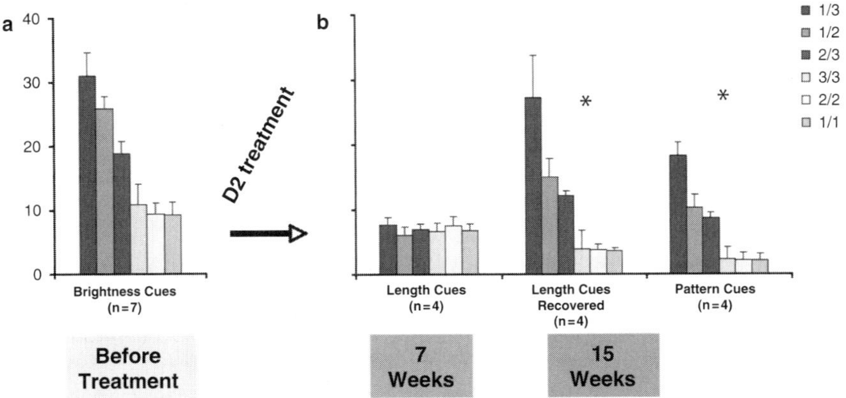

FIGURE 14.4. Behavioral performance before and after treatment of perirhinal cortex with D2 antisense expression vector. The monkeys were first given a unilateral ablation of one rhinal cortex. They were then trained in the sequential color discrimination task, which they learned at a normal rate. They were then exposed to the visually cued reward schedules, in which we observed the pattern of behavior we have come to expect with this task, i.e., they showed more errors as the number of trials remaining before reward increased. Four monkeys were then treated with the expression vector targeting the D2 receptor (two of the monkeys received treatment of the D2 material plus material targeting NMDA receptors, also). These four animals failed to react to the second set of cues until at least 10 weeks had passed. Two control monkeys (with unilateral ablations) reacted to the second set of cues as expected for normal monkeys in that they learned them at the normal rate, i.e., within four testing sessions. This failure to learn the cues is indistinguishable from the effect of a bilateral ablation of rhinal cortex, except that monkeys with bilateral ablation never learn to react to new cue sets (up to 30 weeks).

(Fig. 14.4). The question then was whether the monkeys had learned a new strategy, or whether they had recovered.

Two pieces of evidence suggest that they had recovered. First, all of the animals learned a third cue set at the same rate as normal monkeys learn the first, second or third cue sets. Second, four monkeys, one of which had been treated with the D2 material were treated with D2 targeting material at a second injection operation. All four of these retreated monkeys showed the same learning deficit, i.e., they failed to learn a newly introduced cue set, until another 10 weeks had passed. Again after learning this cue set, all of these animals were able to learn a subsequent cue set at the same rate that normal animals learn cue sets.

Discussion

Our results show that the perirhinal cortex is essential for learning the association of visual cues that predict remaining workload. This information has a profound effect on the behavior of the monkeys. At a time when the operatrial is only mildly demanding and, from the behavioral results in the random cue condition, we know that the animals are capable of performing every trial correctly, the monkeys, in what must be regarded as a 'voluntary' choice of strategy, make many 'errors'. These 'errors' do not indicate an inability to carry out the trials, but more likely signify the aversiveness that accompanies the situation of having to work now, but wait until later to obtain the reward (an effect that is easy to relate to through introspection about our own emotional reactions under similar circumstances).

The single neuron recording results show that the neural responses carrying a reward predicting signal arise through associative mechanisms. This leads us to believe that these neural signals are directly responsible for the behavior. These neuronal signals code for the place in reward schedules. However, none of these individual perirhinal neurons carry the information needed to unequivocally decode the schedule state. Because the encoding effectively exists in a table, the responses of these neurons cannot be simply averaged; each neuron's outputs need to pass through the inverse associative memory, and then combine in a (explicit or implicit) decoding operation. Thus, although the general decoding strategy can be preprogrammed, the specific meanings of the responses must be learned via underlying adaptive or plastic mechanisms.

If we consider that feed-forward visual processing proceeds through the ventral stream to area TE (Ungerleider and Mishkin, 1982), the transformation that we sought appears to occur as the signals pass from area TE medially into the perirhinal cortex. Furthermore, the delay in the onset of responses in perirhinal cortex indicates that the signal arising in area TE seems unlikely by itself to be strong enough to activate perirhinal cortex neurons. The source of the additional drive needed in perirhinal cortex could come from any number of sources. Our behavioral results using the D2 antisense expression vector in perirhinal cortex show that dopamine is needed for the plastic changes in behavior, raising the possibility that the responses of dopamine neurons might be an important source of the driving signal for the perirhinal neurons. Other studies have reported that dopamine neurons seem to respond to surprising, i.e., unexpected, stimuli. The signals carried by dopamine neurons do not seem to have the specificity that is reflected in the schedule-state dependent tuning of the perirhinal neurons. However, through the conjunction of visual inputs from area TE and signals from other sources, perirhinal neurons might construct the needed signal. Whether the selectivity of the perirhinal neuronal responses reflects input from prefrontal cortex (Suzuki and Amaral, 1994), orbitofrontal cortex, entorhinal cortex and thus indirectly from the hippocampus, or other areas requires further study.

The neural correlates of the visually cued reward schedules are seen in the ventral striatum and the perirhinal cortex. Both brain regions have responses that are tightly coupled to the appearance of the cue. However, in both areas the responses depend on the cue's meaning in the task, not the identity of the cue. We have suggested earlier that reward-related responses in the striatum might reflect the outcome of associative processes in perirhinal cortex, or might require an interaction between those two brain regions. Thus, the perirhinal responses are not, strictly speaking, visual, but are visually gated.

There is strong evidence suggesting that dopamine plays an important role in reward-related behaviors (Schultz, 2002, 2004). Learning the meaning of these cues appears to require that the D2 receptor be functioning normally (Liu et al., 2004). At a minimum, our results support a strong role for the D2 receptor in perirhinal cortex for the development of this type of associative memory. There has been a great deal of work focused on the role on NMDA receptors in learning via Long Term Potentiation (LTP). However, several studies suggest that also dopamine through the D2 receptor can play a role in LTP-like processes (Chen et al., 1996; Setlow and McGaugh, 2000; Wang et al., 2004). As yet, there has not been enough focus on the exact role of dopamine in these processes to provide a good account of how the learning we have observed might be mediated by dopamine, especially by the D2 receptor.

Whatever the mechanisms might be, there is a striking transformation in the character of neural signals as information passes from area TE, a brain area that seems to be part of a system organizing visual information so that objects in the environment can be identified, to perirhinal cortex where it is conceivable that the apparent stimulus selectivity could be driven by assignment of predicted reward contingency. Our results have implications for the interpretation of studies of stimulus selectivity in perirhinal cortex. Indeed, whether the apparent stimulus selectivity that we and others have seen in perirhinal cortex would arise from associative experience with previously seen visual stimuli, or even from tentative classification on first viewing will require further experiments in which stimuli are presented in an associative situation and then tested in an episodic memory condition such as in the delayed matching-to-sample task. Until such an experiment is carried out, any firm conclusion about the origin of stimulus selectivity should be avoided, since it is conceivable that the responses of perirhinal neurons and the behavioral effects of perirhinal ablations are related to this property of trying to associate visual stimuli with their predicted reward contingency, rather than being due to some role in the identification of stimulus identity without regard to the reward contingency.

Finally, the use of the DNA expression vector allowed us to study the role of the D2 receptor and compare it to the role of the NMDA receptor. From our experiment we know that the learning of the reward predictive value of cueing stimuli requires the D2 receptor in perirhinal cortex. We also know that knocking down the NMDA receptor to the extent that we did was not adequate to interfere with the learning or performance of the visually cued reward schedules. Whether a more extensive knockdown of NMDA receptors would alter the behavior remains an open question. At this point, the expression vector approach has worked in altering behavior in mice when the striatum was treated (Weiss et al., 1997; Davidkova et al., 1998; Weiss et al., 1999), and in monkeys when the perirhinal cortex was treated (Liu et al., 2004). Work is ongoing to determine the generality of this approach in knocking down other receptors to a sufficient degree to cause a phenotypic change in behavior. Although this type of work is still in its infancy, the present results indicate the large potential of the combination of systems neuroscience techniques with molecular approaches to study higher cognitive function.

References

Akil M, Lewis DA (1993) The dopaminergic innervation of monkey entorhinal cortex. Cereb Cortex 3:533–550.

Akil M, Lewis DA (1994) The distribution of tyrosine hydroxylase-immunoreactive fibers in the human entorhinal cortex. Neuroscience 60:857–874.

Baylis GC, Rolls ET (1987) Responses of neurons in the inferior temporal cortex in short term and serial recognition memory tasks. Exp Brain Res 65:614–622.

Berger B, Trottier S, Verney C, Gaspar P, Alvarez C (1988) Regional and laminar distribution of the dopamine and serotonin innervation in the macaque cerebral cortex: a radioautographic study. J Comp Neurol 273:99–119.

Bowman EM, Aigner TG, Richmond BJ (1996) Neural signals in the monkey ventral striatum related to motivation for juice and cocaine rewards. J Neurophysiol 75:1061–1073.

Bussey TJ, Saksida LM, Murray EA (2002) Perirhinal cortex resolves feature ambiguity in complex visual discriminations. Eur J Neurosci 15:365–374.

Bussey TJ, Saksida LM, Murray EA (2003) Impairments in visual discrimination after perirhinal cortex lesions: testing 'declarative' vs. 'perceptual-mnemonic' views of perirhinal cortex function. Eur J Neurosci 17:649–660.

Chen Z, Ito K, Fujii S, Miura M, Furuse H, Sasaki H, Kaneko K, Kato H, Miyakawa H (1996) Roles of dopamine receptors in long-term depression: enhancement via D1 receptors and inhibition via D2 receptors. Receptors Channels 4:1–8.

Davidkova G, Zhou LW, Morabito M, Zhang SP, Weiss B (1998) D2 dopamine antisense RNA expression vector, unlike haloperidol, produces long-term inhibition of D2 dopamine-mediated behaviors without causing Up-regulation of D2 dopamine receptors. J Pharmacol Exp Ther 285:1187–1196.

Desimone R, Albright TD, Gross CG, Bruce C (1984) Stimulus-selective properties of inferior temporal neurons in the macaque. J Neurosci 4:2051–2062.

Erickson CA, Desimone R (1999) Responses of macaque perirhinal neurons during and after visual stimulus association learning. J Neurosci 19:10404–10416.

Eskandar EN, Richmond BJ, Optican LM (1992) Role of inferior temporal neurons in visual memory. I. Temporal encoding of information about visual images, recalled images, and behavioral context. J Neurophysiol 68:1277–1295.

Gross CG, Rocha-Miranda CE, Bender DB (1972) Visual properties of neurons in inferotemporal cortex of the Macaque. J Neurophysiol 35:96–111.

Gross CG, Bender DB, Gerstein GL (1979) Activity of inferior temporal neurons in behaving monkeys. Neuropsychologia 17:215–229.

Hadfield WS, Baxter MG, Murray EA (2003) Effects of combined and separate removals of rostral dorsal superior temporal sulcus cortex and perirhinal cortex on visual recognition memory in rhesus monkeys. J Neurophysiol 90:2419–2427.

Higuchi S, Miyashita Y (1996) Formation of mnemonic neuronal responses to visual paired associates in inferotemporal cortex is impaired by perirhinal and entorhinal lesions. Proc Natl Acad Sci U S A 93:739–743.

Holscher C, Rolls ET, Xiang J (2003) Perirhinal cortex neuronal activity related to long-term familiarity memory in the macaque. Eur J Neurosci 18:2037–2046.

Insausti R, Amaral DG, Cowan WM (1987) The entorhinal cortex of the monkey: II. Cortical afferents. J Comp Neurol 264:356–395.

Iwai E, Mishkin M (1968) Two visual loci in the temporal lobe of monkeys. In: Neurophysiological Basis of Learning and Behavior. (Yoshi N, Buchwald NA, eds), pp 1–11. Osaka, Japan: Osaka University Press.

Li L, Miller EK, Desimone R (1993) The representation of stimulus familiarity in anterior inferior temporal cortex. J Neurophysiol 69:1918–1929.

Liu Z, Murray EA, Richmond BJ (2000) Learning motivational significance of visual cues for reward schedules requires rhinal cortex. Nat Neurosci 3:1307–1315.

Liu Z, Richmond BJ, Murray EA, Saunders RC, Steenrod S, Stubblefield BK, Montague DM, Ginns EI (2004) DNA targeting of rhinal cortex D2 receptor protein reversibly blocks learning of cues that predict reward. Proc Natl Acad Sci U S A 101:12336–12341.

Mishkin M (1982) A memory system in the monkey. Philos Trans R Soc Lond B Biol Sci 298:83–95.

Miyashita Y, Okuno H, Tokuyama W, Ihara T, Nakajima K (1996) Feedback signal from medial temporal lobe mediates visual associative mnemonic codes of inferotemporal neurons. Brain Res Cogn Brain Res 5:81–86.

Moran J, Desimone R (1985) Selective attention gates visual processing in the extrastriate cortex. Science 229:782–784.

Murray EA, Bussey TJ (1999) Perceptual-mnemonic functions of the perirhinal cortex. Trends Cogn Sci 3:142–151.

Murray EA, Gaffan D, Mishkin M (1993) Neural substrates of visual stimulus-stimulus association in rhesus monkeys. J Neurosci 13:4549–4561.

Nakamura K, Kubota K (1995) Mnemonic firing of neurons in the monkey temporal pole during a visual recognition memory task. J Neurophysiol 74:162–178.

Nakamura K, Matsumoto K, Mikami A, Kubota K (1994) Visual response properties of single neurons in the temporal pole of behaving monkeys. J Neurophysiol 71:1206–1221.

Richfield EK, Young AB, Penney JB (1989) Comparative distributions of dopamine D-1 and D-2 receptors in the cerebral cortex of rats, cats, and monkeys. J Comp Neurol 286:409–426.

Richmond BJ, Sato T (1987) Enhancement of inferior temporal neurons during visual discrimination. J Neurophysiol 58:1292–1306.

Richmond BJ, Wurtz RH, Sato T (1983) Visual responses of inferior temporal neurons in awake rhesus monkey. J Neurophysiol 50:1415–1432.

Rousselet GA, Thorpe SJ, Fabre-Thorpe M (2004) How parallel is visual processing in the ventral pathway? Trends Cogn Sci 8:363–370.

Saleem KS, Tanaka K (1996) Divergent projections from the anterior inferotemporal area TE to the perirhinal and entorhinal cortices in the macaque monkey. J Neurosci 16:4757–4775.

Schultz W (2002) Getting formal with dopamine and reward. Neuron 36:241–263.

Schultz W (2004) Neural coding of basic reward terms of animal learning theory, game theory, microeconomics and behavioural ecology. Curr Opin Neurobiol 14:139–147.

Setlow B, McGaugh JL (2000) D2 dopamine receptor blockade immediately post-training enhances retention in hidden and visible platform versions of the water maze. Learn Mem 7:187–191.

Shidara M, Aigner TG, Richmond BJ (1998) Neuronal signals in the monkey ventral striatum related to progress through a predictable series of trials. J Neurosci 18: 2613–2625.

Sutton RS, Barto AG (1998) Reinforcement Learning: An Introduction. Cambridge, MA: MIT Press.

Suzuki WA, Amaral DG (1994) Perirhinal and parahippocampal cortices of the macaque monkey: cortical afferents. J Comp Neurol 350:497–533.

Tanaka K (1992) Inferotemporal cortex and higher visual functions. Curr Opin Neurobiol 2:502–505.

Tanaka K (1993) Neuronal mechanisms of object recognition. Science 262:685–688.

Ungerleider LG, Mishkin M (1982) Two cortical visual systems. In: Analysis of Visual Behavior (Ingle DJ, Goodale MA, Mansfield RJW, eds). Cambridge, MA: MIT Press.

Wang M, Vijayraghavan S, Goldman-Rakic PS (2004) Selective D2 receptor actions on the functional circuitry of working memory. Science 303:853–856.

Weiss B, Davidkova G, Zhou LW (1999) Antisense RNA gene therapy for studying and modulating biological processes. Cell Mol Life Sci 55:334–358.

Weiss B, Davidkova G, Zhou LW, Zhang SP, Morabito M (1997) Expression of a D2 dopamine receptor antisense RNA in brain inhibits D2-mediated behaviors. Neurochem Int 31:571–580.

Part III

Theoretical Considerations

15

Linking Visual Development and Learning to Information Processing: Preattentive and Attentive Brain Dynamics

STEPHEN GROSSBERG

Department of Cognitive and Neural Systems and Center for Adaptive Systems, Boston University, Boston, MA, USA Grossberg, S

Introduction

Vision research is such a large and vigorous field that it has often advanced through parallel but non-interacting streams of work. Visual development has often been studied separately from adult visual perception. Even studies of adult perceptual learning have often not made contact with other studies of how vision works, and many psychophysical studies have been done without regard to their neurobiological mechanisms. Recently, theoretical advances in understanding the functional organization of visual cortex have begun to synthesize such parallel efforts into a unifying theoretical framework that has disclosed many new issues and, with them, ideas for novel types of experiments. This emerging framework has begun to clarify how the visual cortex autonomously develops, stabilizes its own development, and then gives rise to visual perception in the adult. A rapidly developing cortical model links processes of development in the infant to processes of perception and learning in the adult. This model sheds new light on how cortical circuits can be shaped by environmental statistics, and proposes how the cortex embodies more powerful and subtle computational principles than the Bayesian learning approaches that have recently gained such popularity; e.g., Kersten, Mamassian and Yuille (2004). In particular, the model clarifies the role of attention in learning, and also suggests when attention may not be needed for learning to occur. Cortical computation, I would claim, enables the brain to self-organize in response to ever-changing environmental statistics in a way that the priors and stationary probabilities of Bayesian thinking cannot fully capture. The present article summarizes some of these recent developments.

R. Pinaud, L. A. Tremere, P. De Weerd(Eds.), Plasticity in the Visual System:
From Genes to Circuits, 323-346, ©2006 Springer Science + Business Media, Inc.

This emerging model of visual cortical dynamics, called the LAMINART model (Grossberg, 1999a; Grossberg and Howe, 2003; Grossberg, Mingolla, and Ross, 1997; Grossberg and Raizada, 2000; Raizada and Grossberg, 2001, 2003), suggests how the layered circuits of visual cortex interact to control cortical development and learning, notably how bottom-up, horizontal, and top-down circuits interact to control the formation of perceptual boundaries, or groupings. These studies include analyses of how the cortical layers develop their receptive field properties in a coordinated manner, how grouping and attentional circuits develop within these layers, and how cortical circuits embody statistical environmental constraints that support 3D vision (Grossberg and Seitz, 2003; Grossberg and Swaminathan, 2004; Grossberg and Williamson, 2001). The LAMINART models have hereby clarified how both pre-attentive and attentive feedback interactions may influence cortical development and learning (Grossberg, 1999a, 2003a). Related studies have clarified how learning tunes interactions between boundary circuits and the circuits that control the formation of perceptual surfaces. Indeed, properties of the well-known McCollough effect, or long-term orientation-specific color adaptation, have been traced to such learned boundary/surface interactions (Grossberg, Hwang, and Mingolla, 2002).

Computing with Boundaries and Surfaces

One reason for the disconnect between studies of visual development and learning and of adult visual perception has been insufficient understanding of the functional units that support adult visual perception. During the past two decades, experimental and theoretical evidence have provided increasing support for the prediction that the visual cortex builds percepts of object form using three-dimensional representations of perceptual *boundaries* and *surfaces*, notably representations that can separate figures from their backgrounds and complete the representations of partially occluded objects. It has been proposed that these boundary and surface representations are computed within the interblob and blob streams, respectively, between cortical areas V1 to V4 (see Grossberg (1994) for a review). These representations, in turn, are predicted to be the functional units that project to higher levels of the brain, notably inferotemporal and prefrontal cortex, where they are categorized, or unitized, into object representations. All of these cortical areas and their representations are, moreover, linked with each other through feedback pathways.

Early modeling studies that identified boundaries and surfaces as basic perceptual units and suggested how they are computed by visual cortex were provided by the author and his colleagues; e.g., Cohen and Grossberg (1984), Grossberg (1984, 1987a, 1987b), Grossberg and Mingolla (1985a, 1985b), Grossberg and Todorovic (1988). Since that time, many experiments have lent support to this hypothesis (see Grossberg (1994, 1997) for reviews), and many authors have further modeled these boundary and surface processes;

e.g., Cao and Grossberg (2004); Douglas *et al.* (1995), Finkel and Edelman (1989), Grossberg (1994, 1997), Grossberg, Hwang, and Mingolla (2002), Grossberg and Howe (2003), Grossberg and Kelly (1999), Grossberg and McLoughlin (1997), Grossberg and Pessoa (1998), Grossberg and Yazdanbakhsh, 2004; Heitger *et al.* (1998), Kelly and Grossberg (2000), Li (1998), McLoughlin and Grossberg (1998), Mumford (1992), Pessoa, Mingolla, and Neumann (1995), Somers, Nelson and Sur (1995), Stemmler, Usher and Niebur (1995), and Ullman (1995). A parallel processing stream, through cortical area MT, helps to compute object motion and cues useful for visual navigation. A neural model of motion processing, called the 3D FORMOTION model, has been progressively developed by Chey, Grossberg, and Mingolla (1997, 1998), Grossberg, Mingolla, and Viswanathan (2001), and Berzhanskaya, Grossberg, and Mingolla (2004) but will not be further discussed in this chapter, except to mention one of its key untested predictions: The circuits in cortical areas MT/MST that help to accomplish pre-attentive motion capture are the same ones that carry out attentive motion-direction priming. This prediction is a variant of the *preattentive-attentive interface* prediction that is further discussed herein for the processing of visual form.

Development and Learning in Laminar Circuits for Filtering, Grouping, and Attention

This chapter will first discuss aspects of how the visual cortex generates the perceptual boundaries that go into representations of visual object form. This process is also known as *perceptual grouping*, or the *binding problem*. These perceptual grouping processes already play an important role in how infants perceive the world. For example, neonates appear to perceive a partly occluded object as disjoint parts. The ability to process these fragments as coherent objects via perceptual grouping develops rapidly within the first two to four months of life (Kellman and Spelke, 1983; Johnson and Aslin, 1996; Johnson, 2001).

The LAMINART model clarifies how perceptual grouping circuits develop and learn, thereby giving rise to cortical circuits that are capable of explaining data about adult visual perception. A central issue in cortical development and learning concerns the *stability* of these processes; namely, how do cortical circuits protect previous development and learning from large-scale overwriting and obliteration by the changing statistics of inputs from the world? The same problem arises during adult learning. I call this problem the "stability-plasticity dilemma" since the brain needs to *balance* cortical plasticity and stability (e.g., Grossberg, 1980, 1982): The brain needs to adapt rapidly enough to improve survival chances, yet not just as rapidly erase useful memories in response to changing environmental inputs. This problem has since often been called the problem of "catastrophic forgetting". Catastrophic forgetting does not refer to the controlled refinement and adjustment of circuits in response to fluctuations in environmental statistics. Such refinement is often important for successful behavior. Rather, it refers to large-scale erasure of still useful cortical circuit properties. Most neural models, such as the popular back propagation model (see Grossberg (1988) for a review), indeed all feedforward neural models, experience catastrophic forgetting, unless they arbitrarily shut down their plasticity as time unfolds.

The LAMINART model proposes neural mechanisms that enable developing cortical circuits to dynamically stabilize themselves using properties of their self-organized circuit interactions. Remarkably, the same processes which help to stabilize development seem to control properties in the adult of perceptual grouping, attention, and learning. Many useful implications follow from this observation. One is that the laws of adult perception are strongly constrained by stability constraints on infant development. Another is that even early stages of visual cortical processing actively carry out such "emergent" processes as perceptual grouping and attentional selection.

The LAMINART model further clarifies why visual cortex, indeed all neocortex, is organized into layered circuits. This laminar organization is predicted to realize at least three interacting processes: (1) the developmental and learning processes whereby the cortex shapes its circuits to match environmental constraints in a *stable* way through time; (2) the binding process whereby cortex groups distributed data into coherent object representations that remain sensitive to analog properties of the environment (the property of "analog coherence"); and (3) the attentional process whereby cortex selectively processes important events. The model proposes that the mechanisms which achieve property (1) imply properties of (2) and (3). The LAMINART model also opens a path towards understanding how variations and specializations of these processes operate in other types of neocortex. This modeling perspective begins to unify three fields: infant cortical development, adult cortical neurophysiology and anatomy, and adult visual perception.

The model is called a LAMINART model because it clarifies how mechanisms of Adaptive Resonance Theory, or ART, can be realized within identified laminar cortical circuits. Earlier ART models were devoted to understanding how bottom-up and top-down cortical interactions work together to stably control cortical development and learning during perception and cognition. Although these studies successfully explained and predicted various behavioral and brain data, they did not show how these processes are realized within laminar cortical circuits. Grossberg (1999b) reviews some of these ART concepts and various data that they can explain. The LAMINART model extends these results by proposing how bottom-up, top-down, and *horizontal* interactions work together in *laminar* cortical circuits, and how they unify processes of development, learning, 3D grouping, and attention. LAMINART hereby joins concepts about ART learning and attention with concepts about perceptual grouping. This innovation was introduced in Grossberg (1999a).

Balancing Cortical Excitation and Inhibition: Stability, Intermittency, and Synchrony

Subsequent work on the LAMINART model has clarified how excitatory and inhibitory connections in the cortex can develop stably by maintaining a *balance* between excitation and inhibition in multiple cortical circuits (Grossberg and Williamson, 2001). It is known, for example, that long-range excitatory horizontal connections between pyramidal cells in layer 2/3 of visual cortical areas play an important role in perceptual grouping (Figure 15.1). The model proposes how the laws that govern cortical development enable the strength of these long-range excitatory horizontal connections to be (approximately) balanced against the strength of short-range disynaptic inhibitory interneurons which input to the same target pyramidal cells. These balanced connections are proposed to realize properties of perceptual grouping in the adult. Figure 15.1 summarizes how these balanced connections enable perceptual groupings to form inwardly between pairs, or greater numbers, of inducers in an image (the case of a Kanizsa square is here illustrated), but not outwardly from a single inducer, which would fill the percept with spurious boundaries.

The model also proposes that development enables the strength of excitatory connections from layer 6-to-4 to be balanced against those of inhibitory interneuronal connections to the same layer 4 cells; see Figure 15.2. Due to this balance, the net excitatory effect of layer 6 on layer 4 is proposed to be modulatory. These (approximately) balanced excitatory and inhibitory connections exist within the on-center of an on-center off-surround network from layer 6-to-4. The off-surround cells can strongly inhibit their target cells, even though the on-center cells can only provide excitatory modulation of their target cells.

The model proposes how this layer 6-to-4 circuit functions as a "selection circuit" because it can help to select the groupings that enter conscious attention. Grouping cells in layer 2/3 can activate the layer 6-to-4 selection circuit via excitatory connections from layer 2/3 to layer 6; see Figure 15.3a. When ambiguous and complex scenes are being processed, many possible groupings can start to form using the horizontal connections within layer 2/3. The selection circuit enables the strongest groupings to inhibit weaker groupings via the 6-to-4 off-surrounds of the strongest groupings.

Top-down attention can bias this selection process, and thereby influence which groupings will enter conscious perception. In particular, it is proposed that top-town attentional signals from higher cortical areas, such as area V2, can also activate the layer 6-to-4 on-center off-surround network; see Figure 15.3b. This circuit is called a *folded feedback* circuit because the top-down feedback is "folded" into the bottom-up signal flow from layer 6-to-4. Attention can hereby modulate, or sensitize, cells in the attentional on-center, without fully activating them, because the excitatory and inhibitory signals in the on-center are balanced. Attention can also inhibit cells in the off-surround.

Because both grouping and attention share the same selection circuit, this anatomical arrangement enables attention to influence which groupings are

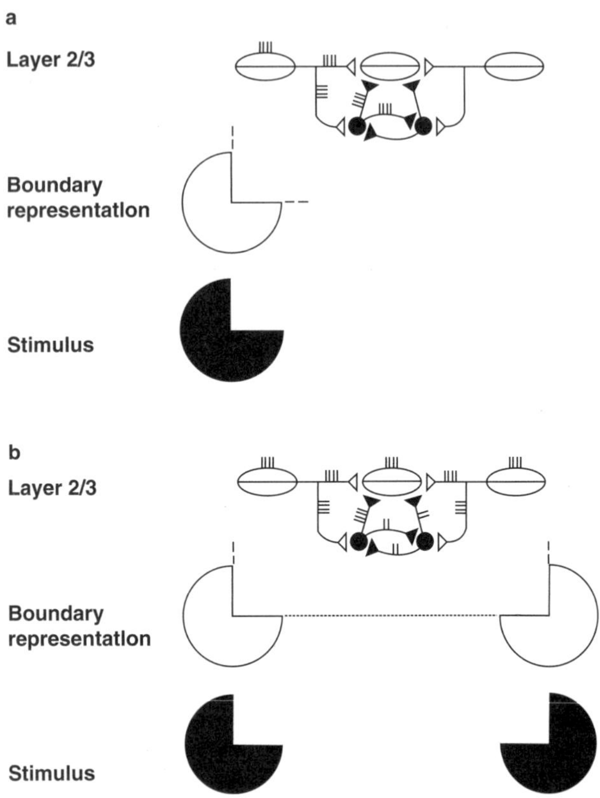

FIGURE 15.1. Schematic of the boundary grouping circuit in layer 2/3. Pyramidal cells with collinear, coaxial receptive fields (shown as ovals) excite each other via long-range horizontal axons (Bosking *et al.*, 1997; Schmidt *et al.*, 1997), which also give rise to short-range, disynaptic inhibition via pools of interneurons, shown filled-in black (McGuire *et al.*, 1991). This balance of excitation and inhibition helps to implement what we call the *bipole property*. **(a)** Illustration of how horizontal input coming in from just one side is insufficient to cause above-threshold excitation in a pyramidal cell (henceforth referred to as the target) whose receptive field does not itself receive any bottom-up input. The inducing stimulus (e.g. a Kanizsa 'pacman', shown here) excites the oriented receptive fields of layer 2/3 cells, which send out long-range horizontal excitation onto the target pyramidal. However, this excitation brings with it a commensurate amount of disynaptic inhibition. This creates a case of 'one against one', and the target pyramidal is not excited above-threshold. The boundary representation of the solitary pacman inducer produces only weak, sub-threshold collinear extensions (thin dashed lines). **(b)** When two collinearly aligned induced stimuli are present, one on each side of the target pyramidal's receptive field, a boundary grouping can form. Long-range excitatory inputs fall onto the cell from both sides, and summate. However, these inputs fall onto a shared pool of inhibitory interneurons, which, as well as inhibiting the target pyramidal, also inhibit each other (Tamas, Somogyi and Buhl, 1998), thus normalizing the total amount of inhibition emanating

selected. This circuit is called the *preattentive-attentive interface* within the LAMINART model. Using this interface, attention can shift the excitatory/inhibitory balance that determines which groupings will enter consciousness. A dramatic example of this influence occurs when attention that is caste on one part of an object can flow selectively along the perceptual groupings that define the entire object. Roelfsema *et al.* (1998) have discovered such a flow of attention along a perceptual grouping during their neurophysiological recordings in macaque area V1. Because of this property, both infants and adults can focus their attention selectively upon whole objects, rather than just random subsets of visual features.

The feedback circuits that govern the grouping and attentional selection processes are predicted to play a key role in helping to stabilize both development and adult learning within multiple cortical areas, including cortical areas V1 and V2. During development, the selection circuit (which itself is developing) helps to prevent the wrong combinations of cells from being co-activated, and thus from being associated, or wired, together. How this is predicted to happen will be discussed further in the next section.

Balanced excitatory and inhibitory connections help to explain the observed variability in the number and temporal distribution of spikes emitted by cortical neurons. Modeling studies have shown how balanced excitation and inhibition can produce the highly variable interspike intervals that are found in cortical data (Shadlen and Newsome, 1998; van Vreeswijk and Sompolinsky, 1998). The LAMINART model suggests that such variability may reflect mechanisms that are needed to ensure stable development and learning by cortical circuits. Given that "stability implies variability," the cortex is faced with the difficult problem that variable spikes are quite inefficient in driving responses from cortical neurons. When one analyses how these balanced excitatory and inhibitory connections generate perceptual groupings, it becomes clear that the grouping circuits automatically have the property of preferentially responding to synchronized inputs. Figure 15.1 illustrates why asynchronously activated cells have a difficult time generating a perceptual grouping, whereas synchronously activated cells do not. According to Figure 15.1a, an asynchronous volley of horizontal signals from a single population of layer 2/3 cells will kill itself off due to balanced excitation and inhibition. According to Figure 15.1b, a synchronous volley from pairs of appropriately positioned cells will initiate grouping. Modeling studies have shown how both perceptual

---

from the interneuron pool, without any individual interneuron saturating. This summating excitation and normalizing inhibition together create a case of 'two-against-one', and the target pyramidal is excited above-threshold. This process occurs along the whole boundary grouping, which thereby becomes represented by a line of suprathreshold layer 2/3 cells (thick dotted line). Boundary strength scales in a graded analog manner with the strength of the inducing signals. [Reproduced with permission from Raizada and Grossberg (2001)].

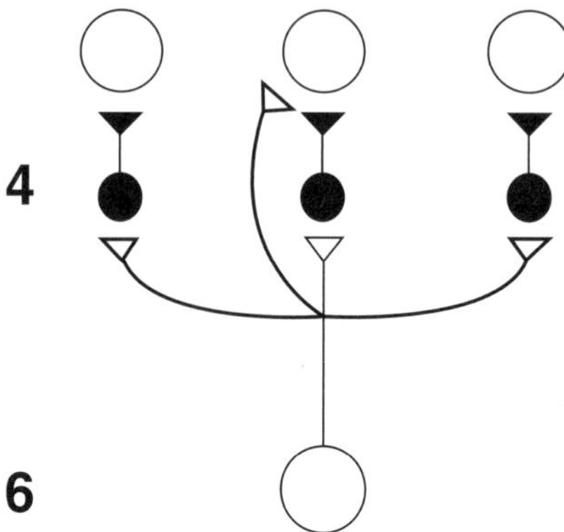

FIGURE 15.2. Schematic of the modulatory layer 6-to-layer 4 on-center off-surround path. Pyramidal cells in layer 6 give on-center excitation to layer 4 spiny stellates in the column above them, but also make medium-range connections onto layer 4 inhibitory interneurons, shown filled-in black (Ahmed *et al.*, 1997; McGuire *et al.*, 1984). These interneurons synapse onto the spiny stellates, creating a 6-to-4 off-surround, and also onto each other (connection not illustrated), thereby helping to normalize the total amount of inhibition (Ahmed *et al.*, 1997). Note that the 6-to-4 off-surround inhibition spatially overlaps with the excitatory on-center, with the consequence that the 6-to-4 excitation is inhibited down into being modulatory, *i.e.* priming or subthreshold (Callaway, 1998b; Stratford *et al.*, 1996). [Reproduced with permission from Raizada and Grossberg (2001)].

grouping and attentional circuits can actively resynchronize signals that have become partially desynchronized (Grossberg and Somers, 1991; Grossberg and Grunewald, 1997; Yazdanbakhsh and Grossberg, 2004). The model hereby discloses a previously unsuspected link between properties of stable development, adult learning, grouping, attention, and synchronous cortical processing.

The Link Between Attention and Learning: The Role of Adaptive Resonance

The solution that ART proposes to the stability-plasticity dilemma is to allow neural representations to be modified only by those incoming stimuli with which they form a sufficiently close match. If the match is close enough, then learning occurs. Precisely because the match is sufficiently close, this learning causes fine-tuning of the existing representation, rather than a radical overwriting. Matching gets started by initially endowing the top-down matching circuits with broadly distributed adaptive weights. Learning prunes these weights and makes them more selective. If the active neural representation does not match with the incoming stimulus, then its neural activity will be inhibited and hence unable to cause plastic changes. The network is designed so that inhibition of the initially active representation enables other representations to win the competition and become active instead. In other words, the network embodies a *search* mechanism that is typically realized by interacting matching and habituative processes (e.g., Carpenter and Grossberg, 1990; Grossberg and Seitz, 2003). Search either gives rise to a new match, thereby allowing learning, or a non-match, causing the search process to repeat until either a match is found or the incoming stimulus selects a totally new representation as a basis for learning.

A key mechanism that implements the matching process is top-down attentional feedback directed to behaviorally relevant sensory stimuli. The ART model predicted that top-down attentional signals exist that are expressed through a modulatory on-center off-surround network (Figure 15.3b), whose role is to select and enhance behaviorally relevant bottom-up sensory inputs (match), and suppress those that are irrelevant (non-match). Mutual excitation between the top-down feedback and the bottom-up signals that they match was predicted to amplify, synchronize, and maintain for a sufficient amount of time the matched neural activity pattern, thereby triggering learned synaptic changes. Thus, attentionally relevant stimuli are learned, while irrelevant stimuli are suppressed and hence prevented from destabilizing existing representations.

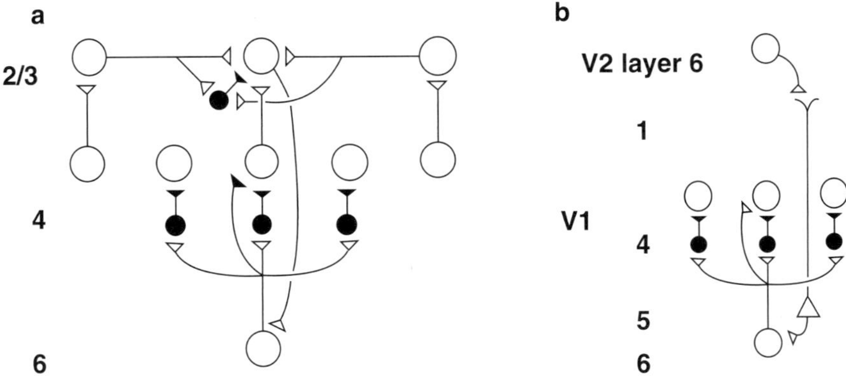

FIGURE 15.3. (a) Connecting the 6-to-4 on-center off-surround to the layer 2/3 grouping circuit: Like-oriented layer 4 simple cells with opposite contrast polarities compete (not shown) before generating half-wave rectified outputs that converge onto layer 2/3 complex cells in the column above them. Like attentional signals from higher cortex, groupings which form within layer 2/3 also send activation into the *folded feedback* path, to enhance their own positions in layer 4 beneath them via the 6-to-4 on-center, and to suppress input to other groupings via the 6-to-4 off-surround. There exist direct layer 2/3-to-6 connections in macaque V1, as well as indirect routes via layer 5. (b) *Folded feedback* carries attentional signal from higher cortex into layer 4 of V1, via the modulatory 6-to-4 path. Corticocotrical feedback axons tend preferentially to originate in layer 6 of the higher area and to terminate in the lower cortex's layer 1 (Salin and Bullier, 1995, p. 110), where they can excite the apical dendrites of layer 5 pyramidal cells whose axons send collaterals into layer 6. Several other routes through which feedback can pass into V1 layer exist. Having arrived in layer 6, the feedback is then 'folded' back up into the feedforward stream by passing through the 6-to-4 on-center off-surround path (Bullier *et al.*, 1996). [Reproduced with permission from Raizada and Grossberg (2001)].

The attentional feedback pathway through the layer 6-to-4 modulatory on-center off-surround network in the LAMINART model is predicted to implement ART matching in cortical laminar circuitry. The ART prediction raises two key questions: First, does top-down cortical feedback have an on-center off-surround structure? Second, is there evidence that top-down feedback controls plasticity in the area to which it is directed?

The prediction that top-down attention has an on-center off-surround characteristic has received a considerable amount of psychological and neurobiological empirical confirmation in the visual system (Bullier *et al.*, 1996; Caputo and Guerra, 1998; Downing, 1988; Mounts, 2000; Reynolds, Chelazzi, and Desimone, 1999; Smith, Singh, and Greenlee, 2000; Somers *et al.*, 1999; Sillito *et al.*, 1994; Steinman, Steinman, and Lehmkuhle, 1995; Vanduffell, Tootell, and Orban, 2000). Based on such data, this conclusion has recently been restated, albeit without a precise anatomical realization, in terms of the concept of "biased competition" (Desimone, 1998; Kastner and Ungerleider, 2001), in which attention biases the competitive influences within the network. Feedback from auditory cortex to the medial geniculate nucleus (MGN) and the inferior colliculus (IC) also has an on-center off-surround form (Zhang *et al.*, 1997). Temereanca and Simons (2001) have produced reported a similar feedback scheme in the rodent barrel system.

The claim that bottom-up sensory activity is *enhanced* when matched by top-down signals is in accord with an extensive neurophysiological literature showing the facilitatory effect of attentional feedback (e.g., Luck *et al.*, 1997; Roelfsema, Lamme and Spekreijse, 1998) but not with models in which matches with top-down feedback cause suppression (e.g., Mumford, 1992; Rao and Ballard, 1999).

Recent data also support the ART claim that top-down feedback controls plasticity. Psychophysically, the role of attention in controlling adult plasticity and perceptual learning was demonstrated by Ahissar and Hochstein (1993). Neurophysiological evidence of Gao and Suga (1998) showed that acoustic stimuli caused plastic changes in the inferior colliculus of bats only when the IC received top-down feedback from auditory cortex. The authors also found that this plasticity is enhanced when the auditory stimuli were made behaviorally relevant, in accord with the ART proposal that top-down feedback allows attended—that is, relevant—stimuli to be learned, while suppressing unattended irrelevant ones. Evidence that cortical feedback also controls thalamic plasticity in the somatosensory system has been found by Nicolelis and colleagues (Krupa, Ghazanfar and Nicolelis, 1999) and by Parker and Dostrobsky (1999). Kaas (1999) reviews these findings.

Another predicted role of these feedback connections is to synchronize the firing patterns of higher and lower cortical areas. Given that "cells that fire together wire together", synchronous firing of this sort would further increase the ability of the mutually excitatory resonant activity caused by ART matching to allow synaptic plasticity and learning to take place. It has elsewhere been shown that variants of the ART and LAMINART models are capable of rapidly synchronizing their activation patterns during both perceptual grouping and attentional focusing (Grossberg and Somers, 1991; Grossberg and Grunewald, 1997; Yazdanbakhsh and Grossberg, 2004). Recent discussions of top-down cortical feedback, synchrony, and how they support ART predictions are given by Engel, Fries, and Singer (2001), Fries *et al.*, (2001) and Pollen (1999).

Learning without Attention: The Role of Pre-attentive Resonance and Synchronization

The hypothesis that attentional feedback exerts a controlling influence over plasticity in sensory cortex does not imply that unattended stimuli can never be learned. Indeed, the LAMINART model has clarified how the stability of early development can be controlled, even before top-down attention may be able to modulate it (Grossberg, 1999a). During development, plastic changes in cortex are driven by stimuli that occur with high statistical regularity in the environment (e.g., Grossberg and Swaminathan, 2004; Grossberg and Williamson, 2001). Given that there is experimental support for the ART prediction that top-down attention plays a matching role which helps to control cortical plasticity, how can we explain other data which, at the outset, seem to contradict this prediction by showing that perceptual learning can occur without attention under certain circumstances; e.g., Watanabe *et al.* (2001)? This issue can be understood by considering the following question: How can pre-attentive groupings form over positions that receive no bottom-up inputs, without destabilizing cortical development and learning?

This is an issue because, as described above, the ART matching rule has three aspects: First, incoming sensory signals that receive matching top-down excitatory feedback should be enhanced; second, non-matching inputs that do not receive excitatory feedback should be suppressed; and third, top-down feedback on its own should be only modulatory, that is, unable to produce above-threshold activity in the lower area in the absence of incoming bottom-up signals. The conceptual challenge is this: If ART matching is needed to stabilize cortical development and learning, and if ART matching requires that suprathreshold activation can occur only where there are bottom-up inputs, then does not the existence of illusory contours contradict the ART matching rule, since such groupings form over positions that receive no bottom-up inputs? Moreover, the horizontal connections that underlie such groupings are known to develop and learn in response to visual inputs, yet do not seem to destabilize cortical development or learning. How is this possible?

Here is where the laminar organization of the visual cortex, as conceptualized by the LAMINART model, offers a parsimonious and elegant solution by using the preattentive-attentive interface circuit that occurs between layers 6 and 4. When a horizontal grouping starts to form in layer 2/3, it also activates the *intra*cortical, interlaminar feedback pathway from layer 2/3 to the modulatory on-center off-surround network from layer 6 to 4. This feedback pathway helps to select which cells will remain active to participate in a winning grouping. But this is the same network that ART requires attention to use when it stabilizes cortical development and learning. In other words, the layer 6-to-4 selection circuit, which in the adult helps to choose winning groupings, is also predicted, during brain development and learning, to ensure that the ART matching rule holds at every position along a grouping, including positions that receive no bottom-up input. Because the matching rule holds, only the correct combinations of cells can "fire together and wire together", and hence stability is maintained. *Intra*cortical feedback via layers 2/3-to-6-to-4-to-2/3 can realize this selection process even before *inter*cortical attentional feedback can develop. This property is sometimes summarized with the phrase: "The pre-attentive grouping is its own attentional prime" (Grossberg, 1999a).

Experiments such as those of Watanabe *et al.* (2001) can be explained by noting that the pre-attentive resonances that support unattended learning are not inhibited by the attentive resonances that are activated by the experimental task. Indeed, in these experiments, a form identification task at the center of gaze attracts attention, whereas unattended perceptual learning of motion direction occurs at peripheral locations. These experiments illustrate that attentive resonance can influence learning at certain locations in the What stream (the form stimulus) even as pre-attentive resonance may influence learning at disjoint locations in the Where stream (the motion stimulus). Said in another way, the LAMINART model predicts that this experiment was cleverly set up so that the inhibitory effects of attention did not suppress some pre-attentive resonances at disjoint positions. Further tests of this hypothesis would systematically vary how much attentive inhibition would be expected to suppress learning in such non-attended locations.

Beyond Bayes: Self-Organizing Feedforward/Feedback and Digital/Analog Decisions

The LAMINART model brings into focus several new problems and proposed solutions thereof about the dynamics of cortical processing, including predicted links between balanced excitation and inhibition, synchrony, resonance, attention, and learning. The LAMINART model embodies a novel way to compute in several other senses as well, which illustrate my assertion that cortical dynamics are not adequately conceptualized by traditional concepts such as Bayesian computing. I claim that cortical computation represents a new type of hybrid between feedforward and feedback computing, and also between digital and analog computing for processing perceptual groupings as well as other types of distributed data. The LAMINART model predicts that these properties allow the fast but stable self-organization that is characteristic of cortical development and life-long learning.

The new hybrid between feedforward and feedback processing works as follows: When an unambiguous scene is processed, the LAMINART model can quickly group the scene in a fast feedforward sweep of activation that passes directly through layer 4 to 2/3 and then on to layers 4 to 2/3 in subsequent cortical areas (see Figures 15.3 and 15.4). This property clarifies how recognition can be so fast in response to unambiguous scenes; e.g., Thorpe et al. (1996). On the other hand, if there are multiple possible groupings in a scene, as in a complex textured scene, then competition among these possibilities due to inhibitory interactions in layers 4 and 2/3 can cause all cell activities to become smaller. This happens because the competitive circuits in the model are *self-normalizing*; that is, they tend to conserve the total activity of the circuit: When some activities get bigger, others must get smaller. This self-normalizing property is related to the ability of the shunting on-center off-surround networks that realize these competitive circuits to realize the property of *contrast normalization*; that is, to process input contrasts over a large dynamic range without saturation (Douglas *et al.*, 1995; Grossberg, 1973, 1980; Heeger, 1992). In other words, these self-normalizing circuits carry out a type of real-time probability theory in which the amplitude of cell activity covaries with the certainty of the network's selection, or decision, about a grouping.

Cell activation amplitude is, in turn, translated into processing speed. Low activation greatly slows down the feedforward processing in the circuit because it takes longer for cell activities to exceed output thresholds and to activate subsequent cells above threshold. In the model, network uncertainty is resolved through feedback: Weakly active layer 2/3 grouping cells feed back signals to layers 6-then-4-then-2/3 to close an intracortical cortical feedback loop that rapidly contrast enhances and amplifies the winning grouping. This is the feedback circuit that embodies the prediction that "The pre-attentive grouping is its own attentional prime", and thus the circuit that is predicted to stabilize cortical development and learning when attention is not available. As the winner is selected, and weaker groupings are suppressed, its cells become more active, hence can again rapidly send the cortical decision to subsequent processing stages.

In summary, the LAMINART circuit behaves like a real-time probabilistic decision circuit that operates in a fast feedforward mode when there is little uncertainty, and automatically switches to a slower feedback mode when there is uncertainty. Feedback selects a winning decision that enables the circuit to speed up again. In all, activation amplitude and processing speed both increase with certainty. The large activation amplitude of a winning grouping is facilitated by the synchronization that occurs as the winning grouping is selected. Bayes may be able to compute probabilities whose values embody variable degrees of uncertainty, but does not have the power to self-time decision-making until contentions are automatically resolved.

The LAMINART circuit also embodies a novel kind of hybrid computing that simultaneously realizes the stability of digital computing and the sensitivity of analog computing. This is true because the feedback loop between layers 2/3-6-4/-2/3 that selects or confirms a winning grouping has the property of *analog coherence* (Grossberg, 1999a; Grossberg, Mingolla, and Ross, 1997; Grossberg and Raizada, 2000; Yazdanbakhsh and Grossberg, 2004); namely, this feedback loop can synchronously store a winning grouping with-

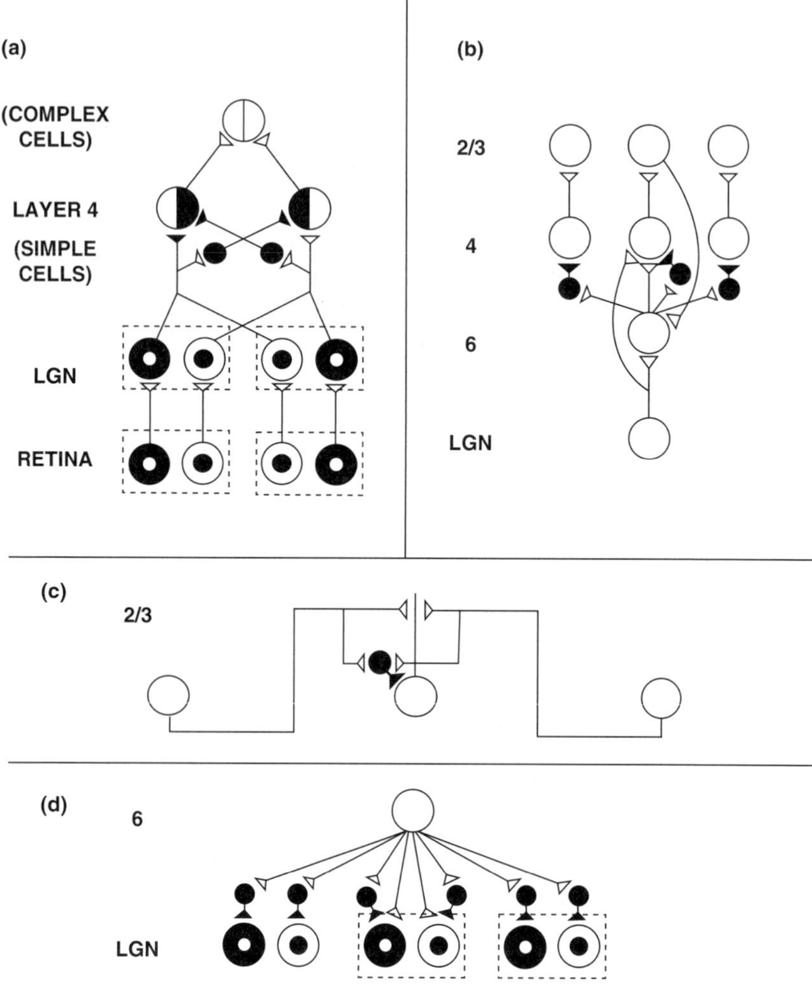

FIGURE 15.4. The adult network of retinal, V1, and lateral geniculate nucleus (LGN) neurons to which the developmental model converges: **(a)** Feedforward circuit from retina to LGN to cortical layer 4. **Retina:** Retinal ON cells have on-center off-surround organization (white disk surrounded by black annulus). Retinal OFF cells have an off-center on-surround organization (black disk surrounded by white annulus). **LGN:** The LGN ON and OFF cells receive feedforward ON and OFF cell inputs from the retina. **Layer 4:** LGN ON and OFF cell excitatory inputs to layer 4 establish oriented simple cell receptive fields. Like-oriented layer 4 simple cells with opposite contrast polarities compete before generating half-wave rectified outputs. Pooled simple cell outputs enable complex cells to respond to both polarities. They hereby full-wave rectify the image. See text for details. **(b)** Cortical feedback loop between layers 4, 2/3, and 6: LGN activates layer 6 as well as layer 4. Layer 6 cells excite layer 4 cells with a narrow on-center and inhibit them using layer 4 inhibitory interneurons that span a broader off-surround. Layer 4 cells excite layer 2/3 cells, which send excitatory feed-

out losing analog sensitivity to amplitude differences in the input pattern. The coherence that is derived from synchronous storage in the feedback loop provides the stability of digital computing, while preserving the sensitivity of analog computation. Bayes may be able to compute the analog values of its probabilities, but does not have the coherence needed to bind them together into emergent perceptually meaningful structures, indeed structures that can spontaneously complete missing information, as in the case of illusory contours.

All of these properties are predicted to be a manifestation of the ability of cortical laminar circuits to stabilize development and learning using the *intra*cortical feedback loop between layers 2/3-6-4-2/3 by selecting cells that fire together to wire together. The same *intra*cortical decision circuit is predicted to help stabilize development in the infant and learning throughout life, as well as to select winning groupings in the adult (Grossberg, 1999a). Thus, properties of perceptual grouping in the adult are predicted to be constrained by the requirements of stable development in the infant. This intracortical circuit can work even before *inter*cortical attentional feedback can develop to also stabilize cortical development and learning. Bayesian learning typically requires that priors and stationary probabilities exist. In contrast, LAMINART circuits are designed to develop and learn in a stable way in response to changing world statistics.

Development of Perceptual Grouping and Learning Circuits

Three types of quantitative modeling studies support the conclusions drawn in this article. These models are more extensively reviewed in Grossberg (2003b). One type of study concerns how horizontal and interlaminar connections develop within cortical layers 2/3, 4, and 6 in cortical area V1, and by extension to V2 and higher cortical regions (Grossberg and Swaminathan, 2004;

---

back signals back to layer 6 cells via layer 5 (not shown). Layer 2/3 can hereby activate the feedforward layer 6-to-4 on-center off-surround network. **(c)** The horizontal interactions in layer 2/3 that initiate perceptual grouping: Layer 2/3 complex pyramidal cells monosynaptically excite one another via horizontal connections, primarily on their apical dendrites. They also inhibit one another via disynaptic inhibition that is mediated by model smooth stellate cells. Together these interactions can realize the "bipole property" which enables groupings to form inwardly across the space between two or more inducers, but not awardly from a single inducer. **(d)** Top-down cortico-geniculate feedback from layer 6: LGN ON and OFF cells receive topographic excitatory feedback from layer 6, and more broadly distributed inhibitory feedback via LGN inhibitory interneurons that are excited by layer 6 signals. The feedback signals pool outputs over all cortical orientations and are delivered equally to ON and OFF cells. See the text for further details. [Reproduced with permission from Grossberg and Williamson (2001)].

Grossberg and Williamson, 2001). These interactions are often cited as the basis of "non-classical" receptive fields that are sensitive to the context in which individual features are found (von der Heydt, Peterhans, and Baumgartner, 1984; Peterhans and von der Heydt, 1989; Born and Tootell, 1991; Knierim and van Essen, 1992). In these modeling studies, it was assumed that receptive fields of individual simple and complex cells in layers 4 and 2/3, respectively, have already substantially developed. Grossberg and Williamson (2001) simulated development of the layer 2/3 horizontal connections that carry out collinear perceptual grouping, and the layer 6-to-4 inhibitory connections that control the preattentive-attentive interface, using the model summarized in Figure 15.4. Grossberg and Swaminathan (2004) simulated how the same laws of cortical development that lead to bipole cell connections for colinear grouping within a single depth can also lead to the development of angle cells (Hedge and Van Essen, 2000; Shevelev, 1998) and disparity gradient cells (Hinkle and Connor, 2001; Thomas, Cumming, and Parker, 2002) that span multiple depths. The different types of cells develop in response to different feature combinations in the visual environment. This study also showed how interactions among the angle cells and disparity gradient cells can contextually disambiguate locally ambiguous visual cues to form unambiguous boundary and surface representations of 3D slanted and curved objects, including percepts of bistable Necker cubes.

The second model (Olson and Grossberg, 1998) investigated the question of how cortical area V1 develops simple cells that respond to different eyes or different orientations at different positions on the retina within the familiar cortical maps of orientation and ocular dominance (Hubel and Wiesel, 1962, 1963, 1968). This organization is called a *map* because cell tuning to orientation and ocular dominance varies in a systematic way as the cortex is traversed in a horizontal direction. Such maps exhibit properties that are called singularities, fractures, and linear zones (Blasdel, 1992a, 1992b; Obeymeyer and Blasdel, 1993). The model showed how these features of cortical maps develop. A number of earlier models also studied how simple cells develop their orientationally tuned receptive fields within maps of orientation and ocular dominance (e.g., von der Malsburg, 1973; Grossberg, 1976a; Willshaw and von der Malsburg, 1976; Swindale, 1980, 1982, 1992; Linsker, 1986a, 1986b; Rojer and Schwartz, 1989, 1990; Durbin and Mitchison, 1990; Obermayer, Ritter and Schulten, 1990, Obermayer, Blasdel and Schulten, 1992; Miller, 1992, 1994; Grossberg and Olson, 1994; Sirosh and Miikkulainen, 1994). The Olson and Grossberg (1998) model showed, in addition, how nearby pairs of simple cells develop that are sensitive to the same orientation but opposite contrast polarities (Liu *et al.*, 1992). Such a model is called a *Triple-O Map* model because it shows how Orientation, Ocular Dominance, and Opposite Contrast Polarities all develop together. Earlier models were either Single-O or Double-O Map models, and many did not represent the dynamics of the cells whose connections were undergoing development. The Triple-O model clarifies how nearby simple cells that are sensitive to opposite contrast polarities could, in principle, cooperate to activate a shared complex cell.

The third model proposed how nearby pairs of simple cells that are sensitive to opposite contrast polarities develop connections to shared complex cells (Grunewald and Grossberg, 1998). In addition to being tuned to position, size, orientation, and pooled contrast polarities, the complex cells in the model, and *in vivo*, are also tuned to binocular disparity, which is a well-known cue to object depth (Julesz, 1971). These complex cell properties help to explain how depth-sensitive perceptual groupings can form over objects that are seen in front of textured backgrounds, and also how figure-ground properties emerge (Grossberg, 1994). A key question for this model concerned how oppositely polarized simple cells, whose activations are *anti-correlated* in time (if a contrast at a given position is dark-to-light, it cannot also be light-to-dark, and conversely), can nonetheless develop connections to a shared complex cell, and thereby become *correlated*.

Grossberg and Seitz (2003) began to unify these three types of models into a more comprehensive model of how cortical development of cells occurs in a coordinated manner across cell layers, guided by signals from the cortical subplate.

Several mechanisms in addition to those summarized in the preceding sections play a role in these models of cortical development. One mechanism causes *antagonist rebounds* to occur between simple cells that are sensitive to opposite contrast polarities but the same positions and orientations. For example, when a simple cell that has been on for awhile in response to a dark-to-light contrast shuts off, an opponent simple cell, that is sensitive to a light-to-dark contrast, briefly turns on. Such rebounds are proposed to be due to the chemical transmitters that carry signals between model cells. These transmitters habituate, or inactivate, when they are released by signals in their pathways, or axons (Abbott, Varella, Sen, and Nelson, 1997; Francis, Grossberg, and Mingolla, 1994; Grossberg, 1972, 1980). Such habituative transmitters play a role in the models of how simple cell maps develop (they help the map to form by preventing the developmental process from getting "stuck" in representations that develop soonest), how complex cells get activated by oppositely polarized simple cells (a rebounding simple cell can get correlated with an active complex cell that was initially activated by a simple cell of opposite contrast polarity), and how bistable percepts can occur in the adult (when two representations activate competitive processes in a balanced way, the one that wins habituates and enables the other to win later), notably percepts of bistable Necker cubes.

Grunewald and Grossberg (1998) also modeled how learned feedback from cortical area V1 to the LGN may carry out a matching process that helps to stabilize the development of disparity tuning in cortical complex cells and, by extension, the cortical map itself; see Figure 15.4d. This V1-to-LGN feedback is homologous to the attentional feedback that is proposed to occur from cortical area V2 to V1 (Figure 15.3b), and by extension other cortical areas as well. These model interactions clarify how complex cells can binocularly match left and right eye image features with the same contrast polarity, yet can also pool signals with opposite contrast polarities, consistent with psychophysical and neurobiological data about adult 3D vision; e.g., Cao and Grossberg (2004) and Grossberg and Howe (2003).

References

Abbott, L.F., Varela, J.A., Sen, K. and Nelson, S.B. (1997). Synaptic depression and cortical gain control. *Science*, **275**, 220–224.

Ahissar, M. and Hochstein, S. (1993). Attentional control of early perceptual learning. *Proceedings of the National Academy of Sciences USA*, **90**, 5718–5722.

Ahmed, R., Anderson, J.C., Martin, K.A.C. and Nelson, J.C. (1997). Map of the synapses onto layer 4 basket cells of the primary visual cortex of the cat. *The Journal of Comparative Neurology*, **380**, 230–242.

Berzhanskaya, J., Grossberg, S. and Mingolla, E. (2004) Laminar cortical dynamics of visual form and motion interactions during coherent object motion perception. Technical Report CAS/CNS TR-2004–011, Boston University. Submitted for publication.

Blasdel, G.G. (1992a). Differential imaging of ocular dominance and orientation selectivity in monkey striate cortex. *Journal of Neuroscience*, **12(8)**, 3115–3138.

Blasdel, G.G. (1992b). Orientation selectivity, preference, and continuity in monkey striate cortex. *Journal of Neuroscience*, **12(8)**, 3139–3161.

Born R.T. and Tootell, R.B. (1991). Single-unit and 2-deoxyglucose studies of side inhibition in macaque striate cortex. *Proceedings of the National Academy of Sciences*, **88**, 7071–7075.

Bosking, W., Zhang, Y., Schofield, B. and Fitzpatrick, D. (1997). Orientation selectivity and the arrangement of horizontal connections in tree shrew striate cortex. *Journal of Neuroscience*, **17(6)**, 2112–2127.

Bullier, J., Hupé, J.M., James, A. and Girard, P. (1996). Functional interactions between areas V1 and V2 in the monkey. *Journal of Physiology (Paris)*, **90**, 217–220.

Callaway, E.M. (1998a). Prenatal development of layer-specific local circuits in primary visual cortex of the macaque monkey. *Journal of Neuroscience*, **18**, 1505–1527.

Callaway, E.M. (1998b). Local circuits in primary visual cortex of the macaque monkey. *Annual Review of Neuroscience*, **21**, 47–74.

Caputo, G. and Guerra, S. (1998). Attentional selection by distractor suppression. *Vision Research*, **38**, 669–689.

Cao, Y. and Grossberg, S. (2004). A laminar cortical model of stereopsis and 3D surface perception: Closure and deVinci stereopsis. Technical Report CAS/CNS TR-2004–007, Boston University. *Spatial Vision*, in Press.

Carpenter, G.A. and Grossberg, S. (1990). ART 3: Hierarchical search using chemical transmitters in self-organizing pattern recognition architectures. *Neural Networks*, **3**, 129–152.

Chey, J., Grossberg, S. and Mingolla, E. (1997). Neural dynamics of motion grouping: From aperture ambiguity to object speed and direction. *Journal of the Optical Society of America*, **14**, 2570–2594.

Chey, J., Grossberg, S. and Mingolla, E. (1998). Neural dynamics of motion processing and speed discrimination. *Vision Research*, **38**, 2769–2786.

Cohen, M.A. and Grossberg, S. (1984). Neural dynamics of brightness perception: Features, boundaries, diffusion, and resonance, *Perception and Psychophysics*, **36**, 428–456.

Desimone, R. (1998). Visual attention mediated by biased competition in extrastriate visual cortex. *Philosophical Transactions of the Royal Society of London*, **353**, 1245–1255.

Douglas, R.J., Koch, C., Mahowald, M., Martin, K.A.C. and Suarez, H.H. (1995). Recurrent excitation in neocortical circuits. *Science*, **269**, 981–985.

Downing, C. J. (1988). Expectancy and visual-spatial attention: Effects on perceptual quality. *Journal of Experimental Psychology: Human Perception and Performance*, **14**, 188–202.

Durbin, R. and Mitchison, G. (1990). A dimension reduction framework for understanding cortical maps. *Nature*, **343**, 644–647.

Engel, A. K., Fries, P. and Singer, W. (2001). Dynamic predictions: Oscillations and synchrony in top-down processing. *Nature Reviews Neuroscience,* **2**, 704–716.

Francis, G., Grossberg, S. and Mingolla, E. (1994). Cortical dynamics of feature binding and reset: Control of visual persistence. *Vision Research*, **34**, 1089–1104.

Fries, P., Reynolds, J.H., Rorie, A.E and Desimone, R. (2001) Modulation of oscillatory neuronal synchronization by selective visual attention. *Science,* **291**, 1560–1563.

Gao, E. and Suga, N. (1998). Experience-dependent corticofugal adjustment of midbrain frequency map in bat auditory system. *Proceedings of the National Academy of Sciences* **95**, 12663–12670.

Grossberg, S. (1972). A neural theory of punishment and avoidance. II. Quantitative theory. *Mathematical Biosciences*, **15**, 39–67.

Grossberg, S. (1973). Contour enhancement, short-term memory, and constancies in reverberating neural networks. *Studies in Applied Mathematics*, **52**, 217–257. Reprinted in Studies of Mind and Brain, S. Grossberg (1982). Kluwer Academic Publishers: Norwell.

Grossberg, S. (1976a). Adaptive pattern classification and universal recoding. I: Parallel development and coding of neural feature detectors. *Biological Cybernetics*, **23**, 121–134.

Grossberg, S. (1976b). Adaptive pattern classification and universal recoding. II: Feedback, expectation, olfaction, and illusions. *Biological Cybernetics*, **23**, 187–202.

Grossberg, S. (1980). How does a brain build a cognitive code? *Psychological Review*, **87**, 1–51.

Grossberg, S. (1982). Studies of mind and brain: Neural principles of learning, perception, development, cognition, and motor control. Kluwer Academic Publishers: Norwell.

Grossberg, S. (1984). Outline of a theory of brightness, color, and form perception. In E. Degreef and J. van Buggenhaut (Eds.); Trends in Mathematical Psychology, (pp. 5559–5586); Elsevier Science: Amsterdam.

Grossberg, S. (1987a). Cortical dynamics of three-dimensional form, color and brightness perception, I: Monocular theory. *Perception and Psychophysics*, **41**, 87–116.

Grossberg, S. (1987b). Cortical dynamics of three-dimensional form, color and brightness perception, II: Binocular theory. *Perception and Psychophysics*, **41**, 117–158.

Grossberg, S. (1988). Nonlinear neural networks: Principles, mechanisms, and architectures. *Neural Networks*, **1**, 17–61.

Grossberg, S. (1994). 3-D vision and figure-ground separation by visual cortex. *Perception and Psychophysics*, **55**, 48–120.

Grossberg, S. (1997). Cortical dynamics of 3-D figure-ground perception of 2-D pictures. *Psychological Review*, **104**, 618–658.

Grossberg, S. (1999a). How does the cerebral cortex work? Learning, attention, and grouping by the laminar circuits of visual cortex. *Spatial Vision*, **12**, 163–186.

Grossberg, S. (1999b). The link between brain learning, attention, and consciousness. *Consciousness and Cognition*, **8**, 1–44.

Grossberg, S. (2003a). How does the cerebral cortex work? Development, learning, attention, and 3D vision by laminar circuits of visual cortex. *Behavioral and Cognitive Neuroscience Reviews*, **2**, 47–76.

Grossberg, S. (2003b). Linking visual cortical development to visual perception. In B. Hopkins and S. Johnson (Eds.); Neurobiology of infant vision, (pp. 211–271), Ablex Press: Stamford.

Grossberg, S. and Grunewald, A. (1997). Cortical synchronization and perceptual framing. *Journal of Cognitive Neuroscience*, **9**, 117–132.

Grossberg, S. and Howe, P.D.L. (2003). A laminar cortical model of stereopsis and three-dimensional surface perception. *Vision Research*, **43**, 801–829.

Grossberg, S., Hwang, S. and Mingolla, E. (2002). Thalamocortical dynamics of the McCollough effect: Boundary-surface alignment through perceptual learning. *Vision Research*, **42**, 1259–1286.

Grossberg, S. and Kelly, F. (1999). Neural dynamics of binocular brightness perception. *Vision Research*, **39**, 3796–3816.

Grossberg, S. and McLoughlin, N. (1997). Cortical dynamics of three-dimensional surface perception: {B}Inocular and half-occluded scenic images. *Neural Networks*, **10**, 1583–1605.

Grossberg, S. and Mingolla, E. (1985a). Neural dynamics of form perception: Boundary completion, illusory figures, and neon color spreading. *Psychological Review*, **92**, 173–211.

Grossberg, S. and Mingolla, E. (1985b). Neural dynamics of perceptual grouping: Textures, boundaries, and emergent segmentations. *Perception and Psychophysics*, **38**, 141–171.

Grossberg, S., Mingolla, E. and Ross, W.D. (1997). Visual brain and visual perception: How does the cortex do perceptual grouping? *Trends in Neurosciences*, **20**, 106–111.

Grossberg, S., Mingolla, E. and Viswanathan, L. (2001). Neural dynamics of motion integration and segmentation within and across apertures. *Vision Research*, **41**, 2521–2553.

Grossberg, S. and Olson, S. (1994). Rules for the cortical map of ocular dominance and orientation columns. *Neural Networks*, **7**, 883–894.

Grossberg, S. and Pessoa, L. (1998). Texture segregation, surface representation, and figure-ground separation. *Vision Research*, **38**, 137–161.

Grossberg, S. and Raizada, R.D.S. (2000). Contrast-sensitive perceptual grouping and object-based attention in the laminar circuits of primary visual cortex. *Vision Research*, **40**, 1413–1432.

Grossberg, S. and Seitz, A. (2003). Laminar development of receptive fields, maps, and columns in visual cortex: The coordinating role of the subplate. *Cerebral Cortex*, **13**, 852–863.

Grossberg, S. and Somers, D. (1991). Synchronized oscillations during cooperative feature linking in a cortical model of visual perception. *Neural Networks*, **4**, 453–466.

Grossberg, S. and Todorovic (1988). Neural dynamics of 1-D and 2-D brightness perception: A unified model of classical and recent phenomena. *Perception and Psychophysics*, **43**, 241–277.

Grossberg, S. and Swaminathan, G. (2004). A laminar cortical model for 3D perception of slanted and curved surfaces and of 2D images: Development, attention, and bistability. *Vision Research*, **44**, 1147–1187.

Grossberg, S. and Williamson, J.R. (2001). A neural model of how horizontal and interlaminar connections of visual cortex develop into adult circuits that carry out perceptual grouping and learning. *Cerebral Cortex*, **11**, 37–58.

Grossberg, S. and Yazdanbakhsh, A. (2004). Laminar cortical dynamics of 3D surface perception: Stratification, transparency, and neon color spreading. Technical Report CAS/CNS TR-2004–002, Boston University. Submitted for publication.

Grunewald, A. and Grossberg, S. (1998). Self-organization of binocular disparity tuning by reciprocal corticogeniculate interactions,. *Journal of Cognitive Neuroscience*, **10**, 199–215.

Hedge, J., and Van Essen, D.C. (2000). Selectivity of Complex Shapes in Primate Visual Area V2. *Journal of Neuroscience*, **20**, RC61 (1–6).

Heeger, D.J. (1992). Normalization of cell responses in cat striate cortex. *Visual Neuroscience*, **9(2)**, 181–197.

Heitger, F., von der Heydt, R., Peterhans, E., Rosenthaler, L. and Kubler, O. (1998). Simulation of neural contour mechanisms: Representing anomalous contours. *Image and Visual Computation*, **16**, 407–421.

Hinkle, D. and Connor, C.E. (2001). Three-dimensional orientation tuning in macaque area V4. *Society for Neuroscience Abstracts*, **286.7**, San Diego, USA.

Hubel, D.H. and Wiesel, T.N. (1962). Receptive fields, binocular interaction and functional architecture in the cat's visual cortex. *Journal of Physiology*, **160**, 106–154.

Hubel, D.H. and Wiesel, T.N. (1963). Shape and arrangement of columns in cat's striate cortex. *Journal of Physiology*, **195**, 215–243.

Hubel, D.H. and Wiesel, T.N. (1968). Receptive fields and functional architecture of monkey striate cortex. *Journal of Physiology* (London), **195**, 215–243.

Johnson, S.P. (2001). Visual development in human infants: Binding features, surfaces, and objects. *Visual Cognition*, **8**, 565–578.

Johnson, S.P. and Aslin, R.N. (1996). Perception of object unity in young infants: The roles of motion, depth, and orientation. *Cognitive Development*, **11**, 161–180.

Julesz, B. (1971). Foundations of Cyclopean Perception. University of Chicago Press: Chicago.

Kaas, J. H. (1999). Is most of neural plasticity in the thalamus cortical? *Proceedings of the National Academy of Sciences USA*, **96**, 7622–7623.

Kastner, S. and Ungerleider, L.G. (2001). The neural basis of biased competition in human visual cortex. *Neuropsychologia*, **39**, 1263–1276.

Kellman, P.J. and Spelke, E.S. (1983). Perception of partially occluded objects in infancy. *Cognitive Psychology*, **15**, 483–524.

Kelly, F. and Grossberg, S. (2000). Neural dynamics of 3-D surface perception: Figure-ground separation and lightness perception. *Perception and Psychophysics*, **62(8)**, 1596–1618.

Kersten, D., Mamassian, P. and Yuille, A. (2004). Object perception as Bayesian inference. *Annual Review of Psychology*, **55**, 271–304.

Knierim, J.J. and van Essen, D.C. (1992). Neuronal responses to static texture patterns in area V1 of the alert macaque monkey. *Journal of Neurophysiology*, **67**, 961–980.

Krupa, D.J., Ghazanfar, A.A. and Nicolelis, M.A. (1999). Immediate thalamic sensory plasticity depends on corticothalamic feedback. *Proceedings of the National Academy of Sciences USA*, **96**, 8200–8205.

Li, Z. (1998). A neural model of contour integration in the primary visual cortex. *Neural Computation*, **10**, 903–940.

Linsker, R. (1986a). From basic network principles to neural architecture. Emergence of spatial-opponent cells. *Proceedings of the National Academy of Sciences*, **83**, 7508–7512.

Linsker, R. (1986b). From basic network principles to neural architecture: Emergence of spatial-opponent cells. *Proceedings of the National Academy of Sciences*, **83**, 8390–8394.

Linsker, R. (1986c). From basic network principles to neural architecture: Emergence of orientation columns. *Proceedings of the National Academy of Sciences*, **83**, 8779–8783.

Liu, Z., Gaska, J.P., Jacobson, L.D. and Pollen, D.A. (1992). Interneuronal interaction between members of quadrature phase and anti-phase pairs in the cat's visual cortex. *Vision Research*, **32**, 1193–1198.

Luck, S. J., Chelazzi, L., Hillyard, S. A. and Desimone, R. (1997). Neural mechanisms of spatial selective attention in areas V1, V2, and V4 of macaque visual cortex. *Journal of Neurophysiology*, **77**, 24–42.

McGuire, B.A., Gilbert, C.D., Rivlin, P.K. and Wiesel, T.N. (1991). Targets of horizontal connections in macaque primary visual cortex. *Journal of Comparative Neurology*, **305(3)**, 370–392.

McGuire, B.A.; Hornung, J.P., Gilbert, C.D. and Wiesel, T.N. (1984). Patterns of synaptic input to layer 4 of cat striate cortex. *Journal of Neuroscience*, **4(12)**, 3021–3033.

McLoughlin, N. and Grossberg, S. (1998). Cortical computation of stereo disparity. *Vision Research*, **38**, 91–99.

Miller, K.D. (1992). Development of orientation columns via competition between on- and off-center inputs. *NeuroReport*, **3**, 73–76.

Miller, K.D. (1994). A model for the development of simple cell receptive fields and the ordered arrangement of orientation columns through activity-dependent competition between ON- and OFF-center inputs. *Journal of Neuroscience*, **14**, 409–441.

Mounts, J.R.W. (2000). Evidence for suppressive mechanisms in attentional selection: Feature singletons produce inhibitory surrounds. *Perception and Psychophysics*, **62**, 969–983.

Mumford, D. (1992). On the computational architecture of the neocortex, II. The role of cortico-cortical loops. *Biological Cybernetics*, 241–251.

Obermayer, K. and Blasdel, G.G. (1993). Geometry of orientation and ocular dominance columns in monkey striate cortex. *Journal of Neuroscience*, **13**, 4114–4129.

Obermayer, K., Blasdel, G.G. and Schulten, K. (1992). Statistical-mechanical analysis of self-organization and pattern formation during the development of visual maps. *Physical Review A*, **45**, 7568–7589.

Obermayer, K., Ritter, H. and Schulten, K. (1990). A principle for the formation of the spatial structure of retinotopic maps, orientation and ocular dominance columns. *Proceedings of the National Academy of Sciences USA*, **87**, 8345–8349.

Olson, S.J. and Grossberg, S. (1998). A neural network model for the development of simple and complex cell receptive fields within cortical maps of orientation and ocular dominance. *Neural Networks*, **11**, 189–208.

Parker, J.L. and Dostrovsky, J.O. (1999). Cortical involvement in the induction, but not expression, of thalamic plasticity. *The Journal of Neuroscience,* **19**, 8623–8629.

Pessoa, L., Mingolla, E. and Neumann, H. (1995) A contrast and luminance-driven multiscale network model of brightness perception. *Vision Research*, **35**, 2201–2223.

Peterhans, E. and von der Heydt, R. (1989). Mechanisms of contour perception in monkey visual cortex. II. Contours bridging gaps. *Journal of Neuroscience*, **9**, 1749–1763.

Pollen, D.A. (1999). On the neural correlates of visual perception. *Cerebral Cortex*, **9**, 4–19.

Raizada, R.D.S. and Grossberg, S. (2001). Context-sensitive binding by the laminar circuits of V1 and V2: A unified model of perceptual grouping, attention, and orientation contrast. *Visual Cognition* (Special Issue on Neural Binding of Space and Time), **8(3/4/5)**, 431–466.

Raizada, R.D.S. and Grossberg, S. (2003). Towards a theory of the laminar architecture of cerebral cortex: Computational clues from the visual system. *Cerebral Cortex*, **13**, 100–113.

Rao, R.P.N. and Ballard, D.H. (1999). Predictive coding in the visual cortex: A functional interpretation of some extra-classical receptive field effects. *Nature Neuroscience*, **2**, 79–87.

Reynolds, J., Chelazzi, L. and Desimone, R. (1999). Competitive mechanisms subserve attention in macaque areas V2 and V4. *The Journal of Neuroscience*, **19**, 1736–1753.

Roelfsema, P.R., Lamme, V.A.F. and Spekreijse, H. (1998). Object-based attention in the primary visual cortex of the macaque monkey. *Nature*, **395**, 376–381.

Rojer, A.S. and Schwartz, E.L. (1989). A parametric model of synthesis of cortical column patterns. *International Joint Conference on Neural Networks*, **2**, 603.

Rojer, A.S. and Schwartz, E. (1990). Cat and monkey cortical columnar patterns modeled by band-pass-filtered 2d white noise. *Biological Cybernetics*, **62**, 381–391.

Salin, P. and Bullier, J. (1995). Corticocortical connections in the visual system: Structure and function. *Physiological Reviews*, **75(1)**, 107–154.

Schmidt, K.E., Goebel, R., Löwel, S. and Singer, W. (1997). The perceptual grouping criterion of colinearity is reflected by anisotropies of connections in the primary visual cortex. *European Journal of Neuroscience*, **9**, 1083–1089.

Shadlen, M.N. and Newsome, W.T. (1998). The variable discharge of cortical neurons: Implications for connectivity, computation, and information coding. *The Journal of Neuroscience,* **18**, 3870–3896.

Shevelev, I.A. (1998). Second-order features extraction in the cat visual cortex: Selective and invariant sensitivity of neurons to the shape and orientation of crosses and corners. *BioSystems*, **48**, 195–204.

Sillito, A.M., Jones, H.E., Gerstein, G.L. and West, D.C. (1994). Feature-linked synchronization of thalamic relay cell firing induced by feedback from the visual cortex. *Nature*, **369**, 479–482.

Sirosh, J. and Miikkulainen, R. (1994). Cooperative self-organization of afferent and lateral connections in cortical maps. *Biological Cybernetics*, **71**, 66–78.

Smith, A.T., Singh, K.D. and Greenlee, M.W. (2000). Attentional suppression of activity in the human visual cortex. *Neuroreport*, **11**, 271–277.

Somers, D.C., Nelson, S.B. and Sur, M. (1995). An emergent model of orientation selectivity in cat visual cortical simple cells. *Journal of Neuroscience*, **15**, 5448–5465.

Stemmler, M., Usher, M. and Niebur, E. (1995). Lateral interactions in primary visual cortex: A model bridging physiology and psychophysics. *Science*, **269**, 1877–1880.

Steinman, B.A., Steinman, S.B. and Lehmkuhle, S. (1995). Visual attention mechanisms show a canter-surround organization. *Vision Research*, **35**, 1859–1869.

Stratford, K.J., Tarczy-Hornoch, K., Martin, K.A.C., Bannister, N.J. and Jack, J.J.B. (1996). Excitatory synaptic inputs of spiny stellate cells in cat visual cortex. *Nature*, **382**, 258–261.

Swindale, N. (1980). A model for the formation of ocular dominance column stripes. *Proceedings of the Royal Society of London, Series B*, **208**, 243–264.

Swindale, N. (1982). A model for the formation of orientation columns. *Proceedings of the Royal Society of London, Series B*, **215**, 211–230.

Swindale, N. (1992). A model for the coordinated development of columnar systems in primate striate cortex. *Biological Cybernetics*, **66**, 217–230.

Tamas, G., Somogyi, P. and Buhl, E.H. (1998). Differentially interconnected networks of GABAergic interneurons in the visual cortex of the cat. *Journal of Neuroscience*, **18(11)**, 4255–4270.

Temereanca, S. and Simons, D.J. (2001). Topographic specificity in the functional effects of corticofugal feedback in the whisker/barrel system. *Society for Neuroscience Abstracts,* **393.6**.

Thomas, O.M., Cumming, B.G. and Parker, A.J. (2002). A specialization for relative disparity in V2. *Nature Neuroscience*, **5**, 472–478.

Thorpe, S., Fize, D. and Marlot, C. (1996). Speed of processing in the human visual system. *Science*, **381**, 520–522.

Ullman, S. (1995). Sequence seeking and counter streams: A computational model for bi-directional information flow in the visual cortex. *Cerebral Cortex*, **5**, 1–11.

van Vreeswijk, C. and Sompolinsky, H. (1998). Chaotic balanced state in a model of cortical circuits. *Neural Computation*, **10**, 1321–1371.

Vanduffel, W., Tootell, R.B. and Orban, G.A. (2000). Attention-dependent suppression of metabolic activity in the early stages of the macaque visual system. *Cerebral Cortex*, **10**, 109–126.

von der Heydt, R., Peterhans, E. and Baumgartner, G. (1984). Illusory contours and cortical neuron responses. *Science*, **224**, 1260–1262.

von der Malsburg, C. (1973). Self-organization of orientation sensitive cells in the striate cortex. *Kybernetik*, **14**, 85–100.

Watanabe, T., Nanez, J.E. and Sasaki, Y. (2001). Perceptual learning without perception. *Nature* **413**, 844–848.

Willshaw, D. and von der Malsburg, C. (1976). How patterned neural connections can be set up by self-organization. *Proceedings of the Royal Society of London, Series B*, **194**, 431–445.

Yazdanbakhsh, A. and Grossberg, S. (2004). Fast synchronization of perceptual grouping in laminar visual cortical circuits. *Neural Networks*, **17**, 707–718.

Zhang, Y., Suga N. and Yan, J. (1997). Corticofugal modulation of frequency processing in bat auditory system. *Nature*, **387**, 900–903.

16

Conclusion: A Unified Theoretical Framework for Plasticity in Visual Circuitry

LIISA A. TREMERE[1], PETER DE WEERD[2]
AND RAPHAEL PINAUD[3]

[1]CROET, OHSU, Portland, OR, USA, [2]Department of Psychology,
University of Maastricht, Maastricht, The Netherlands and
[3]Neurological Sciences Institute, OHSU, Portland, OR, USA

Introduction

In this final chapter, it is our hope to discuss key issues in visual processing focused on the demonstrations and utilities of neural plasticity at each processing station within the ascending visual pathway. In previous chapters, key historical experiments were reviewed that tended to be based on a single methodological approach. Subsequently, these early works were discussed in the context of more recent studies, where advances in understanding visual processing and plasticity have been achieved using multidisciplinary neuroscience approaches. One aim of this final chapter is to use information provided in each chapter to delineate the processes and events that constitute plasticity from those that strongly correlate with the appearance of remodeling, but that do not directly mediate changes in circuitry or in function, referred to herein as plasticity markers.

Technical advances in many areas of neuroscience have greatly facilitated the detection and characterization of central nervous system (CNS) plasticity at the systems, cellular and sub-cellular levels. Electrophysiological characterizations of neural activity and those changes in membrane or neuronal firing behavior that occur with plasticity can be accomplished at the level of whole-cell patch clamp electrophysiology, and through single-unit or multi-electrode recordings in anesthetized and awake animals. Gains have also been made in the areas of anatomy, biochemistry, and molecular and cellular biology, which have been extremely useful to localize and control aspects of where and how plasticity proceeds. These methods for characterizing and understanding plasticity-associated changes are accompanied by a new generation of questions on neural plasticity either as a process that is developmentally regulated

R. Pinaud, L. A. Tremere, P. De Weerd(Eds.), Plasticity in the Visual System:
From Genes to Circuits, 347-355, ©2006 Springer Science + Business Media, Inc.

and controlled by environmental statistics, or induced by pathological conditions such as lesions. The implications of studying these various forms of neural plasticity and plasticity markers are discussed below.

Neural Plasticity

One of the first and possibly essential steps to critically evaluate studies of neural plasticity is to set operational boundaries on the term plasticity, which has historically been used to refer to the potential of a system to undergo change. With regard to visual function, plasticity could apply to many levels of change including behavioral, physiological, cellular and sub-cellular, all of which capture some aspect of alteration and of visual processing. In this book, the term plasticity is intended to refer to neural plasticity, indicating that we are concerned with changes in neuronal circuitry that enable stable alterations in visual processing.

One consistent idea that surfaces when one tries to ensure the identification of plasticity is that plastic events should have temporal criteria such as stability, permanence or reversibility. It has, however, been challenging to achieve agreement on what temporal parameters define one versus another type of plasticity. One example which illustrates the ambiguity in defining plasticity as a function of time is experience-dependent molecular modifications (e.g. Chapter 8). It is of course possible to make absolute reference to units of time, with respect to the duration of a particular set of events. Defining relative time scales, however, especially with reference to an event being slow or fast, or an effect being short- or long-term is a perennial problem in neuroscience. For example, one of the fastest experience-dependent genomic responses in the visual cortex occurs within minutes (to tens of minutes) of stimulus onset. In the context of evaluating molecular events, these phenomena would be considered as fast. Alternatively, when the means of identifying plastic change is whole cell electrophysiology, data captured over minutes is commonly regarded as slow or long-term.

In other scientific fields interested in changes in system performance across time, the context for evaluating temporal events can be established quickly by reference to particular terms such as "large time", that immediately communicate the numerical scale in time, i.e. greater than seconds. However, at present, the field of visual neuroscience lacks a corresponding terminology to establish such temporal context or scale. So, while the demonstration of a stable change continues to be central to how one identifies the occurrence of a plastic event, a specific reference to a temporal border between mechanisms of short versus long-term plasticity and slow or fast events may not be realistic.

In addition to various assumptions regarding the stability of plastic changes, researchers interested in neural plasticity often hold different views on the extent to which they interpret plasticity to be desirable, perhaps because for some, the hope of functional recovery resides within the possibility for physical and functional restructuring. One anatomical location where this issue is, at times, hotly debated is the retina.

Plasticity in the Retina: Implications for Structure and Function

To date, the clearest paradigmatic demonstrations of plasticity in the retina have been shown secondary to insult. These studies have provided evidence that almost all types of retinal cell types can undergo plastic change, even in the adult (Chapters 2-4). Robust rewiring programs however, do not necessarily lead to improved circuit functionality, or even the formation of functional connections. In fact, Robert Marc and colleagues (Chapter 3) have discussed data that shows that certain demonstrations of plasticity do not hold adaptive value, as is the case with microneuromas that follow trauma; these structures clearly show the potential for physically based changes in retinal circuitry, however, it appears that they are not capable of processing meaningful visual information.

Initial gains have been made, nevertheless, towards demonstrating adaptive plasticity in retina. In the detached retina, for example, structural changes appear to enable the retention and redistribution of transduced visual signals into recovering retinal circuitry (Chapter 4). There are also indications for the induction of gene expression programs putatively underlying adaptive reorganization in the absence of injury, in conditions of enhanced visual experience (Chapter 5). All these data strongly suggest that under certain sensory conditions the retina can undergo some forms of adaptive reorganization.

Plasticity in the Thalamus: Altered Organization of Sensory Processing in the Time Domain

The visual thalamus is often regarded as a mere relay for sensory information with little to no involvement in the processing of visual information. By virtue of its anatomical position, the thalamus handles visual information transiting through both ascending as well as descending visual pathways. Work in the somatosensory and auditory systems has suggested that this structure plays a vital role in governing the timing of information transfer in either ascending or descending routes. Thalamic function has also been implicated in the generation of reverberating loops of neural activity, suggested to be a mechanism of gain control for brightness perception (Rossi and Paradiso, 2003).

Interestingly, the same experimental paradigms that trigger plasticity-associated molecular changes in other visual stations including retina and primary visual cortex (V1) do not alter plasticity-associated gene expression in the main anatomical subdivision of the visual thalamus, the lateral geniculate nucleus (LGN; Arckens et al., 2000; Pinaud et al., 2003; Pinaud, 2004). Fast modifications that have been clearly shown for LGN neurons include the rapid modifications of temporal parameters of firing patterns; modifications that are well-suited for modulating gain of sensory signals (Chen and Regehr, 2000; Chen et al., 2002; Chen and Regehr, 2003; Blitz et al., 2004). In other thalamic nuclei, such as the thalamic reticular nucleus, activity-dependent changes appear to be modulated by complex behavioral states, such as attention (Chapter 6).

If the role of the thalamus in sensory processing operates mainly in the time domain, it may be that the expression of plasticity within this structure cannot be identified as changes in spatial relationships of receptive fields (RFs) such as size or amplitude of the excitatory response. Rather, plasticity in the visual thalamus may be expressed within the physical basis of temporal filters thought to reside in this structure.

Cortical Plasticity: Retuning Sensory Filters Beyond the Dynamic Range

Many excellent descriptions currently exist that describe how the RFs of sensory neurons are based upon the interplay of excitatory and inhibitory components. In Chapter 11, on the role of inhibition in visual processing, we reviewed data that shows that pharmacological disinhibition of cortical neurons with GABAergic antagonists can unmask all excitatory inputs to abolish complex network properties such as directional and orientation selectivity of V1 cells (Sillito, 1975, 1977, 1979; Ramoa et al., 1988; Eysel et al., 1998).

These studies raise important questions about the pre-existing response capabilities or dynamic range, versus new response capabilities that result from the induction and implementation of mechanisms of plasticity within sensory neurons. While many cells will respond preferentially to stimuli to which they are "tuned", the tuning of a sensory neuron is constantly refreshed with the on-going stream of synaptic activity. The dynamic nature of this interplay between pre-existing and newly-formed connectivity can make it difficult to assess the occurrence of plasticity within the visual system, as in many cases new responses may only reflect a different use of pre-existing circuitry. It therefore may be argued that, as a function of their dynamic range in processing, visual neurons may constantly express a certain degree of response plasticity. However, given that these changes could conceivably arise from within the dynamic range of that cell, they may not be useful as direct measures of neural plasticity. It is somewhat difficult to conceive of the need for plasticity in a system where all required responses can be achieved from within the dynamic range of that neuron. Moreover, it has been postulated that it is precisely the inability of this system to handle the current sensory load that invokes mechanisms of plasticity. In accordance with this line of thinking, reorganizational pressure and neuronal plasticity may occur when consistent patterns of input impinge upon a sensory neuron beyond the borders of its dynamic range, signaling both inefficient information handling and the requirement for an adaptive change.

Alterations in the electrophysiological signature of population and/or single cell activities were also described in response to the administration of plasticity-associated paradigms such as high-frequency stimulation-inducing long-term potentiation (LTP). An experimental criterion for demonstrating plasticity with these paradigms was that the modification in synaptic efficacy or strength be stable for minutes to hours (and sometimes to days), depending on the nature of the investigation, hence that the modification be "remembered" in the circuit. This criterion raises again the issue of temporal parameters in defining plastic change. Such a demonstration of memory in the circuit is consistent with the requirement that the response unit, be that cell or circuit, move beyond its dynamic range of performance.

If we consider the examples of LTP, we find that the circuit is in fact pushed to express a new balance between pre- and post-synaptic activity, putatively in order to access a previously unattainable range of processing. This new response set is very likely to constitute a new dynamic range for that neuron. Therefore, while it remains valid to state that a plastic system is one that retains the capacity to undergo change, an appropriate operational definition of neural plasticity may be considered to include some demonstration of memory for that change.

Neuromodulator Function and Plastic Change

A second means of altering response output in the visual system has been to couple sensory inputs within normal parameters to neuromodulator application (or depletion) in order to promote re-weighting of the value of its excitatory or inhibitory drive (Chapter 7). In the acute condition, neurotransmitters can influence the biophysical properties of the membrane to alter input resistances, firing thresholds and ion conductances. Neuromodulators can also potently activate and potentiate intracellular biochemical cascades that alter neuronal functioning for subsequent inputs on times scales on the order of minutes to days. Yet another effect of neuromodulatory influence occurs at the genetic level when its biological actions modifies gene expression activity leading to potentially permanent changes in cell performance (for example, see Chapter 7).

In Chapter 7, roles for different neuromodulatory systems, such as the cholinergic, noradrenergic, dopaminergic and serotonergic systems, were reviewed in terms of RF regulation, network excitability and information filtering. The impact of neuromodulatory systems in shaping the excitatory drive into visual system networks appears to extend the dynamic range for sensory processing, enabling the system to respond with greater precision to a wider variety of contexts and conditions.

For example, the discussion in Chapter 7 points to noradrenaline (NA) as a potent neuromodulator that when co-applied during the arrival of incoming sensory information has been shown to facilitate the responses of cells in supragranular layers of V1. Interesingly, NA at V1 cells from the infragranular layers has been shown to primarily depress neural responsiveness. This process illustrates a mechanism to modulate intra- and inter-cortical connectivity by controlling the influence of subcortical output and feed-back connections during visual processing. Importantly, noradrenergic input is tightly regulated by behavioral states such as arousal and attention.

It is also noteworthy to recall that neuromodulatory influences at a single cell have the potential to impact several junctions of sensory processing when evaluated across an entire circuit. For example, the different effects of neuro-modulators on intra- and inter-cortical circuits to cortical information pro-cessing is an idea that closely fits with computational modeling concepts proposed by Grossberg (Chapter 15). In his model, perceptual phenomena such as grouping and binding are closely related to interactions within super-ficial layers, while attention and learning, which have the effect of changing fil-tering properties of neurons in superficial layers, achieve their effect through inter-laminar interactions that originate in layer VI, through which also feed-back from higher-order cortex can influence on-going neural activity.

Activity-Driven Reorganization: Plasticity in Anatomical Connectivity

As briefly discussed above, one interesting ongoing debate in remodeling research is the extent to which altered function depends upon anatomical rewiring versus the formation of new connectivity. Altered anatomical con-nectivity putatively involves morphologically detectable changes, such as the appearance of new synapses or neurites at a given nerve cell, extending into regions that under normal circumstances would not be contacted by that neuron.

The early experiments that tried to characterize neural plasticity, and asso-ciated morphological changes, within sensory systems relied heavily on the use of tract tracing, and histological approaches. It has become clear that CNS connections, in particular in the visual system, can change as a function of experience, learning, and neurochemical conditions. In a classic paper from Volkmar and Greenough (1972), it was shown that the dendritic arbors of cells in the rodent V1 undergo a dramatic enhancement as a function of expe-rience in a complex environment (Volkmar and Greenough, 1972), indicating that a substantial formation of new synaptic contacts is part of a plasticity program in the mammalian CNS. Subsequent studies using anatomical-trac-ing approaches provided definitive evidence for significant morphological rearrangement of cortical connections in a variety of plasticity-inducing par-adigms employing peripheral or central lesions (Danek et al., 1991; Thanos, 1997).

The hypothesis that significant network rewiring and the formation of new anatomical connections are part of a large-scale restructuring program asso-ciated with altered sensory input is supported by studies that involve the inhi-bition of protein synthesis. For example, inhibition of protein synthesis in the rodent V1 by micro-osmotic application of cycloheximide repressed ocular dominance plasticity (Taha and Stryker, 2002). This finding suggests that local protein synthesis is a requirement for experience-dependent structural and functional plasticity to occur in the visual cortex. Thus, gene expression is a requirement for these plastic changes to take place.

One of the fastest growing areas in plasticity research is the characterization of the contributions of individual genes to the plastic response. Several "candidate-plasticity genes" have been implicated in the reorganization of visual connections. A first group of genes encodes proteins that are constitutively expressed, but post-translationally modified as part of the plasticity response, often regulated by visual experience. As discussed in Chapter 8, particular attention has been paid to a few kinases, including MAPK, CaMKII and PKA, whose post-translational modifications, such as phosphorylation, alter their activity and trigger a cascade of events associated with plasticity phenomena, including ocular dominance shifts and learning. A second group of molecules, including a few constitutive transcription factors such as CREB, which are also post-translationally modified, have also been repeatedly implicated in neural plasticity; these molecules have been suggested to integrate modifications in the cell surface, in an activity-dependent manner, with the genomic machinery located in the cell nucleus, as part of a gene expression program aimed at altering the structural properties and organization of neural networks.

A group of molecules that has received increasing attention from researchers interested in experience-dependent plasticity are inducible immediate-early genes (IEGs), which are a class of genes that are activated transiently and very rapidly without the requirement of de-novo protein synthesis; it has therefore been proposed that IEG expression reflects the first inducible genomic response to cell activation (Chapter 8). Due to IEG sensitivity to neuronal depolarization, some IEGs (e.g., c-fos and NGFI-A) have been extensively used as markers for neuronal activation, allowing for the generation of large-scale activity maps, with single-cell resolution (see Chapter 8 and Kaczmarek and Chaudhuri, 1997). This methodology has proven useful to map not only activity, but in some cases, the process of experience-dependent and injury-induced plasticity (Arckens et al., 2000; Temple et al., 2003; Pinaud, 2004). More recently, however, efforts have been placed at trying to pinpoint the individual contributions of IEGs to plasticity changes, rather than simply use them as activity markers. Molecular approaches such as RNA interference, antisense technology, as well as conditional knock-outs, have been used to identify the roles of the proteins encoded by these genes to experience- or injury-induced plasticity in a number of experimental paradigms. Particular attention has focused on IEGs that encode transcriptional regulators as they are well positioned to coordinate activity-dependent waves of gene expression involved in the structural rewiring itself (discussed in Chapter 8).

Characterizing the role of these transcriptional regulators and identifying their target genes, which are commonly referred to as "late genes; LGs" and likely involved in restructuring *per se*, will be the focus of this area of research in the next few years. In fact, the identities of LGs that are regulated by IEGs of interest are slowly but surely emerging; LGs that have been repeatedly implicated in various forms of network plasticity are under tight regulation of some IEGs. Unraveling the identity of molecules that participate in circuit rearrangements and the precise cascade of events and interactions that lead to rewiring will be necessary for understanding the syntax of the plasticity lexicon.

It would appear that researchers from diverse scientific and technical backgrounds have some clear overlap in how they define plastic events. Some of the main criteria used by many in validating an event as plastic are temporal stability, novel performance, the manufacturing, insertion and employment of new components within neural circuitry. At present, the current or future utility of plasticity demonstrated at certain early anatomical levels of visual processing remains difficult to assess. Plasticity in the retina has been virtually shown for most cell types, however, the organized use of these mechanisms as part of a controlled repair process appears to be in very preliminary stages. Thalamic plasticity within the visual system continues to be elusive and reduced when compared to retinal and cortical plasticity. Alternatively, cortical plasticity can be detected in abundance. In this chapter, we have presented an argument that plasticity in visual cortex may result from overloading current sensory filter settings, forcing the visual cortex into compensatory functioning, most probably invoking mechanisms of plasticity to promote adaptive change. Furthermore and finally, we argue for the importance of understanding the role of novel protein synthesis as a new level to which the study of neural plasticity should be focused to uncover the basic mechanisms which enable change in this exquisitely engineered neural system.

References

Arckens L, Van Der Gucht E, Eysel UT, Orban GA, Vandesande F (2000) Investigation of cortical reorganization in area 17 and nine extrastriate visual areas through the detection of changes in immediate early gene expression as induced by retinal lesions. J Comp Neurol 425:531–544.

Blitz DM, Foster KA, Regehr WG (2004) Short-term synaptic plasticity: a comparison of two synapses. Nat Rev Neurosci 5:630–640.

Chen C, Regehr WG (2000) Developmental remodeling of the retinogeniculate synapse. Neuron 28:955–966.

Chen C, Regehr WG (2003) Presynaptic modulation of the retinogeniculate synapse. J Neurosci 23:3130–3135.

Chen C, Blitz DM, Regehr WG (2002) Contributions of receptor desensitization and saturation to plasticity at the retinogeniculate synapse. Neuron 33:779–788.

Danek A, Fries W, Faul R (1991) Fibre divergence in the distal optic radiation: possible basis of functional plasticity in adult primate visual cortex. J Hirnforsch 32: 421–427.

Eysel UT, Shevelev IA, Lazareva NA, Sharaev GA (1998) Orientation tuning and receptive field structure in cat striate neurons during local blockade of intracortical inhibition. Neuroscience 84:25–36.

Kaczmarek L, Chaudhuri A (1997) Sensory regulation of immediate-early gene expression in mammalian visual cortex: implications for functional mapping and neural plasticity. Brain Res Brain Res Rev 23:237–256.

Pinaud R (2004) Experience-dependent immediate early gene expression in the adult central nervous system: evidence from enriched-environment studies. Int J Neurosci 114:321–333.

Pinaud R, Vargas CD, Ribeiro S, Monteiro MV, Tremere LA, Vianney P, Delgado P, Mello CV, Rocha-Miranda CE, Volchan E (2003) Light-induced Egr-1 expression in the striate cortex of the opossum. Brain Res Bull 61:139–146.

Ramoa AS, Paradiso MA, Freeman RD (1988) Blockade of intracortical inhibition in kitten striate cortex: effects on receptive field properties and associated loss of ocular dominance plasticity. Exp Brain Res 73:285–296.

Rossi AF, Paradiso MA (2003) Surface completion: psychophysical and neurophysiological studies of brightness. In: Filling-in: from perceptual completion to cortical reorganization (Pessoa L, De Weerd P, eds), pp 59–80. San Francisco: Oxford University Press.

Sillito AM (1975) The contribution of inhibitory mechanisms to the receptive field properties of neurones in the striate cortex of the cat. J Physiol 250:305–329.

Sillito AM (1977) Inhibitory processes underlying the directional specificity of simple, complex and hypercomplex cells in the cat's visual cortex. J Physiol 271:699–720.

Sillito AM (1979) Inhibitory mechanisms influencing complex cell orientation selectivity and their modification at high resting discharge levels. J Physiol 289:33–53.

Taha S, Stryker MP (2002) Rapid ocular dominance plasticity requires cortical but not geniculate protein synthesis. Neuron 34:425–436.

Temple MD, Worley PF, Steward O (2003) Visualizing changes in circuit activity resulting from denervation and reinnervation using immediate early gene expression. J Neurosci 23:2779–2788.

Thanos S (1997) Neurobiology of the regenerating retina and its functional reconnection with the brain by means of peripheral nerve transplants in adult rats. Surv Ophthalmol 42 Suppl 1:S5–26.

Volkmar FR, Greenough WT (1972) Rearing complexity affects branching of dendrites in the visual cortex of the rat. Science 176:1145–1147.

Index